Zoophysiology Volume 21

Zoophysiology

Volumes already published in the series:

Volume 1: *P. J. Bentley*
Endocrines and Osmoregulation

Volume 2: *L. Irving*
Arctic Life of Birds and Mammals

Volume 3: *A. E. Needham*
The Significance of Zoochromes

Volume 4/5: *A. C. Neville*
Biology of the Arthropod Cuticle

Volume 6: *K. Schmidt-Koenig*
Migration and Homing in Animals

Volume 7: *E. Curio*
The Etology of Predation

Volume 8: *W. Leuthold*
African Ungulates

Volume 9: *E. B. Edney*
Water Balance in Land Arthropods

Volume 10: *H.-U. Thiele*
Carabid Beetles in Their
Environments

Volume 11: *M. H. A. Keenleyside*
Diversity and Adaptation in
Fish Behaviour

Volume 12: *E. Skadhauge*
Osmoregulation in Birds

Volume 13: *S. Nilsson*
Autonomic Nerve Function in the
Vertebrates

Volume 14: *A. D. Hasler, A. T. Scholz*
Olfactory Imprinting and Homing
in Salmon

Volume 15: *T. Mann*
Spermatophores

Volume 16: *P. Bouverot*
Adaption to Altitude-Hypoxia
in Vertebrates

Volume 17: *R. J. F. Smith*
The Control of Fish Migration

Volume 18: *E. Gwinner*
Circannual Rhythms

Volume 19: *J. C. Rüegg*
Calcium in Muscle Activation
A Comparative Approach

Volume 20: *J.-P. Truchot*
Comparative Aspects of
Extracellular Acid-Base Balance

Volume 21: *A. Epple, J. E. Brinn*
The Comparative Physiology of the
Pancreatic Islets

A. Epple · J. E. Brinn

The Comparative Physiology of the Pancreatic Islets

With 36 Figures

Springer-Verlag
Berlin Heidelberg New York
London Paris Tokyo

Professor Dr. August Epple
Department of Anatomy
Jefferson Medical College
Thomas Jefferson University
Philadelphia, PA 19107, USA

Professor Dr. Jack E. Brinn
Department of Anatomy and Cell Biology
School of Medicine
East Carolina University
Greenville, NC 27858, USA

Cover illustration: a modified version of Fig. 7.6, p. 56.

ISBN 3-540-18157-1 Springer-Verlag Berlin Heidelberg New York
ISBN 0-387-18157-1 Springer-Verlag New York Berlin Heidelberg

Library of Congress Cataloging-in-Publication Data. Apple, August. The comparative physiology of the pancreatic islets. (Zoophysiology v. 21) Bibliography: p. Includes indexes. 1. Islands of Langerhans. 2. Physiology, Comparative. I. Brinn, J.E. (Jack E.), 1942– . II. Title. III. Series. [DNLM: 1. Islands of Langerhans – physiology. 2. Physiology, Comparative. W1 Z0615M v. 21 / WK 800 E64c]. QP188. P26E67 1987 596′.01′42 87-26316.

Printed in Germany

Typesetting: K + V Fotosatz GmbH, Beerfelden
Printing and bookbinding: Brühlsche Universitätsdruckerei, Giessen
2131/3130-543210

Preface

As far as we are aware, this is the first attempt to cover the comparative physiology of the pancreatic islets in a monograph. The topics discussed would probably have sufficed to fill about half a dozen monographs, a matter that becomes obvious from a look at the Contents. Hence, we have tried to present the material more in the form of a digest, to emphasize evolutionary perspectives, to point out critical issues, and to identify challenging topics for future research.

This approach required an arbitrary reduction of the number of references, and we therefore join the chorus of recent authors who beg their colleagues for understanding if some of their publications do not appear in the bibliography. Keeping up with the current literature was like fighting one of those monsters that grow a couple of new heads for each one that is cut off. Nevertheless, we hope that we have covered most of the key publications up to the autumn of 1986.

We gratefully acknowledge the advice of many colleagues, and in particular the invaluable criticisms of Robert L. Hazelwood and Erika Plisetskaya. Special thanks are due to the series editor, Donald S. Farner, for his patience and guidance, both of which were fresh proof of his legendary diplomatic skills. Finally, we wish to thank Dr. D. Czeschlik and his staff at the Springer Verlag for their patience and support.

Philadelphia, PA — AUGUST EPPLE
Greenville, NC — JACK E. BRINN

September 1987

Contents

Chapter 1

Introduction

1.1 Landmarks in Islet Research

In the history of islet research, it is easy to identify four distinct phases: an initial one that began with the description of the islets in the rabbit pancreas by Langerhans (1869) and gained momentum with the discovery of pancreatectomy diabetes by von Mehring and Minkowski (1889). Two other key events of this phase were the recognition of the islets of Langerhans as an endocrine tissue (Laguesse 1893), and the demonstration that atrophy of the exocrine pancreas is not followed by diabetes mellitus (Ssobolew 1902). The latter two observations together pointed to the islets as the source of an antidiabetic hormone. MacCallum (1909) proved that this suspicion was correct when he induced diabetes mellitus by pancreatectomy in dogs that had remained normoglycemic after the experimental destruction of their exocrine pancreas, a result that was reconfirmed later when Macleod (1922) compared the insulin content of teleost Brockmann bodies and exocrine pancreas. The initial phase in islet research culminated with the isolation of insulin by Banting and Best (1922), a feat that was followed within a year by the insulin treatment of diabetics (see e.g., Bliss 1982). Banting and Best had succeeded where several other investigators had failed, though it turned out that some of these had been tantalizingly close to their goal (see e.g., Bliss 1982; Volk and Wellmann 1985 a). In retrospect, it is only fair to note that the rapid success of Banting and Best was not just a case of ingenuity, but also of serendipity. At today's state of knowledge, who would expect that a rather crude extract from a mixed exocrine-endocrine gland of another species could be used for the long-term treatment of thousands of patients? Fortunately, in this case, immunology was in its infancy, and side effects of preparations were not yet a matter of too great a concern, either. Had it been necessary for Banting, Best and their coworker, Collip (see e.g., Collip 1923) to abide by modern safety standards, many more diabetics would probably have died before the issuance of a permit for the large-scale application of their early extracts. Only a few years ago, it was shown that many commercial insulins used well into the 1970's contained considerable traces of other peptide hormones, especially vasoactive intestinal peptide (VIP) and pancreatic polypeptide (PP). In one study, antibodies to these substances were found in the blood in 63% of insulin-dependent (Type I) diabetics, but were absent in non-insulin-dependent (Type II) diabetics (Bloom et al. 1979). Another study reports the presence of PP-antibodies in 69% of insulin-dependent diabetics (O'Hare 1983). The possible role of these iatrogenic antibodies in "diabetic" disorders has not been fully established.

During the second phase of islet research, which lasted for about 20 years, the investigations concentrated mainly on insulin and the etiology of diabetes mellitus. The quality of insulin preparations was greatly improved by crystallization (Abel 1926) and the production of the first long-acting protamine zinc insulin (Hagedorn et al. 1936). Also during this phase, the antagonistic relationships between the pituitary and adrenals on one hand, and of the insulin-producing pancreatic B-cells on the other, were discovered (for literature see Volk and Wellmann 1985e). At the same time, however, important data on other aspects of islet research were ignored. A case in point is the demonstration of glucagon by several research groups (Collip 1923; Murlin et al. 1923; Blum 1927; cf. Ferner 1952).

A third phase of avalanche-like progress began with the accidental discovery that alloxan selectively destroys the B-cells (Dunn et al. 1943). This observation made it possible to differentiate the specific symptoms of insulin-deficiency from those of total pancreatectomy. Furthermore, the survival and behavior of the A- and D-cells in the islet organ of alloxan-diabetic mammals indicated that these long-neglected elements were sources of hormones different from insulin. This notion ultimately led to the purification and crystallization of glucagon (Staub et al. 1955), the identification of the primary structure of glucagon (Bromer et al. 1957), and the definitive demonstration of the hormonal status of this peptide by Foà (cf. Foà 1968, 1973). At the same time, also insulin research achieved new and spectacular successes. After 10 years of persistence, Sanger and coworkers identified the primary structure of insulin (Ryle et al. 1955), a landmark in polypeptide research that was followed 12 years later by the discovery that the two peptide chains of insulin are synthesized via a common single-chain precursor (Steiner and Oyer 1967). The identification of this single-chain peptide, called "proinsulin", opened a new era in research on the biosynthesis of messenger peptides, as it now appears that perhaps all peptide hormones and related substances are synthesized via larger precursors. Another recent surprise was the accidental discovery in the chicken (Kimmel et al. 1968) of a third islet hormone, pancreatic polypeptide (PP), which was soon confirmed for mammals (Lin and Chance 1972). Curiously, the cells of origin of PP seemed to be different from a third type of islet cells, the D-cells, whose functional autonomy had been firmly established during the early 1960's (cf. Epple 1963; Fujita 1968). This riddle was resolved when immunohistology showed that the D-cells are the source of a fourth islet hormone, somatostatin (M. P. Dubois 1975; Polak et al. 1975), whereas in mammals and birds the inconspicuous PP-cells occur at the islet periphery and often among exocrine elements (L.-I. Larsson et al. 1974). Somatostatin had been identified previously in the rat hypothalamus as a putative neurohormonal inhibitor of growth-hormone release (Krulich et al. 1968). Almost simultaneously, Hellman and Lernmark (1969) had shown in vitro that the D-cells release a factor that inhibits insulin secretion; however, the identity of this factor with somatostatin was not recognized then. Probably, one of the reasons for this was that the structure of somatostatin remained unknown until after its rediscovery by Brazeau et al. (1973). On the other hand, the idea of hormones common to neurons and islet cells was, perhaps, too revolutionary for the late 1960's, though the coexistence of hypothalamic and islet hormones in multiple endocrine tumors had been established (O'Neal et al. 1968). However, by the mid-1970's, improved immunocytology had created a tidal wave of unexpected data on the distribution of messenger peptides. Since then "islet hormones" have been found in "extra islet" sites; and vice versa, neurosecretions, gastrointestinal hormones and messenger peptides originally known from other tissues have been localized to the islet cells (see Chaps. 2.2, 9.1, 9.1.2 and 9.1.3).

The latest phase in islet research was inevitably ushered in by the enormous recent progress in genetic engineering with the insulin gene being one of the most-studied models. There is no question that, e.g., the cloning of cDNA copies of mRNA's of islet hormones will provide in the near future both vistas and experimental possibilities that surpass the wildest dreams of comparative islet physiologists of a decade ago (see e.g., Steiner et al. 1984; Tager 1984). Perhaps the most frustrating area in islet research remains the molecular basis of insulin action. Although gene sequence data are now available for the insulin receptor, we still do not know how precisely the multiple cellular signals generated by this hormone are mediated. At last, however, the identification of the low-molecular-weight mediators of insulin action appears imminent (Czech 1985; Stevens and Husbands 1985; Saltiel et al. 1986).

1.2 The Islet Organ: New Interpretations of an Old Gland

Only a few years ago, it would have been possible to limit the scope of this book more or less to four hormones, i.e., insulin, glucagon, PP and somatostatin, and to the pancreatic cells (B-, A-, F- and D-cells, respectively) that produce them. Today, however, we are in the midst of a revolution in our understanding of both messenger substances and the cells of their origin (Bloom and Polak 1981; Schwyzer 1982; Krieger 1983), which no longer permits consideration of the islet organ as an isolated entity. Instead, we must interpret it as a member of a large group of endocrine organs and tissues, the secretions of which are mixtures of very heterogeneous substances. It appears possible that these endocrine cells, receptor cells, and neurons all share a common ancestor, that was structurally more or less identical with the "open-type basal-granulated cells" of the gastroderm of hydra (Fujita et al. 1980). The latter elements (Fig. 1.1) occur in minor modifications as "open cells" in the digestive mucosa of many metazoa, including the human (cf. Fritsch et al. 1982; Fujita 1983). Here, their apical microvilli scan the gastrointestinal contents for specific stimuli (e.g., pH, amino acids), which are then answered by exocytotic release of messenger substances in the basal cell region (Fig. 1.2). Figure 1.3 shows that minor modifications of the hypothetical ancestor cells suffice to create a large array of neural, receptor, and endocrine cells, and that boundaries among these categories can only be drawn by arbitrary decisions. Therefore, Fujita (1977, 1980) proposed the term paraneuron for all members of this group that are not "true" neurons, i.e., receptor and endocrine cells. On the other hand, Pearse (1977) proposed the collective term "diffuse neuroendocrine system" for all endocrine neurons and related peripheral endocrine cells. The close relationship between neurons and paraneurons is supported by the distribution of the enzyme "neuron-specific enolase", which may be a specific molecular marker for both groups of cells (Schmechel et al. 1978;

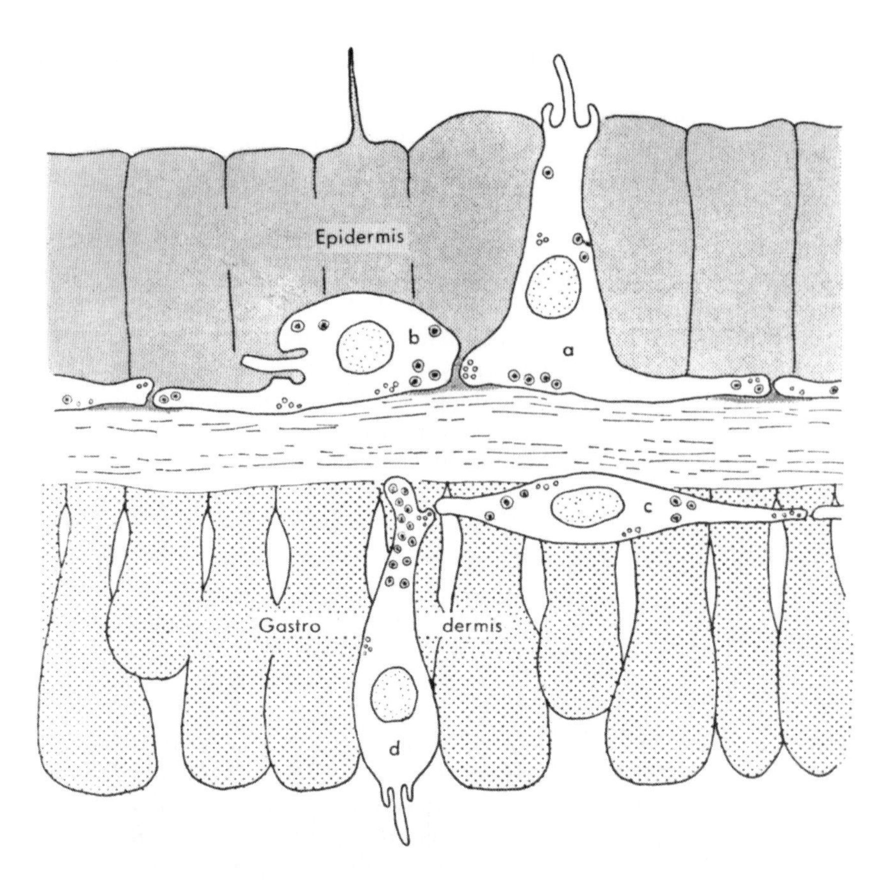

Fig. 1.1. Neuron- and paraneuron-related cells (*a–d*) in epidermis and gastrodermis of *Hydra japonica*. Note the receptor-secretory cell (*d*) in the gastroderm that could represent the ancestor of both "open" and "closed" cells of the endodermal derivatives of the metazoa (Fujita et al. 1980)

Sheppard et al. 1982), with the exception of the parathyroid chief cells (Pearse 1980). Not long ago, it appeared that the endocrine neurons and paraneurons can be delineated easily from other vertebrate endocrines (Epple 1982). However, the discovery of hormonal peptide production in totally different tissues such as, e.g., the myocardium (atrial natriuretic factor: see Cantin and Genest 1986) makes a separation of neurons and paraneurons from other tissues increasingly difficult.

Originally it was thought that all neuron-related endocrine cells are capable of taking up amine precursors and of transforming them into amines. Hence, Pearse (1968, 1969) coined the term APUD cells, the acronym standing for *a*mine *p*recursor *u*ptake and *d*ecarboxylation. Furthermore, it was postulated that all APUD cells are of neural crest origin (Pearse and Polak 1971). Recent progress has made it clear that the original APUD concept, stimulating as it was, requires modification. Thus, as noted by Pearse (1977), both the gonadotrope and thyrotrope cells of the mammalian pars distalis do not have amine "handling" capabilities though they are otherwise typical APUD cells. On the other hand,

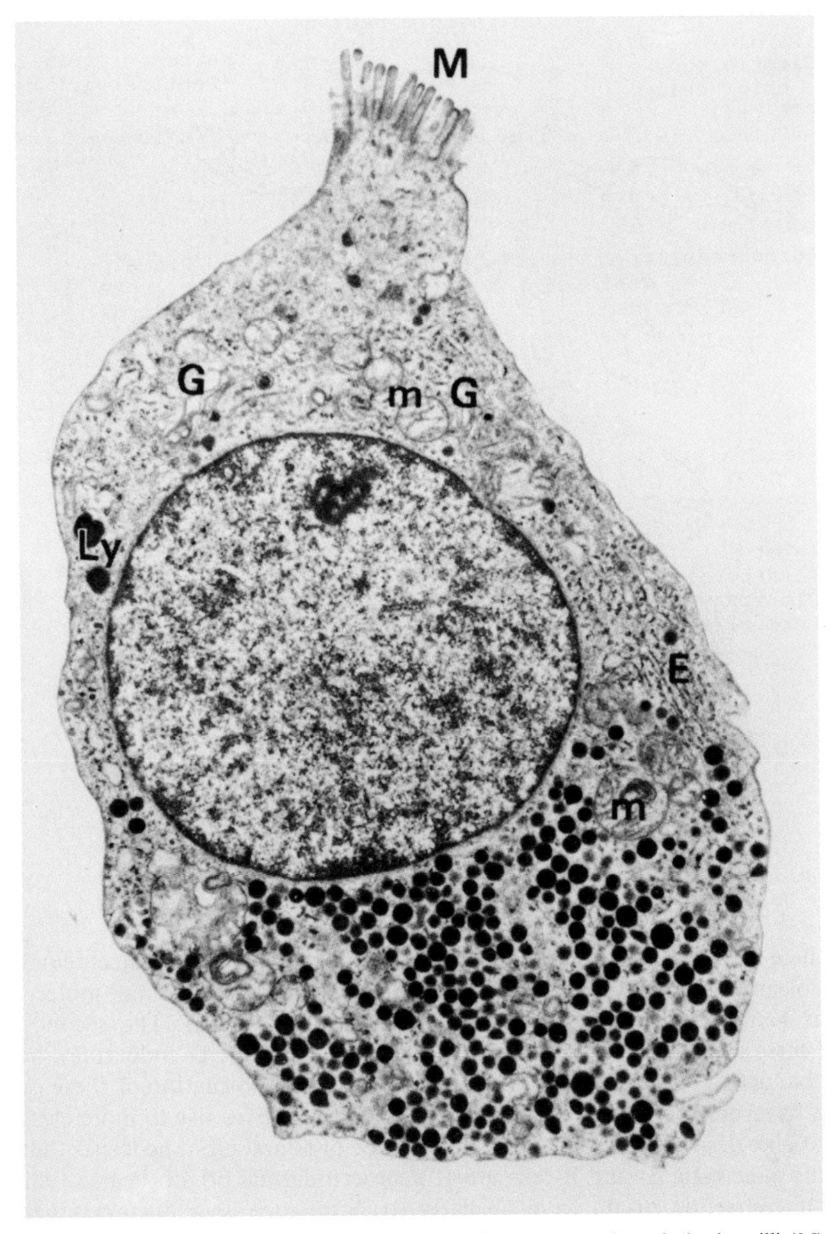

Fig. 1.2. "Open", basal-granulated cell from the human duodenum. Note the apical microvilli (*M*) which receive stimuli from the intestinal lumen, and the basal accumulation of granules at the opposite (capillary) pole of the cell. *G* golgi complex; *m* mitochondria; *E* granular endoplasmic reticulum; *Ly* lysosome (Fujita and Kobayashi 1973)

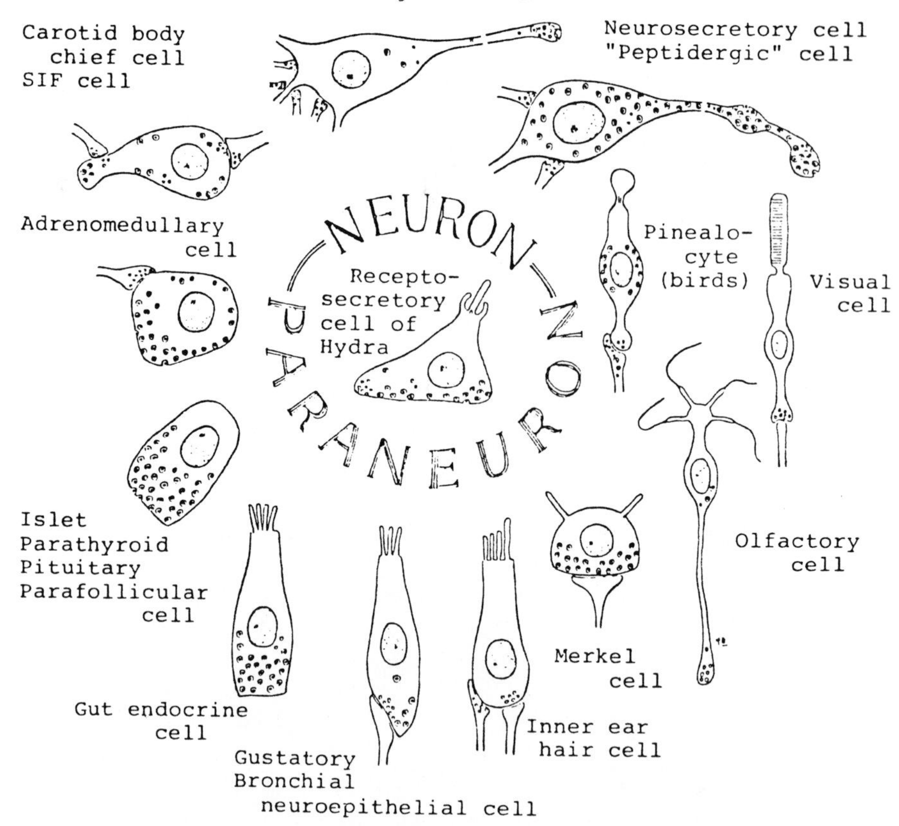

Fig. 1.3. Spectrum of vertebrate neurons and paraneurons (Fujita 1983)

there are presumptive APUD cells in which the amines (catecholamines, indoleamines) are either replaced, or supplemented by other small molecules such as gamma aminobutyric acid (GABA) (see Sect. 9.1.3). The second part of Pearse's hypothesis, the neural crest origin of all APUD endocrines, had to be abandoned. Instead, Pearse (1979) now proposes a formation of these cells from a "neuroendocrine-programmed epiblast", which gives rise to more clearly identifiable descendants such as the neural tube or neural crest; he insists that at least the pancreatic A- and B-cells are of neuroectodermal origin (Pearse 1982, 1984). Of course, an origin from an early structure such as a "neuroendocrine-programmed epiblast" is very difficult, if currently not impossible, to disprove. Therefore, it is not surprising that other authors maintain that the endocrine pancreas is of endodermal origin (Cheng and Leblond 1974; Rutter 1980; Andrew 1982, 1984; Ayer-LeLievre and Fontaine-Perus 1982; Rawdon 1984; Rawdon et al. 1984).

What are the special features of the islet organ? Basically, there are two. The first is its extramucosal location, by which it differs from all related gastrointestinal paraneurons. The second, and most obvious peculiarity of the islet organ is its close association with the exocrine pancreas (Epple and Brinn 1975, 1980, 1986; Epple et al. 1980). However, this association only exists at the gnathostome level, whereas in cyclostomes the ancestors of both pancreas components are often completely separated. Indeed, as will be discussed in Section 2.3, the available evidence suggests that the exocrine pancreas appeared phylogenetically *before* the "follicles of Langerhans", i.e., the ancestral islet organ, which later developed via a different route.

1.3 Terminology

The independent phylogenetic origins of islet organ and exocrine pancreas create terminological difficulties that are compounded by the greatly varying morphology of the gnathostome pancreas (see Sect. 2.3). For clarity we therefore use the following nomenclature:

1. *Islet organ* refers to the entirety of the "follicles of Langerhans" (cyclostomes), or all endocrine cells of the pancreas (gnathostomes).
2. *Endocrine pancreas* is applied for the endocrine cells associated with the pancreas of the gnathostomes. Hence, in this group, "islet organ" is synonymous with "endocrine pancreas".
3. *Islets*, or *islets of Langerhans* or *pancreatic islets* refers to accumulations of endocrine cells that are scattered within the exocrine pancreas of the gnathostomes.

Similar terminological problems exist with respect to the hormones of the islet organ. We follow the traditional terminology which refers to insulin, glucagon, PP, and somatostatin as *the* or *major islet hormones,* although in some species other peptides also appear to be important islet secretions. This is the case in selachians (gastric inhibitory peptide: El-Salhy 1984), holocephalians (glicentin-like substance: Stefan et al. 1981), and the bullfrog (secretin: Fujita et al. 1981 a). Furthermore, it must be noted that gastrin seems to be a pre- and perinatal islet

Table 1.1. Major islet cell types and their major hormones

Cell type	Hormone	Distribution
A-cell	Glucagon	All gnathostomes?
B-cell	Insulin	All vertebrates?
D-cell	Somatostatin(s)	All gnathostomes?
F-cell	PP	All gnathostomes?
Argyrophil cell	?	Adult lampreys
Argyrophil cell	Insulin	Mainly(?) fetal ruminants
Somatostatin cell	Somatostatin(s)	Hagfishes, adult lampreys
Amphiphils	?	Various fishes and amphibians
X-cell	Glucagon-related peptide	Holocephalians
Flame-shaped cell	?	*Lepisosteus*

For further details, see Chapter 4 and Epple and Brinn (1986).

secretion in some species, possibly acting as a pancreatic growth factor (Larsson et al. 1976a). Frequently, peptides with possible or known messenger functions coexist in islet cells with major islet hormones. These substances, referred to as *co-released peptides*, may share a common precursor molecule with *the* islet hormone (e.g., the glucagon-like peptides; see Sect. 11.1); or they may be of apparently independent origin (see Sect. 9.1.2). Small co-released islet substances with known or suspected messenger functions (e.g., amines, GABA, ATP) will be termed *small messenger molecules.* The islet cells will be referred to as indicated in Table 1.1.

Chapter 2

The Evolution of the Islet Organ

2.1 Islet Hormones Before the Islet Organ

Phylogenetically, the islet hormones or closely related substances appeared long before the islet organ. Indeed, there is increasing evidence that all major types of messenger substances (steroids, catecholamines, peptides) are present in prokaryotes (bacteria) and/or unicellular karyotes (for literature see Coupland 1979; Hunt and Dayhoff 1979; Sandor and Mehdi 1979; LeRoith et al. 1983a; Kolata 1984; LeRoith and Roth 1984). However, it has been questioned if this is a truly primitive condition, or the result of secondary gene transfer from higher organisms (LeRoith et al. 1983b), a possibility supported by a recent report of fish to bacterium gene transfer (Bannister and Parker 1985). The functions of the messenger substances in unicellular organisms are largely unknown, although they may be involved in both intra- and extracellular actions. Csaba (1980, 1981) and Josefsson and Johansson (1979) have shown that there are specific receptors for vertebrate messenger substances (peptides and catecholamines) in unicellular organisms; there is no doubt that pheromone-type interactions involving these substances exist in at least some species (Bonner 1971; Kochert 1978; Dunny et al. 1979; Kaiser et al. 1979; O'Day and Horgen 1981). Among the islet hormones so far only insulin- and somatostatin-like material have been described for unicellular organisms. "Insulin" was identified in fungi (*Aspergillus* and *Neurospora*), the ciliate *Tetrahymena pyriformis*, and the bacterium *Escherichia coli* (cf. LeRoith et al. 1983a). "Relaxin", a substance structurally similar to insulin, was also found in *Tetrahymena* (Schwabe et al. 1983), as was "somatostatin" (Berelowitz et al. 1982). Recently, two types of somatostatin-like immunoreactivity (resembling S-14 and S-28, see Sect. 13.1) were also identified in bacteria (*E. coli* and *Bacillus subtilis*) by LeRoith et al. (1985a, b). Until the *production* of these hormones by unicellular species has been proven, these findings must be considered preliminary evidence (Steiner et al. 1984). However, the dicovery of two types of somatostatin also in flowering plants (spinach, *Lemna gibba,* and *Nictiana tabacum*) strongly suggests that the phylogenetic appearance of somatostatin antedates the emergence of multicellular organisms (LeRoith et al. 1985c; Werner et al. 1985).

Among the most primitive metazoa, the coelenterates, messenger substances are formed by both nervous and open-type endocrine cells. Their primitive nerve network contains, besides biogenic amines (cf. Venturini et al. 1984), morphogenetic peptides (Schaller et al. 1982), as well as peptides with known hormonal and/or neurotransmitter functions (Grimmelikhuijzen 1984; S. M. Martin and Spencer 1983). Since islet hormones appear to be absent, at least in *Hydra attenuata* (Grimmelikhuijzen 1984), one wonders about hormones that will be detected in the open-type endocrine cells of the gastroderm (Fujita et al. 1980). This lack of more detailed information is particularly regrettable, since the dual location of messenger peptides (including the islet hormones) in gastrointestinal endocrine cells and neurons seems to be a basic feature shared by both proto- and deuterostomians; and it is noteworthy that Andries and Tramu (1985) differentiated in the midgut of the cockroach (*Blaberus craniifer*) no less than ten different endocrine cell types. The four major islet hormones have been identified in all major groups of proto- and deuterostomians thus far studied (for invertebrates, see Van Noorden 1984). However, it remains to be seen if the ability of the metazoon genome to initiate formation of islet hormones is actually expressed in all forms, or if there are function- and taxon-related omissions.

The functions of the islet hormones in invertebrates are largely unknown, although one can suspect roles in the regulations of growth, development, metabolism, and osmoregulation. The presence of insulin or a very similar substance in insects has been reported by a considerable number of investigators (for literature see Thorpe and Duve 1984; Teller and Pilc 1985). Particularly convincing are the thorough studies of Duve and coworkers (Duve 1978; Duve and Thorpe 1979, 1980; Duve et al. 1979; Thorpe and Duve 1984) in two blowfly (*Calliphora*) species, which suggest that in these insects insulin-like material from neurosecretory brain cells controls carbohydrate metabolism in a manner similar to that of islet insulin (Fig. 2.1) in mammals. Materials with amino acid sequences homologous to insulin have also been identified in the tobacco hornworm (*Manduca sexta*) by Kramer et al. (1982) and in the silkworm (*Bombyx mori*) by Nagasawa et al. (1984); and Mizoguchi et al. (1984). Glucagon-like activity has

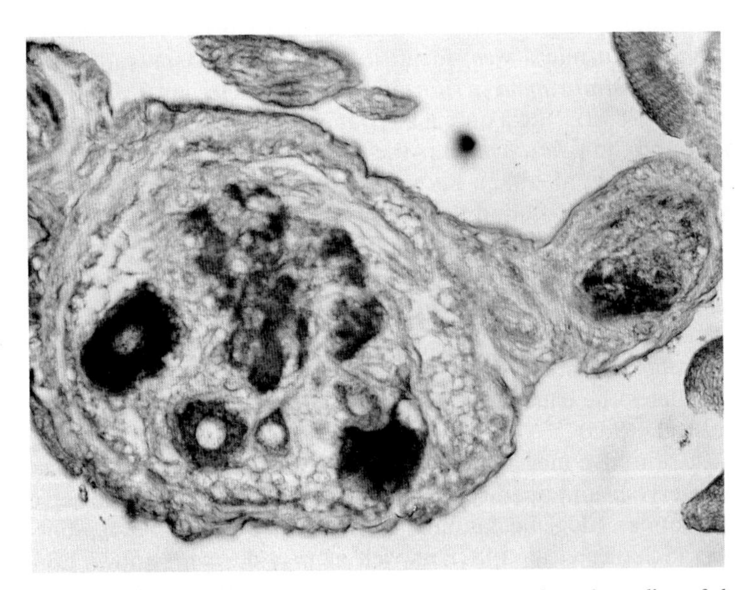

Fig. 2.1. Insulin B-chain immunoreactive neurons on the frontal ganglion of the tobacco hornworm (El-Salhy et al. 1984)

been found in both the nervous and digestive system of insects (cf. Falkmer and Van Noorden 1983; El-Salhy et al. 1984). In the adult form of the mealworm (*Tenebrio molitor*) Gourdoux et al. (1983) observed that gluconeogenesis from amino acids was increased by a glucagon-like activity from nervous tissue (corpora cardiaca), while it was reduced by an insulin-like molecule from the intestine. In the larva of the same species, Teller et al. (1983) identified an insulin-like substance in both head and midgut. Another interesting finding in insects is the presence of insulin-like material in the royal jelly of bees (*Apis mellifera*), which has been reported by three different laboratories (cf. O'Connor and Baxter 1985). Thus, it appears that the presence of insulin in the exocrine secretions of mammals (Countinho et al. 1983; Helgeson et al. 1984) and birds (Turner and Hazelwood 1974) is paralleled in the bee. In addition to insulin and glucagon, also PP- and somatostatin-like material have been reported in insects (Van Noorden 1984).

Both insulin- and glucagon-like activities occur in the gut and "hepato-pancreas" of crustaceans (Maier et al. 1975; Sanders 1983a). However, it must be noted that the name "hepatopancreas" is misleading, since this complex intestinal gland (cf. Gibson and Barker 1979) is not directly homologous with the deuterostomian pancreas and liver. Nevertheless, the secretion of large quantities of an insulin-like substance by an intestinal gland in the lobster (*Homarus americanus*) and the presence of this material in the hemolymph is reminiscent of the vertebrate situation. "Lobster insulin" promotes in vitro glycogenesis in lobster muscle, but has no impact on levels of glucose in hemolymph (Sanders 1983b, c). Similarly, mammalian insulin accelerates hexose transport into the muscle fiber of another crustacean (*Balanus nubilis*) in vitro (cf. Hager and Bittar 1985). On the other hand, Davidson et al. (1971) report that decapode tissue contains a substance which causes hypoglycemia in a mammalian test system. A particularly well-studied protostomian neurosecretion with a probable metabolic function is the crustacean hyperglycemic hormone (also termed eyestalk factor). This peptide is produced in the brain of decapode crustaceans (specifically, the medulla-terminalis-X-organ), from where it is transported to the neurohemal sinus gland, its site of storage and release into the hemolymph. The original suspicion that this peptide could be identical with or related to glucagon was disproven. The hormone is believed to control both resting blood sugar and hyperglycemic responses (Keller et al. 1985; Sedelmeier 1985; Kallen et al. 1986). Somatostatin-like activity has been reported for the nervous system of the isopod *Porcellio dilatatus* (Martin and Dubois 1981), but there seem to be no reports on the occurrence of PP in crustaceans.

Another taxon with both intestinal and "hepatopancreatic" insulin- and glucagon-like material are the molluscs (Gomih and Grillo 1976; Plisetskaya et al. 1978a, b; Banks et al. 1980; Hemminga 1984; Plisetskaya and Joosse 1985). Insulin-like material was localized to intestinal cells in the region of the hepatopancreas of *Mytilus edulis*; however, glucagon was not detected (Fritsch et al. 1976). On the other hand, insulin-, glucagon- und somatostatin-like immunoreactivities were identified in neurons of two species of snails (Schot et al. 1981; Falkmer and Van Noorden 1983; Van Noorden 1984), and it appears that insulin plays a role in mollusc carbohydrate and lipid metabolism (Plisetskaya et al. 1978a, b). Such a role of insulin would be compatible with the presence of insulin-like immunoreactivity in the hemolymph of two slugs (*Ariolymax* and *Phrophysaen*); this insulin-like peptide shows a crossreaction with salmon (*Oncorhynchus kisutch*) insulin (cf. Plisetskaya et al. 1986a). From their own work and a review of the literature, Plisetskaya and Joosse (1985) recently concluded that insulin may have a role in the glycogen metabolism of terrestrial pulmonate gastropods, while the role of this hormone in freshwater pulmonates remains unclear. On the other hand, it has been reported that a somatostatin-like factor stimulates body growth in gastropods (Geraerts 1976; Dogterom 1980).

The only other groups of protostomians studied for the presence of islet hormones are the annelids and the plathelminths. LeRoith et al. (1981) detected an insulin-like material in extracts of both skin region and interior organs of the earthworm, while Sundler et al. (1977) report PP-like immunoreactivity in the nervous system of the same species. In the nervous system of the flatworm, *Dugesia lugubris*, Schilt et al. (1981) identified somatostatin-positive cells. The "somatostatin" of these animals may play an inhibitory role in cellular proliferation (Bautz and Schilt 1986).

Like the protostomians, also the invertebrate deuterostomians possess cells with islet hormones in both the alimentary tract and the nervous system. However, this may not hold for all groups, since Welsch and Dilly (1980) were unable to detect ultrastructurally endocrine cells in the digestive mucosa of the hemichordates. No information on the occurrence of islet hormones in this group seem to be available. In the echinoderms, Falkmer and coworkers (Wilson and Falkmer 1965; Falkmer and Wilson 1967; Davidson et al. 1971) report the presence of insulin and cells producing it in two species of starfish, although the presumed intestinal B-cells show features that are different from their vertebrate counterpart (Chan and Fontaine 1971). Particular attention has been paid to the tunicates which have, besides a variety of other messenger peptides, PP and somatostatin in cells of both the alimentary tract and cerebral complex. Insulin has been localized only to cells of the digestive tract, and glucagon was not detected (for literature see Bevis and Thorndyke 1981; Fritsch et al. 1982; Falkmer and Van Noorden 1983; Van Noorden 1984). *Branchiostoma lanceolatum* ("Amphioxus"), the well-known representative of the cephalochordata, has "open" insulin and glucagon cells in its intestinal mucosa. While the insulin cells may produce one peptide only, the glucagon cells also produce a gastrin-like material (Van Noorden and Pearse 1976; Reinecke 1981). On the other hand, it remains to be seen if a population of aldehyde fuchsin-positive gut cells, which secretes into the gut lumen (Polyakova and Plisetskaya 1976) indeed releases insulin-like material, or if it mereley shows great similarity with another "true" B-cell equivalent. No data on the occurrence of the islet hormones in the CNS of *B. lanceolatum* seem to be available except for an indication of the possible presence of somatostatin by Falkmer and Van Noorden (1983), and a preliminary note on PP (Van Noorden 1984). Data on the functions of the islet hormones in the deuterostomian invertebrates are limited to the observations that glucagon does not affect the stomach epithelium in a tunicate, *Styela clava* (Bevis and Thorndyke 1979), while insulin may be involved in the carbohydrate metabolism of *B. lanceolatum* (Leibson et al. 1976b; Polyakova and Plisetskaya 1976).

2.2 Extrapancreatic Origin of Islet Hormones in Vertebrates

The discovery of islet-hormone-like activities in both nervous and digestive tissues of invertebrates raised the question of an extrainsular occurrence of islet hormones in vertebrates. This suspicion was confirmed in recent studies. However, it remains to be seen if the extra-insular distribution of the other islet hormones is as restricted as it currently seems. The report of insulin in peripheral nerve terminals (Uvnäs-Wallensten 1981) suggests that islet hormones may occur as neuronal or paraneuronal co-secretions which are nondetectable by routine immunocytology.

2.2.1 Insulin

Frequently raised is the question of the absence of B-cells in the intestinal mucosa of the vertebrates even though the islets derive phylogenetically and ontogenetically from the gut (see Sect. 2.3). Actually, there are two largely ignored reports on the occurrence of insulin in the intestine of the rat (Håkanson and Lundquist 1971; Larsson 1977 a), and there is another, possibly related, report on the persistence of insulin activity in the plasma of the pancreatectomized chicken (Colca and Hazelwood 1982). The immunocytochemical demonstration of "open" insulin cells in the intestine of the embryonic and adult shark (*Squalus acanthias*) by El-Salhy (1984) definitely proves that, contrary to a widely held view (cf. Falkmer et al. 1984), the B-cells do not entirely disappear from the gut during the early stages of the islet evolution. Rather, El-Salhy's observations suggest that in the primitive islet organ of the selachians, "... things have not yet been sorted out ..." (see Sect. 4.2). Since Gapp et al. (1986) recently found "open" insulin cells also in the intestine of a turtle (*Pseudemys scripta*), one wonders if insulin cells will be identified in the intestine of additional gnathostomes, especially in the gut of adult *Latimeria chalumnae*, whose islet organ is similar to that of the shark (Epple and Brinn 1975). Insulin-like material has been identified in many mammalian tissues (Perez-Castillo and Blázquez 1980 a), including the adenohypophysis of rat, mouse, and hamster (Hatfield et al. 1981; Budd et al. 1983), the acini and ducts of the parotid gland of the rat (Smith and Patel 1984) and the yolk sac of the rat (Muglia and Locker 1984).

The production of insulin in the nervous system of vertebrates has been a matter of heated debate (see LeRoith et al. 1983 a; Yalow and Eng 1983; LeRoith and Roth 1984; Steiner et al. 1984; Rosenzweig et al. 1985). Rather convincing data in favor of an insulin formation in nerve cells were recently provided by the immunocytochemical demonstration of insulin in cultured neurons from mouse and rat brain (Weyhenmeyer and Fellows 1983; Birch et al. 1984), and by the simultaneous demonstration of C-peptide and insulin-immunoreactivity in cells of the human central nervous system (Dorn et al. 1983 b). This issue is particularly fascinating in the guinea pig. Like a number of other vertebrates (see Chap. 10.1), the guinea pig and relatives (hystricomorphs) seem to have two different insulins. The *pancreas insulin* of the guinea pig is very different from the typical mammalian insulins, but a second one, immunologically similar to the latter type occurs in the brain and other organs of this species (Rosenzweig et al. 1983, 1985; Stevenson 1983). Roth and coworkers suggest that the two insulins of the guinea pig may be phylogenetically ancient and highly conserved products of two different genes (cf. Rosenzweig et al. 1983, 1985). Clearly, this is a radical deviation from the traditional interpretation. It would suggest that the peculiar pancreas insulin of the guinea pig evolved by mutation long before hystricomorphs diverged from the rodent mainline, a view not shared by Steiner's group (Steiner et al. 1984). Insulin-like immunoreactivity has also been described for many regions of the human brain, and for the central nervous system of several rodents, a tortoise (no species given) and the clawed frog by Dorn and coworkers (Bernstein et al. 1980; Dorn et al. 1980b, 1983b), and for cyclostomes by Falkmer et al. (1984). Specific insulin binding in mammalian and avian brain tissues has been found by several laboratories (for literature see Landau et al. 1983; Weyhenmeyer and Fellows 1983; Frank et al. 1985), and radiohistochemical localization of insulin receptors has been reported for the brain of rat (W. S. Young et al. 1980; Van Houten and Posner 1981) and a monkey (Landau et al. 1983). In the fetal rat brain, insulin-binding sites are more evenly distributed and undergo waxing and waning during the development. In the brain of adult rats there is a particularly strong concentration of insulin receptors in the plexiform layer of the olfactory bulb (W. S. Young et al. 1980). The function(s) of neuronal insulin are unknown; both brain insulin and its receptors in the rat are regulated differently from their counterparts in extraneural tissues (Underhill et al. 1982), and the insulin receptors of the brain do not show a correlation with the insulin titers of blood and cerebrospinal fluid (Figlewicz et al. 1986). However, insulin from the blood stream binds specifically in median eminence and other circumventricular organs, the infundibular nucleus and the endothelium of various types of brain vessels (for literature see Landau et al. 1983; Frank et al. 1985). In the pork brain, the receptor affinity constant of human insulin was significantly lower in comparison to pork insulin (Schlüter et al. 1984). Clearly, one of the most urgent tasks for investigation includes the identification of the respective functions of neuronal and circulating insulin in the adult central nervous system. A recent study by Grundstein et al. (1985) supports the concept that insulin may exert its brain action by reducing the glucose utilization in areas implicated in neuroendocrine control.

2.2.2 Glucagon

An extrapancreatic distribution of glucagon and/or its presumed precursor glicentin (see Chap. 11.1) has been reported for both the digestive system and nervous tissues of vertebrates. In addition, large quantities of glucagon were detected immunohistochemically in the smooth muscle of blood vessels and in the myoepithelial cells of sweat glands of the rat by Tanaka et al. (1983), who suggest that these tissues are indeed the original sources of this material. Thus, the extent to which the glucagon immunoreactivity in the extracts of a variety of rat tissues (Perez-Castillo and Blázquez 1980a) can be related to the findings of Tanaka and coworkers remains to be demonstrated. In cyclostomes, intestinal paraneurons of the open type react with antiglucagon, but also with antigastrin and anti-PP (cf. Falkmer and Van Noorden 1983; Van Noorden 1984). In the stomach of the shark (*Squalus acanthias*) cells with respective glucagon- and glicentin-like immunoreactivity were the only two out of 18 gastrointestinal paraneurons that seemed to be of the closed type (El-Salhy 1984). The occurrence of closed-type glucagon cells in the stomach may be a widespread phenomenon among vertebrates, since it has also been reported for amphibians (Fujita et al. 1981a), reptiles (El-Salhy and Grimelius 1981), birds (Yamada et al. 1983; Timson et al. 1979), and mammals (Van Noorden and Falkmer 1980; Falkmer et al. 1981). On the other hand, glucagon or glicentin immunoreactivities have been detected in presumably open-type intestinal cells from chondrichthyes to mammals (Langer et al. 1979; Rombout et al. 1979; El-Salhy et al. 1981; Holmgren et al. 1982; Falkmer and Van Noorden 1983; Andrew 1984; El-Salhy 1984; Rawdon 1984; Rawdon et al. 1984; Sundler and Håkanson 1984). In the extrapancreatic digestive system of mammals, glucagon immunoreactivity has been localized to the submaxillary gland (Perez-Castillo and Blázquez 1980b) and the gastrointestinal mucosa (cf. Falkmer and Van Noorden 1983). Different forms of glucagon-like material (Tominaga et al. 1981) seem to be widespread in the vertebrate CNS. Such material has been reported for the brain of cyclostomes (Falkmer et al. 1984), an elasmobranch (Conlon and Thim 1985), as well as of the clawed frog and a tortoise by Dorn et al. (1983b); and of several mammals, including the human (Conlon et al. 1979; Lorén et al. 1979; Dorn et al. 1980a; Tager et al. 1980; Dorn et al. 1983a, b; Tager 1984; Inokuchi et al. 1986). However, it was not found in the brain of the axolotl (*Ambystoma mexicanum*) although high levels were present in the gastrointestinal tract (Conlon et al. 1985b). No information on the occurrence of glucagon in the CNS of bony fishes and birds seems to be available (cf. Falkmer and Van Noorden 1983).

2.2.3 Pancreatic Polypeptide

Cells with PP-like immunoreactivity have been identified in the digestive tube of cyclostomes (Falkmer and Van Noorden 1983), chondrichthyes (El-Salhy 1984), and all other vertebrates studied so far, except some teleosts and the birds (Larsson et al. 1976b; Alumets et al. 1978a; Langer et al. 1979; Van Noorden and Falkmer 1980; El-Salhy et al. 1981; Fujita et al. 1981a; Hazelwood 1981; Colca and Hazelwood 1982; Wang et al. 1986). Furthermore, PP has been reported for the central and/or peripheral nervous system of cyclostomes (Falkmer et al. 1984), amphibians (Fujita et al. 1981a), the chicken and several mammals (Lorén et al. 1979; cf. Sundler et al. 1983; cf. MacDonald et al. 1985), and for the adenohypophysis of the dog (Fujii et al. 1982). However, it now appears that, at least in mammals, the PP-like activity found in neurons is actually due to a closely related substance, neuropeptide Y (Stjernquist et al. 1983). Furthermore, another peptide of the "PP-family", peptide YY, has been isolated from porcine gut (cf. Emson and de Quidt 1984); this adds one more pitfall to immunological studies with PP (Leduque et al. 1983), which are fraught with difficulties because of the variable structure of the hormone (see Chap. 12.2). Interestingly, PYY has been colocolized with glucagon-related peptides in gut endocrine cells of several species, and in islet A-cells of fetal rats and humans (Ali-Rachedi et al. 1984). The authors indicate a similar result in the pancreas of two species of frogs and in the gut and pancreas of a turtle species, as well as in ratfish, eel, and lamprey species. It is assumed that in the case of the lamprey the authors are referring only to gut glucagon-containing cells. These results extend the observations of Kaung and Elde (1980), El-Salhy et al. (1981), and Falkmer and Van Noorden (1983), who have reported PP immunoreactivity in glucagon-containing cells of frog islets and cyclostome gut. In a recent study on the endocrine cells of the digestive tract of a shark (*Squalus acanthias*) El-Salhy (1984) was able to localize PP and peptide YY to different gastrointestinal cell types. It appears at present impossible to arrive at a clear conclusion as to the extrapancreatic distribution of PP.

2.2.4 Somatostatin

Extrainsular somatostatin or somatostatin-like material has been identified in the digestive system and neurons of all vertebrates studied so far, from cyclostomes to mammals and birds (for literature see Vale et al. 1976; Falkmer et al. 1978, 1984; Jackson 1978; J. A. King and Millar 1979; Bethge et al. 1982; Dupé-Godet and Adjovi 1983; Reichlin 1983a; Conlon et al. 1985a, b; Bennett-Clarke and Joseph 1986; Wright 1986). In a toad (*Bufo marinus*), somatostatin has been localized both in the collecting tubules of the kidney and the urinary bladder (Bolaffi et al. 1980); there is also evidence for its presence in the mammalian kidney (cf. Reichlin 1983a). In mammals, somatostatin occurs in all divisions of the nervous system (central, sensory, sympathetic and parasympathetic, and enteric), in various epithelia of the digestive system, and in endocrine glands. The highest concentrations are found in the hypothalamus (especially the median eminence), but large amounts also occur in the gastrointestinal tract of many species, except for the teleosts (J. A. King and Millar 1979; Gerich 1981). In some CNS neurons, somatostatin coexists with GABA, another inhibitory messenger substance (Oertel et al. 1982). Furthermore, it has been found together with NE in sympathetic fibers, with an unidentified catecholamine in cultured neural crest cells; and in pheochromocytomas (Hökfelt et al. 1977; Wu et al. 1983; Maxwell et al. 1984). In the adrenal medulla, it is one of the numerous, recently discovered "co-peptides" (see Chap. 8.1.1). A fraction of the parafollicular (C-)cells of the mammalian thyroid contains somatostatin in addition to calcitonin, which occurs in all C-cells (cf. Kameda et al. 1984). Quéré et al. (1985) demonstrated somatostatin-like immunoreactivity in the anterior pituitary of the camel (*Camelus dromedarius*). Appreciable quantities of somatostatin are also present in the cerebrospinal fluid of monkey species in which there is a distinct diel rhythm with the lowest values occurring during the day (cf. Reichlin 1983b). The functions of neuronal somatostatin seem to be largely those of neurotransmission and/or neuromodulation, except for the fractions that are released into the cerebrospinal fluid and the hypophysial portal vessels. In the latter case, somatostatin is a potent inhibitor not only of growth hormone secretion, but also of thyrotropin release (cf. Reichlin 1983a). In the thyroid C-cells, an inhibitory autocrine effect (see Chap. 9.3) has been suggested, whereas in the urinary system an antagonism to antidiuretic secretions of the neurohypophysis appears likely (cf. Reichlin 1983a). Since it is probable that several forms of somatostatin with differing spectra of action exist in most vertebrates (see Chap. 13), the precise picture of the somatostatin effects at the above sites may be much more complex than is indicated in this summary.

2.3 Early Interactions in Islet Evolution: Islet Organ, Protopancreas, Liver, and Their Vascular Links

The evolution of organs can often be retraced by careful evaluation of ontogenetic and comparative data (Osche 1982). This is also the case with the islet organ (Fig. 2.2) in which the early evolutionary stages may be summarized as follows (Epple and Brinn 1980; Epple et al. 1980): (1) transformation of an open-type receptor-secretor of the intestinal mucosa into a closed-type cell; (2) formation of submucosal follicles by multiplication of closed-type cells; (3) migration of the follicles to an extramural location. Steps (2) and (3) are apparently recapitulated in the ontogeny of the lampreys (Fig. 2.3).

The transition from this "cyclostome stage" (Epple 1969) to the gnathostome situation in which islet organ and exocrine pancreas form a mixed gland has long been a mystery, for which several possibilities have been suggested (Epple and

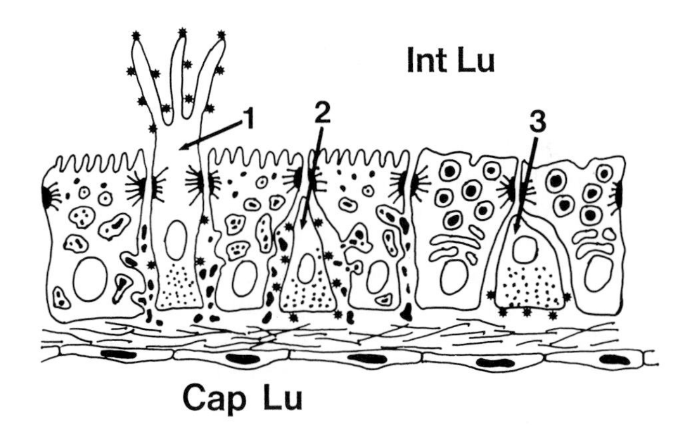

Fig. 2.2. Hypothetical stages in the evolution of "closed" endocrine cells from "open" receptor-secretors of the gastrointestinal mucosa. (1) "Open" cell (*second from left*) with receptor sites (*asterisks*) concentrated on modified microvilli, which extend into the intestinal lumen (*Int Lu*). (2) "Intermediate" receptor-secretor (*center*), located between absorptive cells. This cell responds to stimuli carried by pinocytotic vesicles of surrounding absorptive cells to the intercellular spaces below the tight junctions. (3) "Closed" cell (*second from right*) with receptors concentrated at the capillary pole. This cell responds to blood-borne stimuli from the capillary lumen (*Cap Lu*). Its location between exocrine cells (large, apical granules) indicates that it no longer requires indirect luminal stimuli from absorptive cells (Epple et al. 1980)

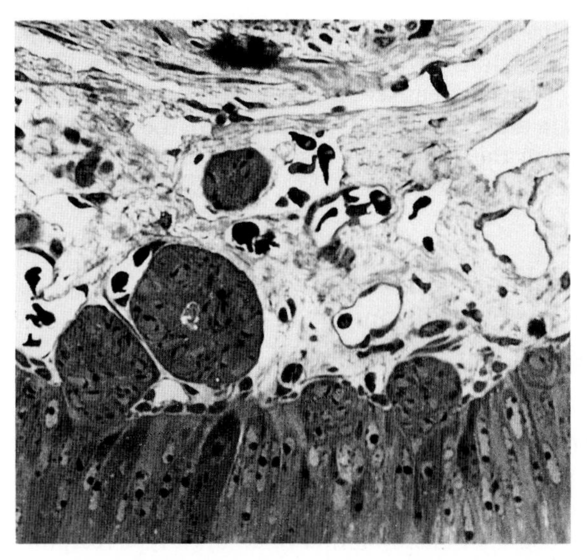

Fig. 2.3. Islet follicles in various stages of formation, budding from the epithelium of an intestinal diverticulum. Ammocoete stage of the Southern Hemisphere lamprey, *Geotria australis* (Hilliard et al. 1985)

Lewis 1973). It appears that the main obstacle in pertinent considerations has been the more or less tacit assumption that the Myxinidae and/or the adult northern lampreys (Petromyzontidae) represent a reasonably good model of the early vertebrates. This may not be so. The Myxinidae seem to be both primitive and secondarily simplified creatures, which, in addition, acquired several specializations (for literature see Hardisty 1979, 1982). The Petromyzontidae, on the other hand, may be of somewhat better use for attempts at phylogenetic reconstructions, provided one recognizes that the postmetamorphic stage has special features, including the disappearance of the bile ducs (for literature Youson 1981 a, b)! Unfortunately, this makes them poor models for studies on the pancreatic phylogeny. Even the larvae (ammocoetes) of the Petromyzontidae seem to have a simplified digestive tube in which one or two diverticula at the esophageal-intestinal junction have been reduced to small caeca (Yamada 1951; Hilliard et al. 1985). It appears that the most primitive islet-pancreas association is displayed by the ammocoetes of a southern lamprey, *Mordacia mordax.* As shown by Strahan and Maclean (1969), this species has a large diverticulum at the esophageal-intestinal junction which consists of deep folds of zymogen cells, justifying the name "protopancreas" (Fig. 2.4). We recently became aware of a further study by Maclean (1965) that thoroughly analyzes the protopancreas and its vascular relations with the intestine, islet organ, and liver. With respect to the evolution of the blood supply of the islet organ, Maclean's (1965) study is in full agreement with our conclusions (Epple and Brinn 1980). Based on Maclean's work, and our data (Hilliard et al. 1985) on the islet organ of the third type of extant lampreys (*Geotria australis*), we propose that the interrelations between intestine, exocrine pancreas, islet organ and liver evolved as follows (Fig. 2.5):

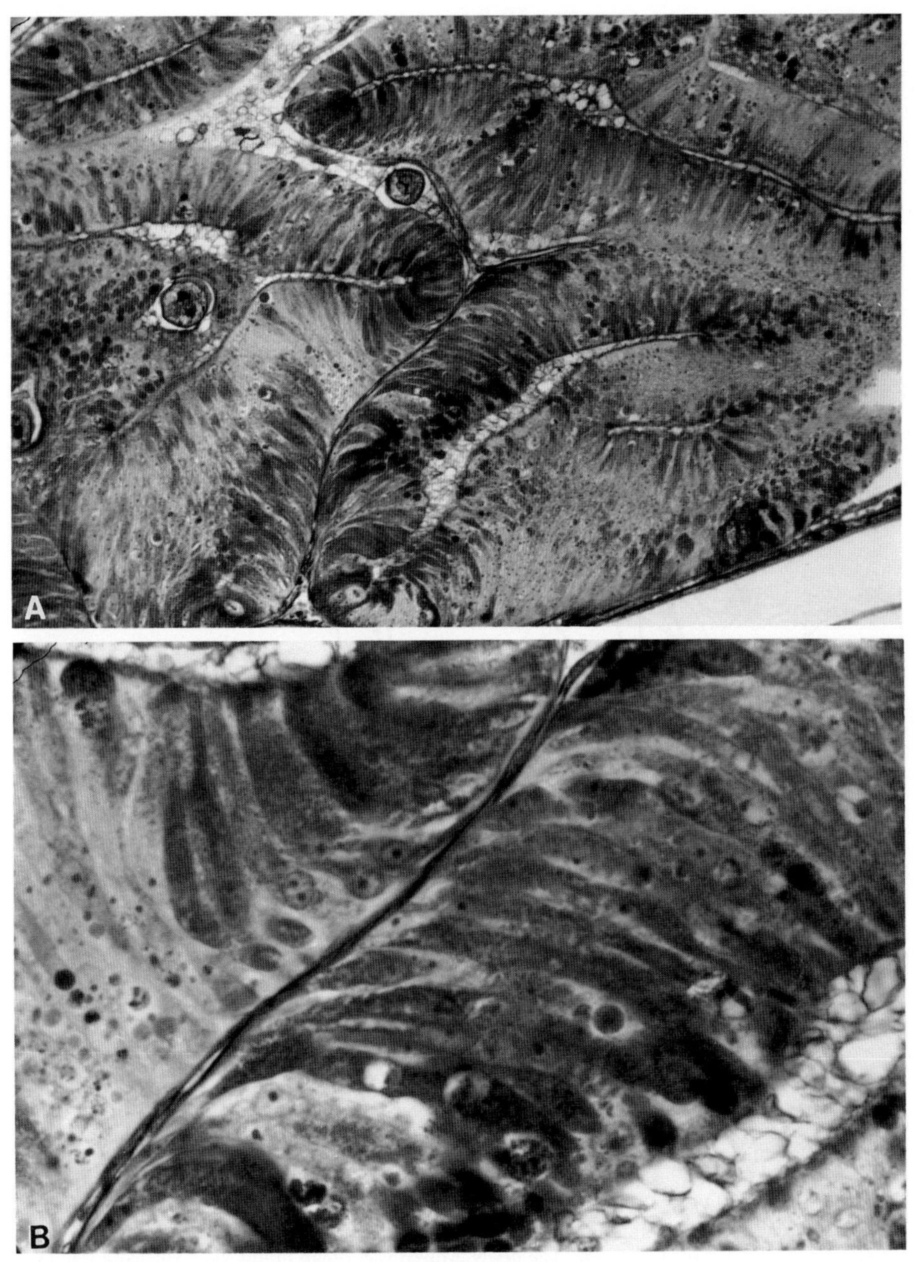

Fig. 2.4. Protopancreas of the ammocoete of *Mordacia mordax*. Aldehyde fuchsin-trichrome method of Epple (1967). *A* Cross-section at low magnification. Note the lining of acinus-like folds by high columnar epithelium; ×240. *B* Higher magnification reveals aldehyde fuchsin-positive granulation of varying density, and other cellular inclusions; ×1900

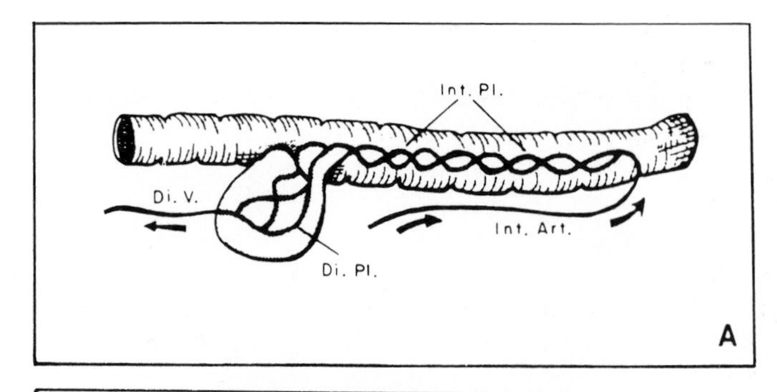

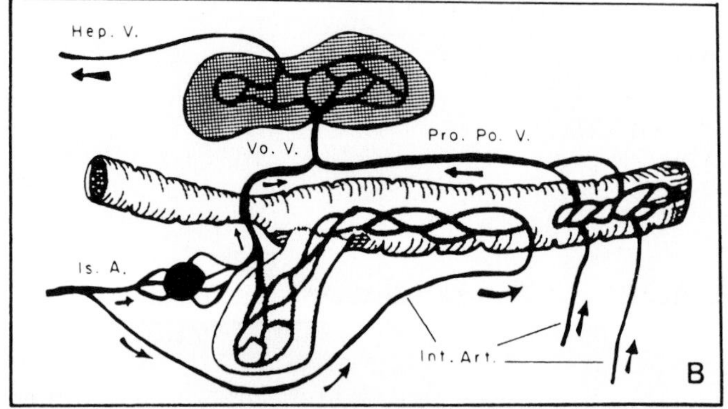

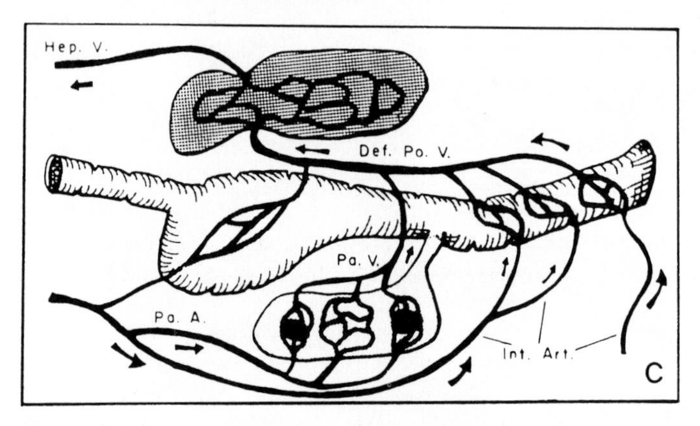

Fig. 2.5. Schematic reconstruction of phylogenetic stages of the vascular connections between islet organ, pancreas, and liver. *A Branchiostoma* stage. A secretory diverticulum receives blood from the intestinal plexus (*Int. Pl.*), which is continuous with the diverticular plexus (*Di. Pl.*). The ancestors of the islet cells reside in the intestinal mucosa (not shown). *Int. Art.* intestinal artery; *Di. V.* diverticular vein. *B* Larval *Mordacia* stage. The newly evolved islet organ drains via a vorticose vein (*Vo. V.*) into a provisional hepatic portal vein (*Pro. Po. V.*), which carries blood from the lower intestinal region. The protopancreas is still supplied by blood from the upper intestinal plexus. *Hep. V.* hepatic vein; *Is. A.* islet artery; *Int. Art.* intestinal artery. *C* Gnathostome stage. Both islets and exocrine pancreas receive arterial blood, and drain via pancreatic vein(s) (*Pa. V.*) into the definitive hepatic portal vein (*Def. Po. V.*) (of the adult lampreys and gnathostomes). The latter has evolved from the intestinal plexus and replaces the provisional portal vein, which has disappeared. No vascular connection between intestine and pancreas. For clarity, hepatic artery and biliary drainage are omitted (Epple and Brinn 1986). *Hep. V.* hepatic vein; *Pa. A.* pancreatic artery; *Int. Art.* intestinal arteries

1. At the stage of *Branchiostoma lancelatum* ("Amphioxus"), the intestine forms a secretory diverticulum at its junction with the esophagus. This structure is supplied by a vascular portal system that carries venous blood from the intestine. The diverticulum is the ancestor of the exocrine pancreas, and the portal system forms a communication link by which metabolites and/or hormones from the intestine affect the secretions of the diverticulum.

2. The next stage is seen in the larva of the lamprey genus *Mordacia*, in which the diverticulum has evolved into the deeply folded "protopancreas". In addition, both follicles of Langerhans and the liver have appeared. The upper two-thirds of the intestine is drained by a venous network that, as in *Branchiostoma lancelatum,* forms a portal system with the diverticulum (protopancreas). However, the blood then passes via a vorticose vein to a provisional hepatic portal vein, which also carries blood from the lower third of the intestine to the liver. Consequently, the blood from the upper intestine passes through a double portal system. The follicles of Langerhans (forming a compact islet organ), which are supplied by arterial blood, drain directly into the vorticose vein. Since their secretions enter the circulation *downstream* from the protopancreas, there is no direct functional link between the two organs. As in all vertebrates, the liver receives already at this stage the islet hormones in higher blood concentrations than any other organ, with the possible exception of the pancreatic "halo" regions (see Chap. 6).

3. At the gnathostome stage, a well-developed extramural exocrine pancreas enters into various anatomical associations with the islet organ (now "endocrine pancreas"). The latter can be assigned to three basic types (Fig. 2.6). Furthermore, both pancreas components, which are now supplied by arterial blood, are drained via veins that join the definitive hepatic portal vein.

From the preceding, the following picture arises: (a) exocrine and endocrine pancreas evolve independently and with independent blood supplies. Consequently, local interactions between both pancreas tissues, as postulated on account of insulo-acinar portal systems (see Chap. 6), must be phylogenetically secondary phenomena; (b) in contrast to the protopancreas, the early islet organ is supplied by arterial blood, and thus receives signals that are carried by the general (systemic) circulation; (c) at the earliest known stages of their evolution, the future endocrine pancreas and the liver are connected by an insulo-hepatic portal system that persists in all groups of vertebrates (Epple and Brinn 1980, 1986).

However, three questions arise: (1) How and "why" did the persistent relationship between the two pancreatic tissues in the gnathostomes develop? (2) Why do the protopancreas of *Mordacia* and the homologous diverticula of *Geotria* (cf. Hilliard et al. 1985) disappear during the transformation to the feeder stage? (3) Why are the diverticula (secondarily) reduced in the *larval* Petromyzontidae?

An answer to the first question is suggested by our findings in larval *Geotria* (Hilliard et al. 1985) in which one of two diverticula at the esophageal-intestinal junction "carries" the common bile duct at its tip. Obviously, it is not difficult to imagine that in the ancestor of the gnathostomes the developing diverticulum carries instead the islet tissue away, since exocrine pancreas, liver, and islet organ are formed in close local association at the esophageal-intestinal junction of the

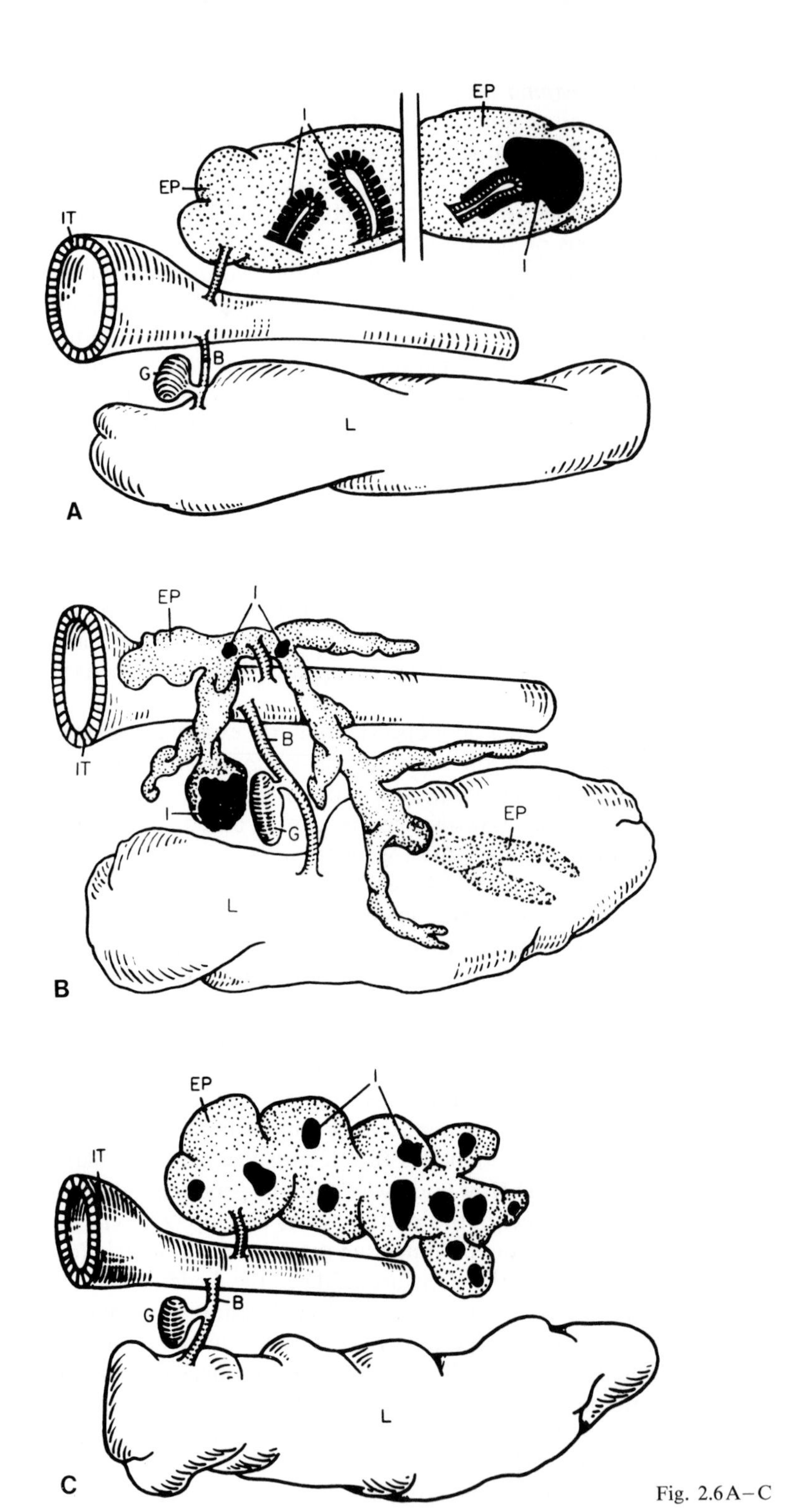

Fig. 2.6 A–C

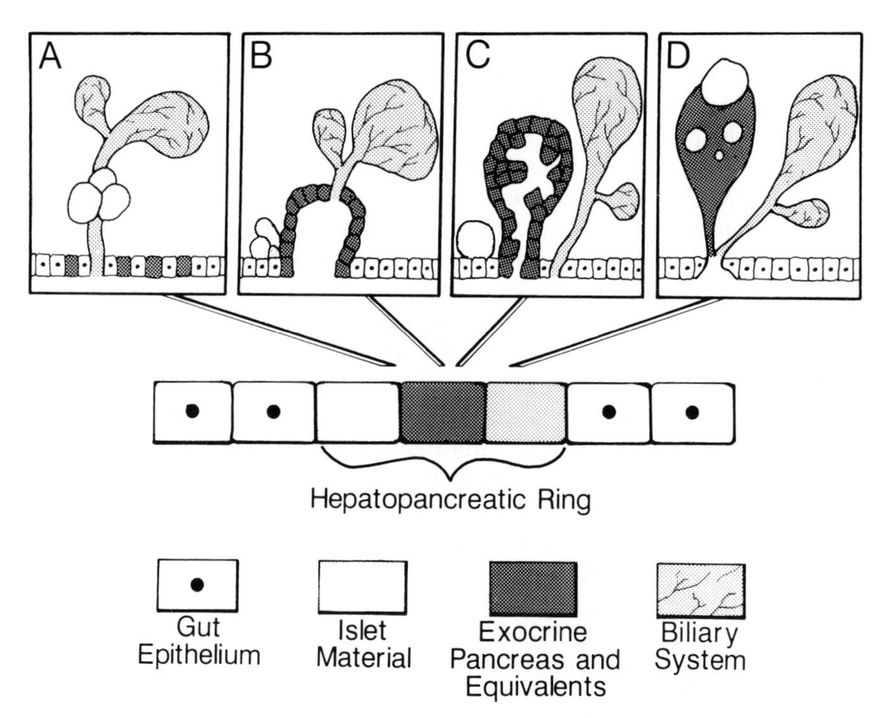

Fig. 2.7. Topographic variations of the derivatives of the hepatopancreatic ring (for literature see Hilliard et al. 1985; Epple and Brinn 1986). *A* Hagfish, and caudal islet of northern lampreys. Islet follicles associated with bile duct; exocrine pancreas equivalents dispersed in gut epithelium. *B* Similar to larval *Geotria.* Islets form from gut (esophagus) and diverticulum (presumptive exocrine pancreas precursor). Biliary system associated with diverticulum. *C* Older (?) larval *Mordacia.* The three descendents of the hepatopancreatic ring are not associated. Note that the diverticulum is replaced by the protopancreas. *D* Basic gnathostome situation. Exocrine pancreas and islets develop separately from liver. However, note that in some forms (particularly actinopterygians) there is an obviously secondary mingling of pancreatic and hepatic tissues that is easy to understand for phylogenetic and ontogenetic reasons

◀

Fig. 2.6. Three basic types of the gnathostome pancreas (Epple 1969). *A* Chondrichthyan pancreas. It seems that all forms possess a well-developed, compact extramural pancreas. *Left*: in many elasmobranchs the majority of islet cells (*dark*) surround small ductules. Possibly these ductules are the "prototype" of the ductules that give rise to islet cells under hyperplastic and tumorous conditions in the mammalian pancreas. *B* Actinopterygian pancreas. In all major groups, though not all taxa, the pancreas is rather diffuse and often mingles with liver tissue. In many taxa, the islet tissue occurs as one or two major accumulations (principal islets), plus a number of small "accessory" islets. *C* Tetrapod pancreas. Although basically of a compact type, the pancreas of many forms has long, fine extensions covering the mesenteries (not shown here). The islets vary greatly with the different groups; in mammals, they tend to be more homogeneous in size and distribution than in other groups. Usually, they are more frequent and larger in the pancreas tail.
L liver; *EP* exocrine pancreas; *I* islet tissue; *G* gall bladder; *B* bile duct; *IT* intestinal tract

lamprey. We propose that minor developmental variations in this region, which corresponds to the hepatopancreatic ring (Ferner 1952) of the higher vertebrates (Fig. 2.7), easily explain the morphological differences shown in Fig. 2.6.

The advantages of the association of the islet organ with the extramural pancreas have been pointed out previously (Epple and Brinn 1975): (1) The extramural location allows an increase of the islet tissue volume without interference with intramural structures (e.g., digestive glands). (2) Because the exocrine pancreas uses the portal vein and its tributaries as a trellis, the associated islets are guaranteed rapid, direct delivery of their secretions to the liver.

During metamorphosis, the disappearance or reduction of the protopancreas of *Mordacia* and of the homologous diverticula of *Geotria* occurs simultaneously with, and most likely for the same reason as the loss of the exocrine liver function: the switch from larval food, algae and similar organisms, to the easily digestible diet of the adult lamprey, blood and tissue remnants, allows a reduction of the exocrine pancreas secretion, and a complete suspension of the exocrine liver functions. As for the probable secondary absence of diverticula (or protopancreas) in the larval Petromyzontidae (Yamada 1951; Maclean 1965), no obvious explanation seems available. Perhaps unknown dietary changes are involved. Admittedly, this discussion avoids the hagfishes. However, they seem to have left the vertebrate "main line" long before the lampreys (cf. Hardisty 1982) and are thus of little relevance for the considerations in Sect. 2.3.

Chapter 3

Ontogeny of the Gnathostome Pancreas Tissues

Whereas the islet organ of the cyclostomes develops directly from the epithelium of the intestine or bile duct, the ontogenies of both components of the pancreas of gnathostomes are closely linked. Therefore, we consider here the embryonic development of the islet organ, now the "endocrine pancreas", together with that of the exocrine pancreas. The older literature on this topic has been sumerized by Bargmann (1939); more recent reviews have been written by Falkmer and Patent (1972), Pictet and Rutter (1972), Epple et al. (1980), Rutter (1980), and Epple and Brinn (1986). It is generally held that the gnathostome pancreas develops from dorsal and ventral outgrowths (anlagen) of the small intestine. According to Siwe (1926), the ventral material disappears in the elasmobranchs very early, leaving the dorsal anlage to form all of the adult pancreas. In the "higher" gnathostomes there seems to be a basic pattern, consisting of one dorsal and two ventral anlagen. Whereas in all cases the dorsal anlage probably persists, the ventral anlagen may (a) fuse, (b) totally disappear or (c) develop only unilaterally as, for example, in the human. In mammals, the derivatives of the dorsal and ventral anlagen fuse almost indistinguishably, although the duct system and differences in cell populations of the adult pancreas reflect the respective origins (see Chap. 4.3). The chicken pancreas develops from three anlagen, a dorsal one appearing at 72 h of incubation and two ventral ones at 96 h (Przybylski 1967). The dorsal anlage gives rise to the third and splenic lobes while the dorsal and ventral lobes are derived from the two ventral anlagen (Dieterlen-Lièvre 1970). Two anlagen give rise to the definitive pancreas in the urodele amphibian, *Ambystoma opacum*, but the primordia do not completely fuse as in mammals; it is only at their overlapping ends that they form an isthmus. Consequently, the two pancreatic ducts of this species retain their independence rather than fusing to form main and accessory ducts as in mammals (Frye 1958). In another salamander, *Eurycea bislineata*, there are two ventral anlagen that appear slightly before the dorsal bud. Both make contact without fusing; however, the dorsal and right ventral anlagen do fuse (Frye 1962). Although little is known of the pancreatic development in other "lower" species (Bargmann 1939), the above examples suffice to show that there is no consistent relation between the embryonic anlagen and the structure of the adult gnathostome pancreas. The lobation of the latter seems to depend on the growth pattern of the anlagen, and their local interactions with the developing organs of the abdominal cavity, particularly liver (see Chap. 4.2) and adipose tissue (see e.g., Baron 1934).

The sequence of appearance of the islet cell types during the embryonic development has been approached in the ultrastructural studies of Like and Orci (1972), Pictet and Rutter (1972) and Pictet et al. (1972). It had long been thought that B-cells were the first endocrine cells in the pancreatic primordia (Bencosme

1955), but the above-cited studies suggested that A-cells are the first to differentiate (9 weeks, human; 12 days, rat), followed by D-cells (10 weeks, human; 15 days, rat) and finally B-cells (10.5 weeks, human; 17 days, rat). Some caution is necessary, however, in the ultrastructural identification of fetal islet cells. Rhoten and Hall (1982) have pointed out that reptilian islet secretory granules undergo remodeling and attain their adult form at different times during development. Therefore, because of its sensitivity and larger sampling area, immunocytochemistry may be a more accurate method of identifying developing islet cells. Not surprisingly then, recent studies indicate that insulin-, glucagon- and somatostatin-immunoreactive cells appear simultaneously in the human pancreas as early as at 8 weeks gestation (P. M. Dubois et al. 1975; Chayvialle et al. 1980). At the same time, glicentin is also detectable (Leduque et al. 1982), and it appears that this substance precedes glucagon in the fetal A-cells (Stefan et al. 1982b). PP-reactive cells were found at 10 weeks gestation (Paulin and Dubois 1978; Chayvialle et al. 1980), the time at which Like and Orci (1972) first found A-cells by EM criteria. The very conspicuous islets of fetal ruminants consist largely, though not exclusively, of a peculiar type of insulin cells. At least in fetal lambs these cells are argyrophilic with the Grimelius technique. The predominance of these cells may be connected with the specialized metabolism of fetal ruminants, since they disappear quickly after birth when "regular" islets are formed (cf. Falkmer et al. 1984). The argyrophilic insulin cell is probably not restricted to fetal ruminants. In smaller numbers, it may occur in the pancreas of other species since Ferner (1952) showed that in the embryonic human pancreas argyrophilic cells give rise to B-cells. Possibly, it is also identical with the stem cell of argyrophilic insulinomas (Creutzfeldt 1985), one type of amphiphil seen in the shark *Scyliorhinus canicula* (Epple 1967), and the argyrophilic islet cell of adult lampreys (see Chap. 4.1).

In the lizard, *Anolis carolinensis*, Rhoten and Hall (1982) observed the simultaneous appearance of all four islet cell types at 4 days post-oviposition, others have shown that D-cells occur in the chick pancreas along with A- and B-cells at approximately 90 h of incubation (Andrew 1977; Macerollo 1977). On the other hand, Kaung (1981), also using immunocytochemistry, reported that B-cells are the first to appear in islets of *Rana pipiens*, followed by A-, F-, and finally D-cells. In the embryos of an ovoviviparous shark (*Squalus acanthias*) which were about half of the size of these animals at birth, El-Salhy (1984) found three out of the five islet cells present in the adult (see Chap. 4.2), i. e., A-, B- and D-cells, PP- and GIP-cells being absent at this stage. Clearly, the available data on the ontogeny of the individual islet cells are insufficient to identify possible trends or rules.

An open question in islet ontogeny concerns the mechanisms that cause the great phylogenetic variability in the associations between exocrine and endocrine pancreas. We have pointed out that it is likely that, in the ancestors of the gnathostomes, the protopancreas (or its diverticular equivalent) carried the early compact islet organ away from the intestine (see Chap. 2.3). As shown by Siwe (1926) (see also Figs. 4 to 7 in Bargmann 1939), this situation seems to be reflected in the first stages of development of the pancreas of many gnathostomes. According to Siwe (1926), the growing dorsal anlage often carries a large, compact accumulation of islet tissue away from the intestine; the degree of subsequent mingling of the two types of pancreatic tissues determines the extent to which islet

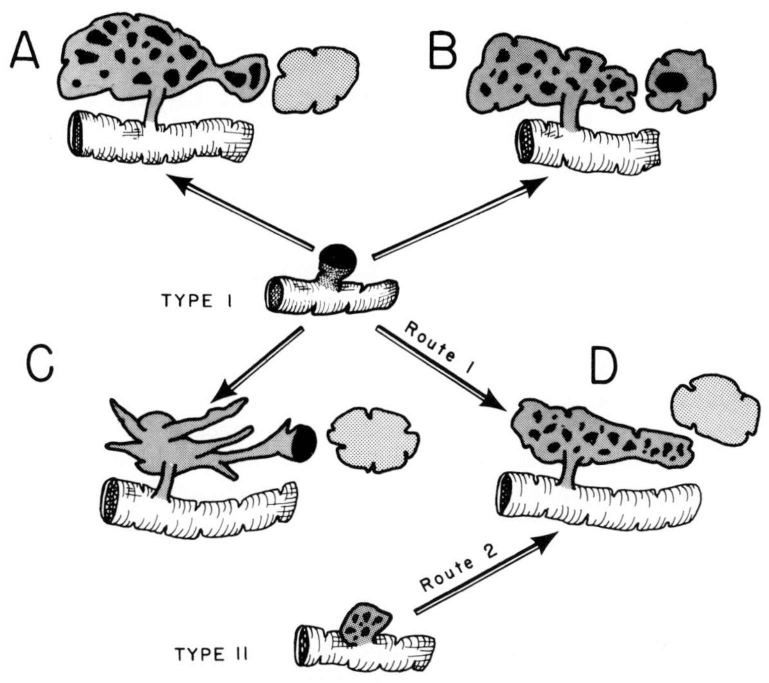

Fig. 3.1. Development of dorsal anlage-derived pancreas tissues in different vertebrates. *Type-I* anlage (with compact endocrine material) gives rise to the third and splenic lobe of the galliform pancreas (*A*), large pancreas regions and the intrasplenic islet of certain *Squamata* (*B*), large pancreas regions, and usually (?) the Brockmann bodies of the Actinopterygii (*C*), and the total pancreas or pancreas tail regions (Muridae) of certain mammals (*D*). However, the configuration shown in *D* may also develop from the *Type-II* anlage with early diffuse islet tissue formation (e.g., guinea pig, human, roe, deer, cow) (Epple and Brinn 1986).
Exocrine pancreas: *shaded* (left); islet tissue: *black*; spleen: *stippled* (right)

tissue splits into smaller aggregations, i.e., islets (see Fig. 3.1). It must be added that the early large islet mass is absent in some species, and that smaller islets also develop independently and probably in all three anlagen, if present. Thus, we suggest that, as in the case of the exocrine pancreas, the structural variations of the gnathostome islet organ are in most cases the result of growth mechanics. However, the small size of the mammalian islets may be related to some functional advantages (see Chap. 6).

Chapter 4

Comparative Cytology of the Islet Organ

As shown in Table 1.1, it is generally held that the cyclostomes have at least two types of islet cells, while the gnathostomes have at least four. However, additional and obviously independent cell types have been found in some species, particularly among the elasmobranchs. Thus, we must expect in these species further, perhaps totally new, islet hormones. On the other hand, it is difficult, if not impossible, to decide if islet cells, originally present, have disappeared secondarily. This question arises with the new cell types in *Petromyzon marinus* (Brinn and Epple 1976), and with the large number of islet cells in the elasmobranchs (see below). Currently, perhaps the most critical issue is the F-cell, the identification of which depends strongly on the particular batch of anti-PP antiserum used (see e.g., Sect. 4.2, Reptilia). The importance of careful evaluation of immunocytochemical data, which has been emphasized many times (see e.g., Falkmer and Van Noorden 1983; and below Sect. 4.2), need not be discussed here in detail. Possible pitfalls are illustrated by (1) staining of neuropeptide Y neurons with anti-PP (Stjernquist et al. 1983), (2) unspecific binding of immunoglobulins to somatostatin and glucagon cells through a non-antigen-antibody mechanism, mediated by the C_{1q} fraction of complement (Buffa et al. 1979; see also Grube and Aebert 1981), and (3) interference by intracellular amines with the reaction between antibodies and peptide hormones (Polak and Buchan 1980). An equally critical issue is the possible existence of intermediate exocrine-endocrine cells (Gerlovin 1976; Yaglov 1976; Eusebi et al. 1981). If such elements should be shown to exist indeed, they could have considerable implications for diabetes research. After all, the suspected transformation from exocrine to endocrine pancreas cells led long ago to Laguesse's "balancement hypothesis" (see Bargmann 1939), the revival of which should spur attempts to induce neoformation of B-cell from acinar tissue. However, the view that at least in normal pancreas – the mixed exocrine-endocrine cells are artifacts of poor fixation (Brinn 1973) is reiterated here. The structural lability of anoxic islet cells and the slow penetration by aldehydes, particularly poor cross-linkers such as formaldehyde, yield spurious and often unacceptable results in EM studies. The absence of intermediate cells has also been emphasized by others in work on well-fixed pancreas of humans (Like and Orci 1972; Laito et al. 1974), rabbit (Bencosme 1955), and amphibian species (Frye 1958). Forssmann (1976) compared immersion and perfusion fixation of horse islets, and noted that intermediate cells were present in the former and absent in the latter. Another source of confusion may arise with the presence of macrophages. Such cells, containing varying amounts of endocrine and exocrine granules, were seen in the islets of older sea bass (*Dicentrarchus labrax*) by Carillo et al. (1986). Finally, it cannot be excluded that neoplastic islet cells sometimes fuse with neighboring exocrine or endocrine cells (cf. Bosman et al. 1985).

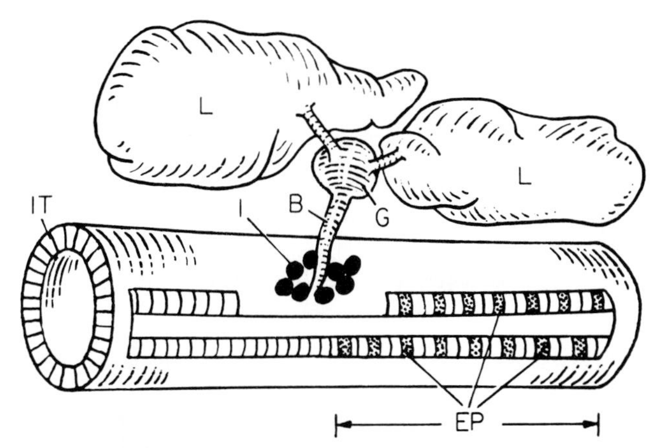

Fig. 4.1. Islet follicles (*I*) surround the lower bile duct (*B*) in Myxinidae. All islet tissue is concentrated in this region and totally separated from the presumptive equivalents of the exocrine pancreas (*EP*), which occur in the intestinal mucosa (Epple 1969). *L* liver; *G* gall bladder; *IT* intestine

4.1 Agnatha

The extant cyclostomes are the only survivors of the "agnathan" stage of vertebrate evolution (Hardisty 1979, 1982). The islet organ of the Myxinidae, which lack a larval stage, has many B-cells and a varying number of somatostatin cells throughout the life cycle. In all species studied, they surround the lower bile duct as a single compact mass of endocrine follicles or cords (Fig. 4.1), though isolated islet cells may occur in the epithelium of the bile duct (Östberg et al. 1975, 1976; cf. Hardisty and Baker 1982). An interesting difference between species of *Myxine* and *Eptatretus* is the scarcity of somatostatin cells and the high incidence of colloid cysts in the follicles of the former (cf. Winbladh 1976a, b) which may reflect differences in metabolism. The extant lampreys (holarctic Petromyzontidae, with several genera; and two different forms from the southern hemisphere, *Geotria australis* and *Mordacia* spec.; see Hardisty 1979; Potter 1980), all have an ammocoete stage, and in both the northern lampreys and *Geotria*, the only specific islet cell during early life is the B-cell (cf. Hardisty and Baker 1982; Hilliard et al. 1985). *Mordacia* has not yet been studied in detail. However, at the beginning of the metamorphosis, there appear numerous somatostatin cells (Elliott and Youson 1986) with staining properties that differ from those of D-cells (Brinn and Epple 1976). In addition, a Grimelius-positive silver cell was detected in *Petromyzon marinus* (Van Noorden, Brinn and Epple, unpublished) and has been confirmed for postlarval *Geotria australis* (Hilliard et al. 1985). Although it is tempting to interpret the silver cell as A-cell (Epple and Brinn 1980), further clarification is needed since the anti-glucagon antisera so far applied react only with gut cells of the lamprey (Falkmer and Van Noorden 1983). On the other hand, the identity of the lamprey silver cells with the argyrophil B-cells (Falkmer et al. 1984) also cannot now be excluded. Additional ultrastruc-

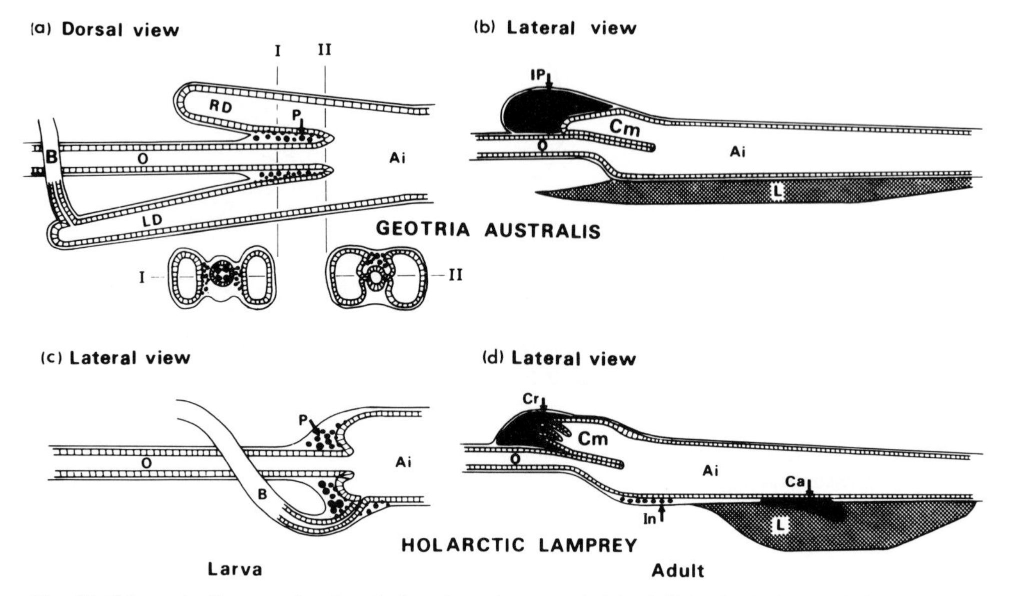

Fig. 4.2. Schematic diagrams showing the location of pancreatic islet follicles in the larvae (*left*) and adults (*right*) of *Geotria australis* (*top*) and a holarctic lamprey (*bottom*). Location of follicles are depicted in *black*. N. B. For clarity, the blood vessels, sinuses and typhlosole and the liver and gall bladder of the larvae are not illustrated (Hilliard et al. 1985)

turally distinct cell types have been found in the final phase of the life cycle of *Petromyzon marinus* (Brinn and Epple 1976). The existence of these cells in fasting, prespawning *Petromyzon marinus* is difficult to explain. Considering that the parasitic, adult lamprey has "simplified" its digestive system, we have suggested that these cells may merely reflect a phylogenetically earlier and more complex islet organ (Epple et al. 1980). The structure of the islet organ of adult lampreys differs between the Petromyzontidae and the two southern forms. In the former, there are three aggregations of islet tissue (Fig. 4.2). In the two southern genera, there is only one compact accumulation near the esophago-intestinal junction (cf. Barrington 1972; Hardisty and Baker 1982; Hilliard et al. 1985).

In conclusion, only insulin and somatostatin have been localized to the cyclostome islets, but the detection of one or more additional islet hormone(s) in adult lampreys can be expected.

4.2 Gnathostomes

Chondrichthyes. The cartilagenous fishes are divided into two groups, the elasmobranchs (sharks, skates and rays) and the holocephalians (ratfishes). They are the phylogenetically oldest group of living vertebrates with a combined exocrine-endocrine pancreas (Fig. 2.7A). Their islet cells, which have a close relationship with ducts of the exocrine components of the gland, frequently form an outer layer around ductules. However, islet formation varies in different species (Thomas 1940), and it is particularly pronounced in the holocephalians (Fujita 1962; Patent and Epple 1967). The islet organs of the selachians often resemble the early stages of development of mammalian islets (cf. Bargmann 1939). On the other hand, selachians possess an unusually large number of tinctorially different islet cells (Epple 1967), which is matched by their ultrastructural heterogeneity (Kobayashi and Syed Ali 1981). Sekine and Yui (1981), who applied four antisera in a ray (*Dasyatis akajei*), found, in addition to the expected islet cell types (A-, B-, D- and F-cells), a population of cells with immunoreactivity to both somatostatin and PP. In a species of shark (*Squalus acanthias*), El-Salhy (1984) identified immunocytologically five different islet cells with antisera to glucagon, insulin, somatostatin, PP, and GIP. The holocephalians have four major islet cells (A-, B-, D- and X-cells), in addition to smaller numbers of pancreatic PP, GIP and enkephalin cells (Falkmer et al. 1984). Glucagon, insulin and somatostatin have been shown in the A-, B-, and D-cells, respectively. The X-cells, which comprise about 50% of all islet cells, represent a puzzle. While the ratfish A-cells react with both C- and N-terminal anti-glucagon, the X-cells react only with N-terminal anti-glucagon sera. However, after proteolytic digestion of the sections, they become reactive with C-terminal anti-glucagon (see also Chaps. 11.1 and 11.2). From these findings, plus the lack of a glicentin reaction, Stefan et al. (1981) conclude that the X-cells are similar, but not completely identical with the L-cells of the mammalian digestive mucosa, which contain glicentin and other glucagon-related substances. The striking cytological difference between the islet organs of the elasmobranchs and the holocephalians is difficult to explain. Falkmer et al. (1981, 1984) suggest that the holocephalian islet organ is more primitive, which, however, is unlikely (see Sect. 4.3).

Latimeria chalumnae. The islet-pancreas system of this "living fossil" is surprisingly similar to that of the selachians. Unfortunately, no detailed analysis of the islet cell types has been reported so far (Epple and Brinn 1975).

Actinopterygii (Including Polypteridae). In this group there is a strong tendency to form large accumulations of islet tissue, called principal islets. Together with a more or less complete rim of exocrine tissue, or invading strands of acini, the principal islets form "Brockmann bodies" (Fig. 2.7B). The exocrine tissue varies from compact to "diffuse" (cf. Epple 1969; Epple and Brinn 1975). Often, it invades the liver, but islets are rarely found in this location. One exception to this rule are the Polypteridae, in which the liver contains the bulk of their islet tissue in the form of a large intrahepatic Brockmann body (Epple and Brinn 1975; Mazzi 1976). On the other hand, almost all exocrine tissue of *Lepisosteus osseus* is scattered within the liver, except for a small amount that surrounds the aggregated islet tissue along the common bile duct (Epple and Brinn 1975).

There has been little new information on islet cells in the primitive groups of bony fishes since the study of Epple and Brinn (1975). Ultrastructurally and by tinctorial methods, the islets of polypterids (*Calamoichthys calabaricus*) and holosteans (*Amia calva* and *Lepisosteous osseus*) contain five cell types, but there is no information on those of the chondrosteans. PP immunoreactivity in *Lepisosteous osseus* islets (Van Noorden and Patent 1978) and histological data (Brinn and Epple, unpubl.) suggest that four of the cell types of the polypterids and holosteans may correspond to the four main types of other gnathostomes.

In teleosts, the first electron-microscopic definitions of four granule-containing islet cells were given by Brinn (1973, 1975) in *Ictalurus catus*, *I. punctatus*, and *I. nebulosus*, and Thomas (1975) in *Limanda limanda*. A precise identification of three of these cells (A-, B-, and D-cells) by immunocytochemistry was reported by Johnson et al. (1976). Klein and Van Noorden (1978), using correlative light-microscope immunocytochemistry and electron microscopy confirmed that somatostatin-positive cells are indeed D-cells. Van Noorden and Patent (1978) subsequently demonstrated PP immunoreactivity in the islets of pike (*Esox lucius*), swordtail (*Xiphophorus helleri*), eel (*Anguilla anguilla*), catfish (*Ictalurus nebulosus*), and *Gillichthys mirabilis*, suggesting that the fourth cell type detected by electron microscopy is the F-cell. Recently, this was confirmed by Abad et al. (1986) for the sea bream, *Sparus auratus*. Combining microscopic immunocytochemistry and electron microscopy, Klein and Van Noorden (1980) corrected an earlier report of two populations of A-cells in islets of *X. helleri* by showing glucagon and PP immunoreactivity in the two cells. A PP-like immunoreaction was also detected in the Brockmann bodies of anglerfish, *Lophius americanus*; it seems to be due to a newly discovered member of the PP family, termed aPY (Noe et al. 1986). Islet cells with PP-like material were also described for catfish, *Ictalurus punctatus* (Johnson et al. 1982), the cyprinid, *Barbus conchonius* (Rombout and Taverne-Thiele 1982), and the sea bass, *Dicentrarchus labrax* (Carrillo et al. 1986). In addition to typical immunoreactivity in the four main cell types, the Brockmann body of the rainbow trout (*Salmo gairdneri*) shows a co-localization of glucagon and GIP in A-cells (Wagner and McKeown 1981). These authors noted that the GIP reaction could be blocked with glucagon; however, when used in radioimmunoassays the same antiserum was not blocked by glucagon. Therefore, the significance of this finding is uncertain. Langer et al. (1979) were able to detect PP immunoreactivity unequivocally only in the Brockmann body of one of the three teleosts they examined. It was present in *Pelmatochromis pulcher*, but absent in *Helostoma temmincki* and *Idus idus*. Considering the results with the snake *Thamnophis sirtalis* (see below), one is inclined to explain the negative data in this study with technical problems.

The usefulness of species-specific antisera in immunohistological studies is underlined by findings of Wang et al. (1986) in a Pacific salmon (*Oncorhynchus kisutch*). In this species, A-, B-, D- and F-cells were identified in both the Brockmann body and smaller islets. However, species-specific antisera against salmon insulin and salmon somatostatin (S-25) were more effective than antisera against the corresponding mammalian hormones. An interesting difference in the distribution of F-cells was noted by Stefan and Falkmer (1980) in the daddy sculpin (*Cottus scorpius*), a species which has two Brockmann bodies, one adjacent to the spleen and the other near the pylorus. The former contains A-, B-, and D-cells, but no F-cells, whereas the latter contains A-, B-, D-, and F-cells. The A-cells are less numerous in the juxtapyloric islet. Similar observations in the mammalian pancreas have been related to the embryonic diverticula from which the islets are derived (see Sect. 4.3).

In the actinopterygians there emerges a general pattern of islet architecture in which B- and D-cells usually occupy and mingle in the central region of the islets with A-cells being more and F-cells most peripherally located. This pattern holds for many "higher" vertebrates. Falkmer and Östberg (1977) have summarized the relative percentages of cell types in the fish islets, with A-cells being 25–40%, B-cells 30–50%, and D-cells 15–20%. F-cells were not generally recognized by immunocytochemistry at the time of that review. In recently performed point counting on islets of *Ictalurus punctatus*, we found general agreement with percentages of Falkmer and Östberg except for D-cells, for which we found somatostatin-positivity to comprise approximately 38% of the endocrine-cell volume (McNeill et al. 1984). No quantifications have been reported for F-cells in bony fishes.

Dipnoi. The extant lungfishes belong to two very distantly related families. The four African species (genus *Protopterus*) and the South American *Lepidosiren paradoxa* form the Lepidosirenidae, while the Australian *Neoceratodus forsteri* is the only living representative of the Ceratodontidae. The pancreas of all lungfishes is enveloped by folds of the gastrointestinal tract, a peculiar situation that has been thoroughly analyzed in *N. forsteri* by Rafn and Wingstrand (1981). The islet organ differs morphologically greatly between both families. In *Protopterus* ssp., it contains a Brockmann body-like islet, while in *N. forsteri* it resembles the mammalian type with smaller scattered islets. Thus, in the lungfish pancreas we find features of both actinopterygians and tetrapods. Because no recent reports have updated the tinctorial and ultrastructural descriptions A-, B- and D-cells in *Protopterus annectens* (Brinn 1973; Epple and Brinn 1975), the precise number of islet cells is unknown.

Amphibia. The pancreas of all amphibians seems to be basically of the compact tetrapode type (Fig. 2.7D), although considerable variations occur (Penhos and Ramey 1973; Epple and Brinn 1975). Information about urodelan and apodan islets is at best sketchy. By ultrastructural criteria Epple and Brinn (1976) reported four islet cell types in the urodele, *Necturus maculosus.* The secretory granules of the A-, B-, and D-cells in this species very closely resemble those described in the apodan, *Ichthyophis kohtaoensis* (U. N. Welsch and Storch 1972). A further combined immunocytochemical-electron microscope study of the islets of the bullfrog, *Rana catesbeiana* (Tomita and Pollock 1981) showed A-, B-, D-, and F-cells with granules almost identical to those in *Necturus maculosa.* In the bullfrog, the hepatic process of the pancreas has small islets containing numerous A-, B-, and F-cells, and a few D-cells, whereas the duodenal process has larger islets containing predominantly B-cells, scattered A- and D-cells, and very few F-cells. In the same species, Fujita et al. (1981a) find secretin cells as a regular component of the islet organ. El-Salhy et al. (1982) report the four main cell-types in two other anurans, *Bufo orientalis* and *Rana temporaria*, but find no regional differences in cell type distribution. The anurans have only 7–9% of the islet volume as D-cells, and approximately 30% each of A-cells, B-cells, and F-cells. Kaung and Elde (1980) and El-Salhy et al. (1982) describe the co-localization of glucagon and PP in a small group of cells. One wonders if these dually staining cells correspond to the amphiphils identified by Epple (1966a, b, 1967). Frye (1964) observed metamorphic changes in *Rana clamitans* islets, including decreased staining of B-cells with pseudoisocyanine. Kaung (1983), who studied changes in the B-cell population of *Rana pipiens* by immunocytochemical and morphometric methods during metamorphosis, also reports decreased staining intensity as well as a decrease in the B-cell population, suggesting that some degeneration of B-cells occurs in metamorphosis.

The data summarized here on amphibians show the necessity for more studies with modern techniques. The great variability of the islet organ in taxa with species of differing life styles (see e. g., Epple 1966b) suggests many surprises in future research.

Reptilia. Although the reptiles occupy a key-phylogenetic position, the islets of the four extant orders (Chelonia: turtles, tortoises; Crocodilia: crocodiles, alligators; Squamata: lizards, snakes; Rhynchocephalia, with one living species: *Sphenodon punctatus*) are poorly known. The older literature has been reviewed by Gabe (1970), Miller and Lagios (1970), Falkmer and Patent (1972), Falkmer and Östberg (1977), Epple et al. (1980), and Falkmer and Van Noorden (1983), whereas the most recent data are given by Gapp and Polak (1983), Buchan (1984), Rhoten (1984), and Agulleiro et al. (1985).

In turtles, crocodilians, *Sphenodon punctatus*, and some Squamata, the islets seem to be scattered over the entire pancreas, which is basically of the compact type (Penhos and Ramey 1973). The islets reported in turtles are mostly small, with a trend towards larger islets in caudal pancreas region of *Pseudemys scripta* (Gapp and Polak 1983; Agulleiro et al. 1985). The islet tissue of the Crocodilia is dispersed, and some larger islets occur. In many Squamata, there is a trend to form large islet accumulations in the pancreas region near the spleen, and in some snakes (Elaphidae and Crotalidae) and in a lizard (*Varanus exanthematicus*) there are even intrasplenic islets (Hellerström and Asplund 1966; Buchan 1984; Dupé-Godet and Adjovi 1983). Probably the latter represent islet tissue that was carried on the tip of the dorsal pancreas anlage (see Chap. 2.3), before it was "trapped" by the developing spleen. In the turtles, crocodilians, and lizards so far studied immunocytochemically, A-, B-, D- and F-cells have been identified with the respective antisera. However, in a total of 11 species of snakes, PP-immunoreactivity was found only in the islets of two Boidae, but not in the species of Colubridae, Elaphidae, and Crotalidae investigated (Buchan 1984). In the turtle, *Pseudemys scripta*, F-cells are absent in the "splenic lobe", but present in the pancreas head (Gapp and Polak 1983; Agulleiro et al. 1985). On the other hand, the distribution of F-cells in the islets of a crocodilian (*Alligator mississippiensis*) showed no regional variation (Buchan et al. 1982). Interestingly, the alligator F-cells reacted with anti-avian PP but not anti-bovine or anti-human PP. As the authors suggest, this could reflect the relatively close phylogenetic relationship between crocodilians and birds, since Lance et al. (1984) found that PP from the alligator is similar in sequence to that of the chicken. F-cells were also detected with anti-bovine PP in a lizard (*Anolis carolinensis*), on which they are scattered as single cells in the exocrine tissue but rare in the islets (Rhoten and Smith 1978;

Rhoten and Hall 1981). F-cells reacting to anti-bovine PP, as well as species differences in distribution of cell types, have been reported in three additional species of lizards by El-Salhy and Grimelius (1981) and El-Salhy et al. (1983). In *Chalcides ocellatus* the A-, B-, and D-cells constitute a greater volume in the dorsal lobe, while that of F-cells is 20 times greater in the ventral lobe (El-Salhy et al. 1983). On the other hand, *Mabuya quinquetaeniata* also has more F-cells in its ventral lobe but shows equal distribution of the other three cell types (El-Salhy and Grimelius 1981). The failure of Buchan (1984) to identify F-cells in most of the snake species in her material is possibly due to technical reasons, or the batch of antibody used. Although this author failed to detect PP-immunoreactivity with antisera raised to whole molecules of avian or bovine PP, she obtained a positive reaction in the boid snakes with a C-terminal anti-bovine PP. Rhoten (1984), on the other hand, obtained PP-staining with anti-bovine PP in the pancreas of *Thamnophis sirtalis*, a species that gave a negative result in Buchan's study. Rhoten (1984) also notes that the F-cells are the least numerous cell type in the islets of this species, with A- and B-cells dominating, and D-cells amounting to about 15%. Particularly intriguing is the transient occurrence of pancreatic GIP and gastrin cells during the hibernation period in the lizard, *Uromastix aegyptia* (El-Salhy and Grimelius 1981).

In conclusion, the reptiles show a great variability of islet structure. However, it appears likely that all species have A-, B-, D- and F-cells.

Aves. The pancreas of birds, which consists of three or four well-defined lobes (cf. Bargmann 1939; Guha and Ghosh 1978), contains two basic types of islets: large "dark" (A-) islets consisting of A- and D-cells; and smaller "light" (B-) islets, consisting mainly of B-cells, although often a few D- or A-cells are found at the periphery. In addition, varying with the species, some mixed islets may occur (for literature see Epple et al. 1980; Iwanaga et al. 1983). In the galliform birds, the dark islets are restricted largely to the third and splenic lobes (i.e., descendents of the dorsal anlage), while the light islets are distributed throughout the pancreas (Mikami and Ono 1962; Smith 1974). However, this does not hold for other species such as the duck (*Anas platyrhynchos*), in which dark islets also occur in the ventral pancreas (Svennevig 1967). The identity of the A-, B- and D-cells has been confirmed by several immunocytochemical studies (Rawdon and Andrew 1979; Iwanaga et al. 1983; Falkmer and Van Noorden 1983; Andrew 1984); and by radioimmunoassay, the concentrations of glucagon, insulin, and somatostatin correspond to the distribution of these cell types (Weir et al. 1976; Hazelwood 1984). Glicentin has been localized to the A-cells; and somatostatin-28 (see Chap. 13) to the D-cells (Iwanaga et al. 1983). Immunocytochemistry has shown that whereas F-cells are dispersed as single cells or small groups in the exocrine tissue of all four lobes of the chicken pancreas, they are less common at the periphery of the islets (Larsson et al. 1974; Alumets et al. 1978a; Rawdon and Andrew 1979; Andrew and Rawdon 1980; Iwanaga et al. 1983). However, the majority of the F-cells seems to be located in the derivatives of the ventral anlagen (Bonner-Weir and Weir 1979; Andrew 1984; Tomita et al. 1985).

In conclusion, the birds are unique in having two different types of islets. Like many actinopterygians and some reptiles, they possess large islet accumulations close to the spleen. All four major cell types are found in all regions of the pancreas; the A- and D-cells occur mainly in the derivatives of the dorsal anlage, while the F-cells tend to be more common in the derivatives of the ventral anlagen.

Mammalia. The pancreas of this group represents the typical tetrapode type with relatively small, but widely scattered islets. The pancreas tissues of prototherian (egg-laying) mammals appear to have been totally neglected; also little has been done with regard to the islet cytology of metatherians (marsupials) since the study of the opossum (*Didelphis virginiana*) by Munger et al. (1965). These authors, who identified A-, B-, and D-cells by ultrastructural criteria, also identified the E-cell, first reported by T. B. Thomas (1937), as containing glycogen and having secretory granules almost as large as acinar zymogen granules. Larsson et al. (1976b) found F-cells predominantly in the duodenal lobe islets of the opossum pancreas, with a few extra-insular F-cells also in the splenic portion. By light microscopy, White and Harrop (1975) studied the islets of four marsupials, the red kangaroo (*Macropus rufus*), the gray kangaroo (*Macropus major*), the euro (*Macropus robustus*), and the brush-tailed possum (*Trichosurus vulpecula*). Their methods allowed them to only distinguish B-cells and non-B-cells accurately, but interestingly, the kangaroos had B-cell populations of only 8–15% of the islet whereas that of the possum had 53% B-cells. In a recent immunohistologic study on *Trichosurus vulpecula*, Reddy et al. (1986) identified A-, B-, D- and F-cells, and confirmed a high percentage of A-cells.

The islets of eutherian (placental) mammals have been far more extensively studied than those of any other group of vertebrates. Because it is beyond the scope of this discussion to enumerate the many publications, the reader is directed to reviews by Lacy and Greider (1972), Falkmer and Östberg (1977), Erlandsen (1980), and Munger (1981). As a group, the eutherians have a remarkably uniform islet cytology. The A-cells are generally found in greater concentration in the splenic end of the pancreas. In fact, they are virtually absent from the duodenal portion of the dog pancreas (Munger et al. 1965). In contrast to the predominance of A-cells in the islet tissue of birds and reptiles, eutherian A-cells are far less numerous. In a very comprehensive morphometric analysis of the rabbit pancreas, Sato and Herman (1981) report that the A-cells constitute only 7.7% of the islet. Baetens et al. (1979) reported that islets in the dorsal lobe of the rat pancreas contain 28% A-cells and that ventral lobe islets contain less than 2% A-cells. With the exception of the horse, which has centrally located A-cells (A. Forssmann 1976), these cells are often peripheral to a core of B-cells. The A-cell granules of rats, dogs, pigs, and guinea pigs are known to contain glicentin in addition to glucagon (Ravazzola et al. 1979).

B-cells comprise the most numerous cell type in the eutherian islet, the usual average being 60–80%. Sato and Herman's (1981) analysis of the rabbit pancreas yielded 86% B-cells. The percentage composition of B-cells in the dorsal and ventral islets of rats are 66% and 74%, respectively (Baetens et al. 1979).

The D-cells are usually situated between A- and B-cells. Sato and Herman (1981) report that rabbit islets contain 2.2% D-cells, while Baetens et al. (1979) found 4% of the islet cells in both dorsal and ventral lobes of the rat to be D-cells.

The F-cell was first described by EM criteria in the duodenal portion of the dog pancreas by Munger et al. (1965). When PP was localized in islets, the cell containing the peptide was designated as the PP-cell (L.-I. Larsson et al. 1974; L.-I. Larsson et al. 1976b), but upon the discovery that the dog F-cell contained PP immunoreactivity (W. G. Forssmann et al. 1977; Greider et al. 1978), the term "F-cell" has been adopted for most PP-containing cells. Orci et al. (1976) were the first to point out the nonrandom distribution of the F-cells in the rat pancreas, an observation that was confirmed by L.-I. Larsson et al. (1976a), who studied PP distribution in the pancreas of 11 eutherians in addition to man and the opossum. Strictly quantitative procedures were not used, and some interspecific variation was noted; the trend is that F-cells are located mostly in the duodenal portion of the gland, both as insular and extra-insular components. Baetens et al. (1979) quantified F-cells in the rat to be 2% of the islet in the dorsal and 20% in the duodenal portion. In a combined immunocytochemical and ultrastructural study, Fiocca et al. (1983) equated a cell population in the head of the dog pancreas with F-cells. However, PP-reactive cells in the remainder of the pancreas differed from those in the head by having smaller electron-dense granules. According to Solcia et al. (1985), these cells comprise about 15% of the pancreatic PP-cells of the human, and they are identical with D_1-cells. The differences between the two populations of PP-reactive cells are so far unexplained, but they may be related to the presence of different co-secretions (see Chap. 9).

Since recent descriptions of human islets have been presented by Lacy and Greider (1979), Erlandsen (1980), Munger (1981), A. Clark and Grant (1983), Orci (1983), and Volk and Wellmann (1985b), the following discussion will entail mainly new data, especially on the A- and F-cells.

Human A-cells, like those of many other mammals, have electron-dense secretory granules averaging 300 nm in diameter. Many of these granules display a very dense core surrounded by varying amounts of less dense material. Bussolati et al. (1971), while studying some cytochemical characteristics of human A-cell granules in the EM, noted that phosphotungstic acid stained the granule core whereas the less dense surrounding material reacted with silver. The authors suggested that the dense core contains glucagon and the peripheral material is of unknown composition. Using an antibody specific for glicentin, Ravazzola et al. (1979) showed that pancreatic and gastric A-cells were both reactive. With the greater resolution of the protein A-gold technique, it was later found that the A-cell granule core contains glucagon and that the surrounding material contains glicentin (Ravazzola and Orci 1979, 1980), thus confirming the earlier observations of a two-component granule by Bussolati et al. (1971). The presence of glicentin in the pancreatic A-cell granule explains the difficulty of many early workers to obtain glucagon-specific antibodies and the interference by GLIs in glucagon assays. The human B-cell conforms to the general description of most mammalian B-cells (Munger 1981). The D-cell, which is usually situated between the A- and B-cells, appears to be more abundant in fetal and neonatal than in adult islets (see below).

Malaisse-Lagae et al. (1979) found that a distinct lobe from the posterior duodenal portion (head) of the human pancreas is particularly rich in PP-reactive cells. When the volume densities of all four islet cell types are averaged for the entire pancreas, the PP-cell is the second most frequent cell type after the B-cell (Stefan et al. 1982a).

The F-cell is the dominant cell type in the PP-rich lobe, with volume densities ranging from 76% to 89% in five pancreases from persons ranging in age from 9 to 80 years (Malaisse-Lagae et al. 1979). Pancreata from a 33-week fetus and two infants demonstrated F-cell volume densities varying from 43% to 63%, suggesting an age-related increase of F-cell volume. In this study, the B-cell volume densities varied from 9% to 32% in the PP-rich lobe, contrasting with more familiar values of 48% to 85% in the remainder of the gland. A-cell volume densities averaged about 1% in the PP-rich lobe of all specimens and ranged from 11–24% in the tail and body. Islets derived from both dorsal and ventral anlagen in the fetus and infants had high D-cell volume densities (17–24% ventral, 37% dorsal) and drastically reduced values of 3–6% in older cadavers (1–7% ventral, 3–7% dorsal). A separate analysis further supports the effect of age by demonstrating that islets of older subjects (≥54 years) contained a higher volume of F-cells than islets of younger subjects (≤45 years) (Stefan et al. 1982a). Therefore, age appears to affect the volume of D- and F-cells. However, it must be noted that the latter two studies did not differentiate between both populations of PP-cells, i.e., F- and D_1(?)-cells.

Possible sex-related differences were noted in the study of Stefan et al. (1982a) wherein the average total F-cell volume from five male subjects (age 29–72 years) was significantly greater than that of 13 females (age 16–80 years).

There are striking cytological differences between the islets in the two "basic" types of diabetes mellitus. Type-I diabetes, which develops usually during childhood or adolescence ("juvenile diabetes"), is characterized by *insulin deficiency.* At the onset of the disease, the B-cells try to keep up with the demand for insulin, until they are finally exhausted and disappear. According to Gepts and LeCompte (1985) there is simultaneous neoformation of two types of islets, one consisting of A- and D-cells, and another one consisting of PP-cells. The etiology of this disease appears to be multiple; recently there has been strong emphasis on the role of viral (Craighead 1985) and autoimmune mechanisms (cf. Lernmark 1985; Arquilla and Stenger 1985; Unger and Foster 1985). The disease is fatal, unless insulin therapy is given (see Chap. 10.4). Type-II diabetes develops usually in post-adolescent persons ("maturity onset diabetes"), the Pima and Papago Indians being a notable exception. In these tribes, Type-II diabetes develops frequently before adulthood. Type-II diabetes is accompanied by disturbed glucose tolerance and hyperglycemia, due to an *inappropriate B-cell response* to changes in glycemia, and a decrease in peripheral response to insulin. Dietary excess may be a factor in the initiation of this disease; obesity and peripheral insulin resistance often seem to precede hyperglycemia (cf. Efendic et al. 1980; Unger and Grundy 1985; Volk and Wellmann 1985c, d). In contrast to Type-I diabetes, B-cells are present and persist, and the defective insulin response to hyperglycemia may be alleviated by oral antidiabetic drugs (cf. Reaven 1984). For more information and literature on both types of diabetes mellitus, see Katzeff et al. (1985), and Unger and Foster (1985).

In the most definitive immunocytochemical-morphometric analysis of the diabetic pancreas, including four Type-I (insulin-dependent) and eight Type-II (non-insulin-dependent)diabetics, Rahier et al. (1983) reported a decrease in total endocrine cell volume caused by B-cell loss, which was restricted mostly to the PP-poor region of the pancreas. Furthermore, these authors noted no change in the absolute mass of A-, D- and F-cells in Type-I diabetics, indicating that the loss of B-cells has no effect on the volume of the other islet cell types. In the Type-II diabetics, on the other hand, the A-cell mass is increased, while the total volumes of the B-, D- and F-cells remain unchanged. Caution is advised in the interpretation of this type of study, however, because the volume density of a cell population may not accurately reflect the true cell number (numerical density), particularly if a group of cells has undergone hypertrophy. Yet, if borne out by additional data, these results indicate important cytological differences between the islets of human diabetics and some of the animal models of this disease.

4.3 Islet Structure: Phylogenetic Trends

Two frequent trends emerge in the gnathostomes above the chondrichthyan level:

1. Large islets, when close to the spleen, develop from the dorsal anlage. Examples are the Brockmann bodies of the teleosts (cf. Epple 1969; Falkmer and Östberg 1977; Epple et al. 1980), and the splenic lobes and/or intrasplenic islets of certain reptiles (cf. Dupé-Godet and Adjovi 1983; Buchan 1984; Rhoten 1984) and birds (cf. Epple et al. 1980; Epple and Brinn 1986). Furthermore, the increased number or larger size of islets in the caudal pancreas of many tetrapods seems to reflect their origin from the dorsal anlage (Fig. 3.1).
2. A-cells tend to be more frequent in pancreas regions derived from the dorsal anlage, while F-cells tend to be more frequent in the derivatives of the ventral anlagen (cf. Orci 1983; Andrew 1984).

It must be emphasized that these trends do not permit formulation of a *law*. Falkmer and Van Noorden (1983) suggest that the juxtapyloric Brockmann body of *Cottus scorpius* (see above) originates from ventral anlage; F-cells are scattered throughout the pancreas of the chicken (Iwanaga et al. 1983); and the pancreas of the guinea pig produces all four major islet hormones (Baskin et al. 1984) although, according to Siwe (1926), it originates exclusively from the dorsal anlage.

There is no rule as to the phylogenetic distribution of large and small islets. In the Polypteridae, as well as in the most "advanced" teleosts, large principal islets occur, although such accumulations are often absent in "intermediate" groups (such as the Chondrostei and *Amia calva*). In the two families of lungfishes, two totally different types of islets are found in otherwise very similar "intra-intestinal" pancreata. Wide variations in islet structure have also been described for various reptiles (Gabe 1970). These facts must be considered when the possible existence of "insulo-acinar" portal systems is discussed (Chap. 6).

A comparative survey of islet cytology of the gnathostomes does not support the notion that islet composition becomes more complex during the evolution of "higher" vertebrates (cf. Falkmer et al. 1984). As previously pointed out, both groups of chondrichthyes have a surprisingly large variety of islet cells, perhaps a total of eight in the selachians. Possibly this is a reflection of phylogenetic trial-and-error, in which the exocrine pancreas carries a "sampling" of intestinal paraneurons, the usefulness of which in the extramural location is ultimately decided by selection pressures. Be this as it may, cytological simplification rather than diversification seems to be the phylogenetic trend of the gnathostome islets. Clearly, this conclusion is incompatible with a stepwise evolution of the islet complexity which is supposed to have occurred in three stages (cf. Falkmer and Van Noorden 1983; Falkmer et al. 1984): two-hormonal islets in cyclostomes, three-hormonal islets in holocephalians, four-hormonal islets in elasmobranchs and "higher" vertebrates. However, also proper application of cladistic principles (Atz et al. 1980) and the recent, pertinent literature show that the holocephalians and selachians are sister groups that belong to a side branch of vertebrate evolution (Rosen et al. 1981). This makes it impossible to place their islet organs as successive stages in the main line of vertebrate evolution. Furthermore, if the presence of "open" insulin cells in the intestine is indeed of taxionomic (hierarchic) value, as proposed by Falkmer and coworkers, their presence in the elasmobranchs (El-Salhy 1984) reverses this author's sequence of islet evolution, i.e., it makes the elasmobranchs more primitive than the holocephalians, which do not have gastrointestinal insulin cells (cf. Falkmer et al. 1984). This would agree with Fujita's (1962) conclusion that the islet organ in the holocephalians is intermediate between that of the selachians and higher vertebrates. Thus, both the immunohistological data and phylogenetic evidence suggest that the elasmobranch pancreas represents a reasonably good early gnathostome model (cf. Epple 1969).

Chapter 5

The Exocrine Pancreas

Before the discussion of the controversial topic of insulo-acinar portal systems (Chapter 6), we examine briefly the structure and organization of the exocrine pancreas.

We are not aware of a recent comparative account of this gland. Some relevant information is found in the chapters of Neubert (1927), Zimmermann (1927) and in a monograph edited by Go et al. (1986). A phylogenetic survey (Epple and Brinn, unpublished) suggests that the gross morphology of the exocrine pancreas is not important for its function. One critical and yet unanswered question concerns the nature of the small ductules connected with the islets (see also Fig. 2.7A), which seem to give rise to both exocrine and islet cells, particularly under pathological conditions (Lazarus and Volk 1962; Pour et al. 1982; Bjorenson 1985). We find these ductules in seemingly increased numbers close to hyperplastic islets of old rats (Epple, unpubl.), and at one pole of the Brockmann body of the catfish, *Ictalurus punctatus* (Brinn, unpubl.). Are the cells of the small ductules identical with the intercalated ducts and centroacinar cells of the functioning exocrine duct system, or are they undifferentiated "reserve tissue" for

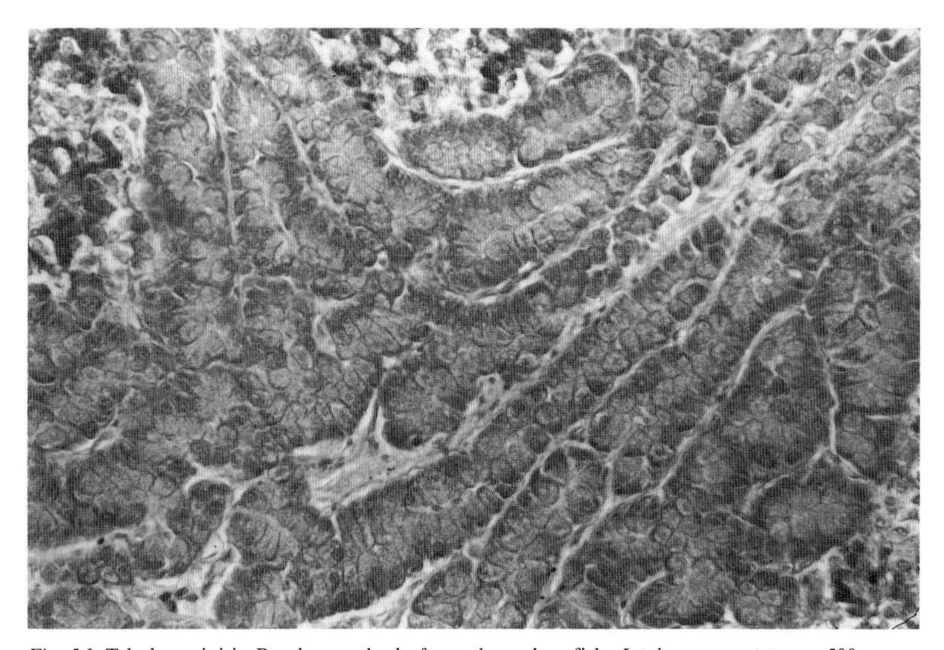

Fig. 5.1. Tubular acini in Brockmann body from channel catfish, *Ictalurus punctatus;* ×500

the islets and/or the exocrine pancreas? And, of course, what is their possible significance for tumor formation?

The mammalian pancreas is comprised of many alveoli, which are formed by 20 to 50 acinar cells. Each alveolus is drained by an intercalated duct that joins increasingly larger ducts, ultimately draining into the small intestine. The pyramidal-shaped acinar cells segregate the exocrine environment from the stromal one by tight junctions on their apico-lateral surfaces, and the lateral plasmalemma demonstrates gap junctions, suggestive of electrical coupling (see below). This "typical" compound tubulo-alveolar organization in mammals and birds differs from that in many of the lower vertebrates. M. Miller and Lagios (1970) point out the tubular nature of the reptilian pancreas; Kobayashi (1969) describes tubular acini in *Xenopus laevis*. Teleosts also have tubular acini (Fig. 5.1) inside the principal islet (Brockmann body).

There has been a suggestion that the smallest functional unit of the mammalian pancreas is larger than a typical acinus (Gorelick and Jamieson 1981). Findlay and Petersen (1983) studied the transfer of intracellularly injected Lucifer Yellow CH between living exocrine cells of mouse pancreas fragments and found dye-coupled acinar units of 110 and 230 cells. There was no evidence of duct-cell involvement in dye transfer, but presumably it was mediated via gap junctions between acinar cells.

Chapter 6

Insulo-Acinar Interactions

The possible functional principle behind the existence of endocrine islets has led to the postulate that their intrapancreatic dissemination is a parallel of the distribution of the Leydig cells in the testis (Ferner 1957), and that the islet vessels form an insulo-acinar portal system.

There is no question that the acini in the immediate neighborhood of the islets often differ from acini in more distant locations (Fig. 6.1), and at least in the duck, this can be seen already during the later embryonic life (Svennevig 1967). This phenomenon, dubbed "zymogen mantle" or "halo" (Jarotzky 1899), has been studied in particular in mammals and birds. The following options appear possible: (1) Zymogen mantles are histological artifacts; (2) they are due to an effect of islet cells that is mediated via gap junctions; (3) they are a paracrine effect of islet secretions which diffuse into the interstitium of the exocrine pancreas; (4) they result from a vascular link between islets and acini, i.e., an insulo-acinar portal system; (5) they are due to neurosecretions (Fujita et al. 1982) that are released in the islets and affect the exocrine tissue via routes (3) or (4).

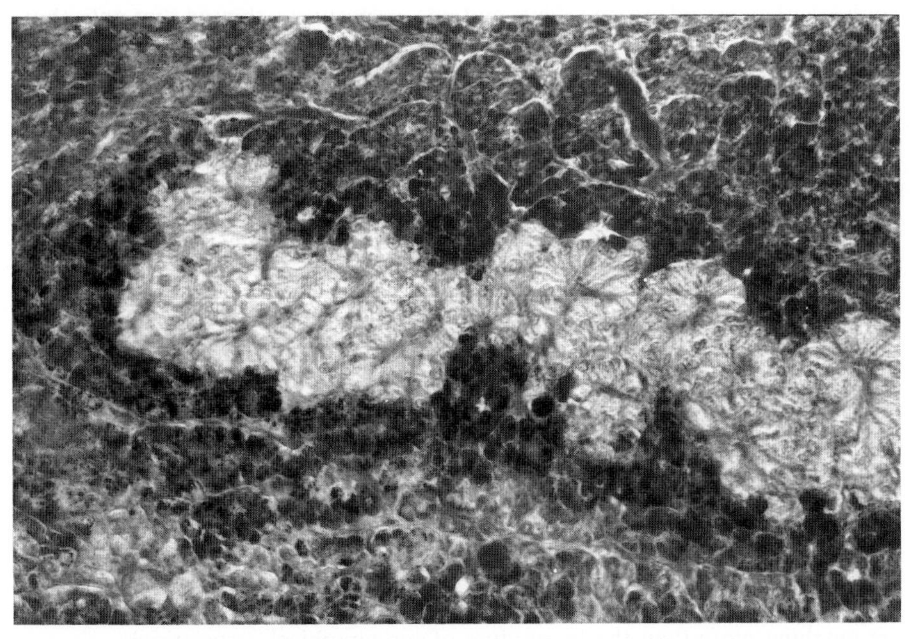

Fig. 6.1. Zymogen mantle surrounding A-islet of a pullet. Note in some regions the presence of giant exocrine "granules" that developed by confluence of regular granules. Aldehyde fuchsin-trichrome (Epple 1967); ×500

Two observations make it unlikely that the zymogen mantles are histological artifacts. Malaisse-Lagae et al. (1976) found different enzyme concentrations between peri-insular and tele-insular acini. Putzke and Said (1975) observed that pilocarpine-stimulated release of zymogen granules was retarded in acini immediately around the islets. Option (2) exists, but the functional significance of the gap junctions between the endocrine and exocrine cells (Alumets et al. 1978b; Bendayan 1982; Agulleiro et al. 1985) remains obscure. A paracrine effect of islet secretions probably exists in the chicken, since in pullets manipulations of the diet caused an enormous progressive enlargement of the zymogen-mantles, far beyond the probable vascular drainage of the islets. In this case, the "mantle cells" showed the histological picture of excessive storage (Epple 1968a). A vascular link between islets and exocrine cells with a specific function has been described or postulated by several authors (for literature see Bonner-Weir and Weir 1979; Henderson et al. 1981; Iwanaga et al. 1983; Ohtani 1983; Syed Ali 1984; Rooth et al. 1985; Trimble et al. 1985; Williams and Goldfine 1985), and the islets have been interpreted as a "neuro-paraneural control center of the exocrine pancreas" (Fujita et al. 1981b). In essence, such postulated insulo-acinar portal systems consist of one (sometimes more) afferent islet arteriole(s) that give(s) rise to a capillary network within the islet. This network then continues some distance into the exocrine tissue before it is drained by collecting venule(s). Alas, there are some problems with the functional interpretation of such systems. First of all, insulo-acinar systems must be restricted to certain taxa. In lampreys, there is no direct vascular link between the islet organ and the exocrine pancreas equivalent (see Chap. 2.3). In teleosts, Lange (1984) found no evidence of such systems. Secondly, studies with microspheres in rabbits and rats, respectively, showed that only about 20% (Lifson et al. 1980, 1985) or even less (Jansson and Hellerström 1983) of the total pancreas blood passes through the islets. This would allow only a limited number of acini to become exposed to concentrated ("portal") islet secretions. Furthermore, in the rat a large amount of islet blood is directly drained by collecting venules (Bonner-Weir and Orci 1982). In the cat (Syed Ali 1982) and in the chicken (Iwanaga et al. 1983) a considerable amount of arterial blood also flows directly to the exocrine capillaries. Thus, in the species studied in more detail only a limited quantity of islet blood can reach the acinar cells. We have previously pointed out that evolutionary pressure should have created a more even intrapancreatic distribution of islets, if insulo-acinar systems were indeed a general functional principle (Epple et al. 1980). We may add here that in some species very wide pancreatic regions are totally devoid of islets. This is the case in many bony fishes, particularly the Polypteridae and *Lepisosteus osseus* (Epple and Brinn, unpubl.), and it has also been reported for turtles (Girone 1928). From the preceding (see also Chap. 4.3) it is clear that insulo-acinar portal systems (1) are not a general phenomenon of gnathostomes, and (2) only involve a limited number of exocrine acini in the species so far thoroughly analyzed. This does not rule out that, in evolutionary terms, they may have secondarily developed into units of functional significance for some species. However, there is no proof yet that they are anything more than the by-product of topographical associations.

The nature of the islet secretions involved in the zymogen-mantle phenomenon is unknown. Since such mantles developed around both A- and B-islets of pullets (Epple 1968a), one is inclined to guess that one or more substance(s) different from insulin and glucagon play(s) a role. The F-cells are rather evenly scattered between the exocrine acini of the chicken (Iwanaga et al. 1983); therefore, PP may not be involved either. This conclusion is consistent with the absence of an effect of PP on rat pancreas tissue in vitro (Louie et al. 1985). Using isolated acini of the guinea pig pancreas, Pandol et al. (1983) found no specific effect of glucagon, insulin, and PP on amylase output, an observation at variance with a stimulatory effect of insulin seen in rat and mouse (Williams and Goldfine 1985). In essence, this leaves, of the four major islet hormones, somatostatin, by sheer exclusion. The D-cells are widely spread in the islets of all pancreas regions (see Chap. 4.2), and putative somatostatin receptors on acinar-cell membranes have been reported (Taparel et al. 1983; Esteve et al. 1984; Sakamoto et al. 1984); furthermore, there is indeed some experimental support for the possible role of somatostatin as an exocrine inhibitor. Kawai et al. (1982a) found that the perfused dog pancreas extracts 50–80% of somatostatin while removing less than 21% of the insulin and glucagon. On the other hand, pancreatic acini from rats and mice have high-affinity insulin receptors, and there seem to be local differences in the distribution of these receptors (cf. Mössner et al. 1984; Williams and Goldfine 1985). Finally, it remains to be seen if the "insular neurosecretion" (Fujita et al. 1982) affects the exocrine cells (see Chap. 7); and it is conceivable that islet secretions other than the major hormones are involved (see Chap. 9.1). For instance, TRH appears to inhibit pancreatic enzyme secretion (Gullo and Labo 1981; Komiya et al. 1984).

Henderson (1969) raised the question: "why are there islets of Langerhans?", going on to propose that the dispersion of small islets throughout the exocrine pancreas serves to distribute islet hormones efficiently to the surrounding exocrine cells. Recently, Henderson et al. (1981) reviewed the support for this hypothesis, implying that the exocrine pancreas is a target organ of islet hormones. While not denying the likelihood of a trophic action of systemic titers of islet hormones (especially in the light of the morphogenetic actions of insulin in the embryo/fetus, see Chap. 10.3), we suggest, however, that the breakup of the islet mass into small, dispersed groups of cells may have advantages other than those proposed by Henderson. The first target organ of the islet hormones is the liver, and in mammals, which depend on tightly regulated blood-glucose levels, fast and efficient delivery of islet hormones to the liver is essential for maintenance of blood-glucose levels. Small islets serve that function better than large aggregates such as the Brockmann body, since blood traverses small islets much faster than large ones; and as a consequence, a quantum of islet hormone is out of the islet and on its way to the liver much more quickly. This anatomic arrangement also eliminates low-pressure sinks and slower, percolating blood flow, as one would expect in a large islet. However, the small, dispersed islets can still function as a collective unit, i.e, as the "islet organ".

The direct arterial blood supply to the islet organ need not have any implications about an insulo-acinar portal system either. Rather, it ensures the delivery to the B-cells of blood-glucose levels comparable to those going to the brain, unmodified by exocrine cells. The islet organ is then able to monitor accurately circulating glucose levels. A further advantage of a direct arterial blood supply to the islet organ and the subsequent centrifugal flow is that the islet (a) is protected from enzyme leakage from exocrine cells, and (b) receives a more effective O_2 supply.

The never-ending saga of the endocrine-exocrine pancreas interactions would not be complete without a new twist. In addition to differences in the amylase content of the dorsal and ventral pancreas regions (Malaisse-Lagae et al. 1983), there may also be a difference in the insulin and glucagon release from these regions (Leclercq-Meyer et al. 1985 b).

Chapter 7

Nervous Regulation of Islet Functions

7.1 Pathways

Despite a recent surge in interest, the neural regulation of the endocrine pancreas is poorly understood. Possibly, islet innervation is important for the pulsatile release of islet hormones. Pulsatile release of hormones, as opposed to tonic delivery, may have advantages such as increased efficiency at target organs (thus "saving" hormone), or prevention of receptor desensitization (for literature see Weigle et al. 1984; Samols et al. 1986). Another function of islet innervation may be an overriding control in the integration of islet responses. The vast majority of new data on islet innervation, which comes from studies in mammals, already reveals considerable inter- and intraspecific differences (Gerich and Lorenzi 1978; Woods and Porte 1978; R. E. Miller 1981; Smith and Madson 1981; Palmer and Porte 1983; Porte and Woods 1983; Rohner-Jeanrenaud et al. 1983; Smith and Davis 1983; Steffens and Strubbe 1983; Luiten et al. 1984; Jeanrenaud 1985). Although comparative studies are in their infancy, the available information shows that it is beforehand impossible to identify phylogenetic or functional patterns of islet innervation (Epple et al. 1980; Buchan 1984). Perhaps the extremes are best illustrated by the cyclostomes which totally lack an islet innervation and the teleosts in which islet cells show particularly intimate contacts with neurons (Epple and Brinn 1975).

It appears possible that the nervous control of the islets involves three efferent pathways: (1) hypothalamic neurosecretions; (2) insular neurosecretions; (3) innervation of islet cells. The question of insulotropic hypothalamic hormones deserves serious attention since hypothalamic extracts with both insulinotropic (Lockhart-Ewert et al. 1976; Hill et al. 1977; Moltz et al. 1979; Knip et al. 1983; Palmer and Porte 1983; Wood et al. 1983; Jeanrenaud 1985) and glucagonotropic (Moltz and Fawcett 1983) effects have been identified in mammals. Furthermore, insulin receptors may well exist in the brain of all vertebrates, from lamprey to mammals (Leibush 1983), and at least in some brain regions of the latter they can be reached by circulating insulin (for literature see Landau et al. 1983; Porte and Woods 1983; Frank et al. 1985; Haskell et al. 1985). This raises the question of a neuroendocrine-feedback system between islet cells and CNS (cf. Melnyk and Martin 1985) which, depending on the species, may involve up to five efferent pathways, and a number of hormonal afferent components (Fig. 7.1). If Van Houten and Posner's (1981) proposed feedback system for blood-borne polypeptide hormones (Fig. 7.2) can be confirmed in its essential features, then the comparative endocrinologists will be faced with new, exciting questions. For example: Is a humoral paraneuron-CNS feedback via circumventricular organs the phylo-

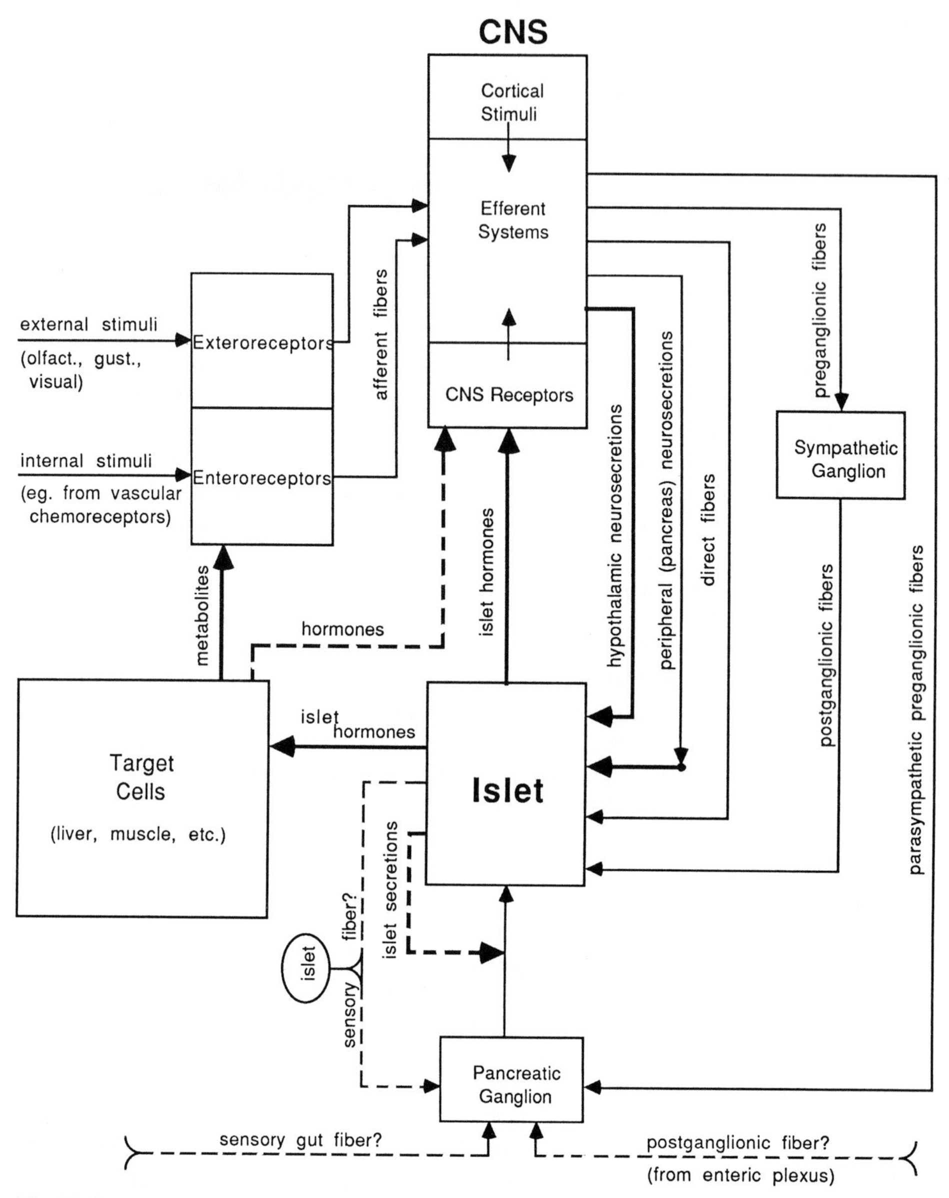

Fig. 7.1. Summary of known and suspected neural pathways of islet regulations. Humoral pathways: *heavy lines*; nerve fibers: *thin lines*; strongly hypothetical connections: *interrupted lines*

genetically original form of neuroendocrine interactions? Is, perhaps, "efferent" brain secretion the *basic* route of CNS-islet communication, which is only supplemented by insular neurosecretion and/or direct innervation whenever evolutionary pressures call for it? And if so, is the probably secondary absence of islet innervation (Hahn von Dorsche et al. 1976) in the spiny mouse (*Acomys*

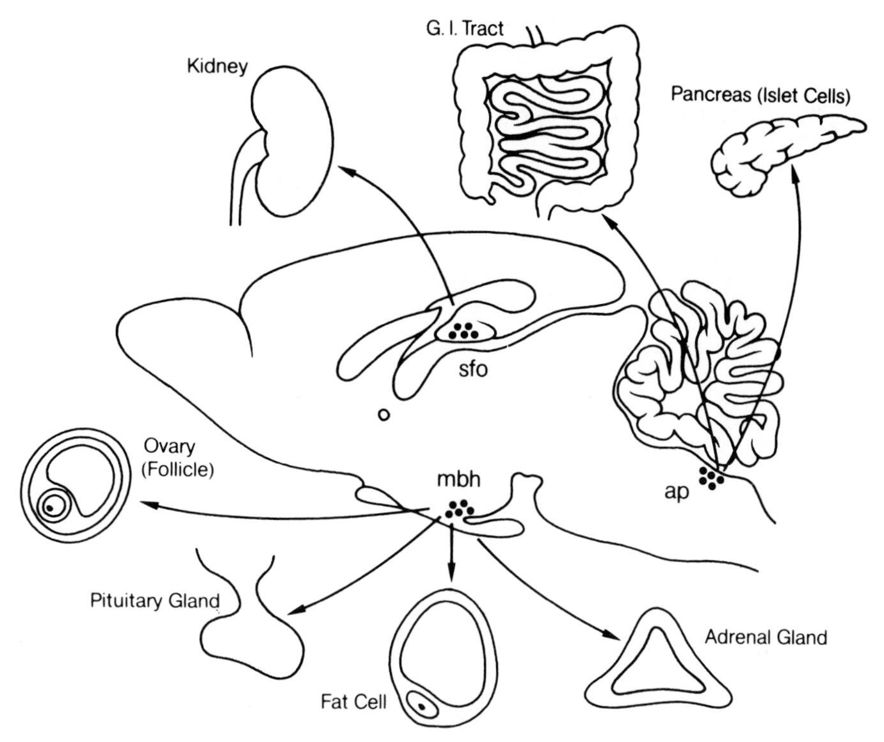

Fig. 7.2. Diagram of locations of some circumventricular organs, depicting their functions in simplified form. The medial basal hypothalamus (*mbh*), containing the median eminence, is involved in regulating rates of glandular secretion, fat deposition and mobilization, and food intake. The subfornical organ (*sfo*) is linked closely to the regulation of water and salt balance, drinking behavior, and blood pressure. Functions of the organum vasculosum of the lamina terminalis (ovlt – not shown) are similar to those of the sfo. The area postrema (*ap*) may be related to vagal functions influencing pancreatic hormone secretion and gastrointestinal/autonomic activity. Specific binding sites for a variety of blood-borne polypeptide hormones occur in all these regions. The role of circumventricular hormone receptors may be to mediate direct feedback effects of blood-borne hormones on metabolic/regulatory functions subserved by these regions (Van Houten and Posner 1981)

cahirinus) an example of selection in which environmental factors allowed a reduction of the CNS control to the basic humoral link? In addition, important clinical questions arise: What is the role of hypothalamic insulotropic secretions in (1) metabolic disorders, and (2) the maintenance and function of transplanted islets?

Insular neurosecretions (Fig. 7.3) have been identified by Fujii et al. (1980) and Fujita et al. (1982). These authors, who find at least four different types of nerve terminals in the pericapillary spaces of the dog islets, also describe nerve endings with VIP in the same location in the mink (*Mustela vison*) and a snake (*Elaphe quadrivirgata*). Beforehand, the targets of these presumed insular neurosecretions remain to be identified. Obvious candidates include endocrine and exocrine pancreas, blood vessels, and the liver. Insular neurosecretions may well be just one example of more widely spread peripheral neurosecretions since

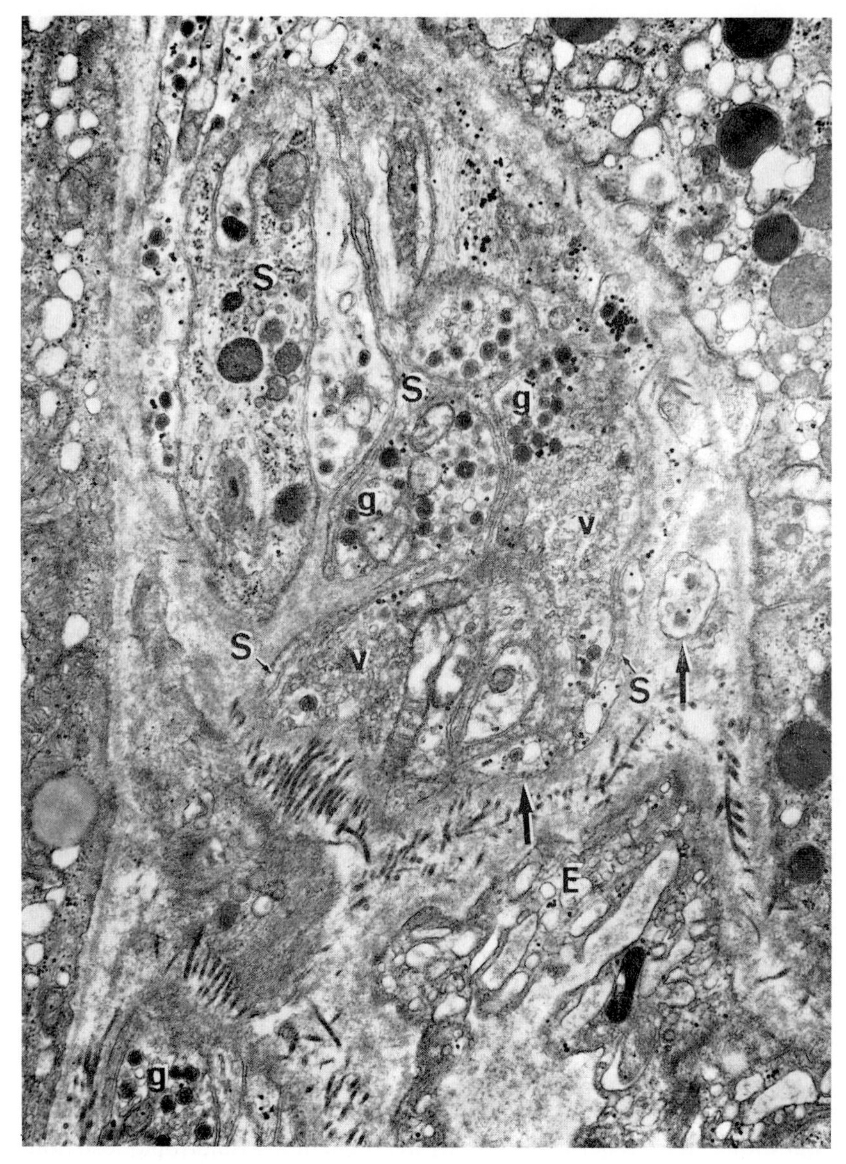

Fig. 7.3. Probable insular neurosecretion in the racer snake (*Elaphe quadrivirgata*). Note the nerve bundle in the intracellular space of this islet. Endocrine cells are seen on the *right* and *left* sides. A blood capillary with a characteristically vesiculated endothelium (*E*) is shown at the bottom. Axons containing large, endocrine-like granules (*g*) and small, synaptic-like vesicles (*v*) are invested by the cytoplasm of Schwann cells (*S*). Axons devoid of the Schwann-cell coverage and facing towards the blood capillary are indicated by arrows (Fujita et al. 1980)

similar axon endings were seen in the poison gland of another snake, *Rhabdophis tigrinus* (Fujita et al. 1982).

Another "unconventional" question of islet innervation comes with the realization that Langley's (1921) concept of an enteric, third division of the peripheral nervous system is well justified. Suffice it to point out that (1) there are many more neurons in the two plexus of the digestive tract than preganglionic fibers that innervate them (Gershon 1981; Gershon et al. 1983); (2) in the enteric nervous system peptides appear to be the predominant form of transmitters (see e.g., Lundberg et al. 1982; Nilsson 1983; Smith and Davis 1983; Polak and Bloom 1984). Since the exocrine pancreas is a derivative of the digestive epithelium, one wonders (a) if the intrapancreatic and other pancreas-related ganglia are an offshoot of the enteric plexus (cf. Stagner and Samols 1985b), and (b) if there are direct connections between pancreatic and intestinal ganglion cells. Such connections, if confirmed, could have considerable functional implications for both the exocrine and endocrine pancreas. Clearly, the ontogenetic origin of the pancreas favors the idea that the pancreatic ganglia are equivalents of the enteric plexus. Pancreatic ganglia are still found after vagotomy in mice (Honjin 1956), and fetal rat pancreas contains neurons after 10 days of culture. In the latter case, the neurons specifically innervate the islets (Brinn et al. 1977). Furthermore, rhythms of hormone release from the isolated, perfused dog pancreas (cf. Stagner and Samols 1985a) strongly resemble the well-known intrinsic rhythms of the enteric nervous system. Thus, one wonders if it is justifiable to apply the term "parasympathetic" for intrapancreatic neurons.

In mammals, the highly complex pathways of central nervous control of islets can be divided as follows:

1. Sensory systems (receptor cells plus associated afferent neurons, central and peripheral chemosensory neurons).
2. Central nervous control systems.
3. Efferent pathways via sympathetic, "direct" and parasympathetic fibers, and possibly also neurosecretions.

Sensory inputs from the hepatic portal vein, liver, intestine, and pancreas, as well as gustatory, olfactory, and visual stimuli, and hypnotic suggestions may all induce insulin secretion (Russek and Racotta 1980; Charles 1981; R. E. Miller 1981; Woods et al. 1981; Oomura and Niijima 1983; Rohner-Jeanrenaud et al. 1983; Smith and Davis 1983; Steffens and Strubbe 1983; Curry 1984). On the other hand, cortical signals reduce the insulin secretion and increase glucagon release at the beginning of strenuous exercise or fight-or-flight responses, before blood-borne signals affect the islet cells (Unger 1983). Furthermore, glucose and insulin sensors occur in several brain regions (Havrankova et al. 1983; Oomura 1983; Porte and Woods 1983; Van Houten and Posner 1983). In mammals, the messages from the digestive system are conveyed via the VIIth, IXth, Xth, and splanchnic nerve to the brainstem, and it is noteworthy that 75–90% of all fibers in the abdominal vagus, and about 50% of all splanchnic fibers are afferent. The first station of integration is apparently the Nucleus tractus solitarius, while the main center for all neural processes related to the islet organ seems to be located in the hypothalamus. The latter is responsible for the mediation of the early phase of secretion of insulin, glucagon, and PP at the beginning of a meal, which occurs before metabolites stimulate the islet cells. Furthermore, hypothalamic control also seems to prevent an exaggerated insulin response during the second, postprandial phase of insulin release. On the other hand, hypothalamic defects can lead to hyperinsulinemia, which in turn may induce hyperphagia and obesity (Oomura and Niijima 1983; Porte and Woods 1983; Rohner-Jeanrenaud et al. 1983; Shimazu 1983; Steffens and Strubbe 1983; Jeanrenaud 1985). In the genetically obese (fa/fa) rats the related messages seem to involve the cholinergic system (Rohner-Jeanrenaud and Jeanrenaud 1985). Parasympathetic fibers supplying the mammalian pancreas originate in the lower medulla (dorsal motor nucleus of the vagus, the nucleus ambiguous and the lateral reticular nucleus), whereas the preganglionic sympathetic fibers originate in the ventral horns of the cervical, thoracic, and upper lumbar regions of the spinal cord. In addition, Luiten et at. (1984) describe in the rat fibers originating in the intermediolateral regions of T_3-L_2. It is generally held that the preganglionic sympathetic fibers synapse with the postganglionic neurons in the coeliac ganglion (Laughton and Powley 1979; Weaver 1980; Powley and Laughton 1981; Rohner-Jeanrenaud et al. 1983; Steffens and Strubbe 1983). However, there are also "unorthodox" direct fibers from the spinal cord to the islets (Luiten et al. 1984). Figure 7.1 summarizes the essential features of the neural pathways involved in the insulin secretion, as outlined above. It must be emphasized that this is a preliminary model, and that it pertains only to an as yet unknown number of mammals. No such detailed data are available from nonmammalian vertebrates.

The efferent innervation of the pancreas is considerably more complex than previously thought. The notion that, for example, in exercise the sympathetic nervous system affects the islets by alpha-adrenergic inhibition of insulin secretion and beta-adrenergic stimulation of glucagon secretion (cf. Porte and Woods 1983; Unger 1983; Schuit and Pipeleers 1986) probably requires some modification because of the co-release of catecholamines and messenger peptides (cf. Smith and Davis 1983). Ultrastructurally, there are at least four basic types of nerve endings in the pancreatic islets. The first is classically held to be cholinergic and contains a predominance of clear 30–50 nm vesicles and a few 80–90 nm vesicles with moderately electron-dense cores. The second is adrenergic and contains two populations of 30–50 nm vesicles, one being clear and the other having a small dense core (Fig. 7.4). Both cholinergic and adrenergic endings have been observed in the islets of most gnathostomes, including fish. A third type, found in the islets of certain primates, contains mainly large dense-cored vesicles approximately 10 nm in diameter (Forssmann and Greenberg 1978). The fourth type, which was observed in large numbers in teleost islets (Klein 1971; Brinn 1973, 1975), contains small clear vesicles and large dense-cored vesicles, measuring 150–200 nm in diameter. Similar large granules were also seen in a snake (Trandaburu and Calugareanu 1966; see also Fig. 7.3). It is almost certain that these are peptidergic, but there is no question that some, perhaps all of the "cholinergic" and "adrenergic" fibers also contain messenger peptides (cf. P. H. Smith and Davis 1983). Based on Burnstock's criteria (Burnstock 1972), Brinn (1975) suggested that the nerves with large granules could be purinergic, but with more recent evidence (Bloom and Polak 1981) they are now thought to be primarily pep-

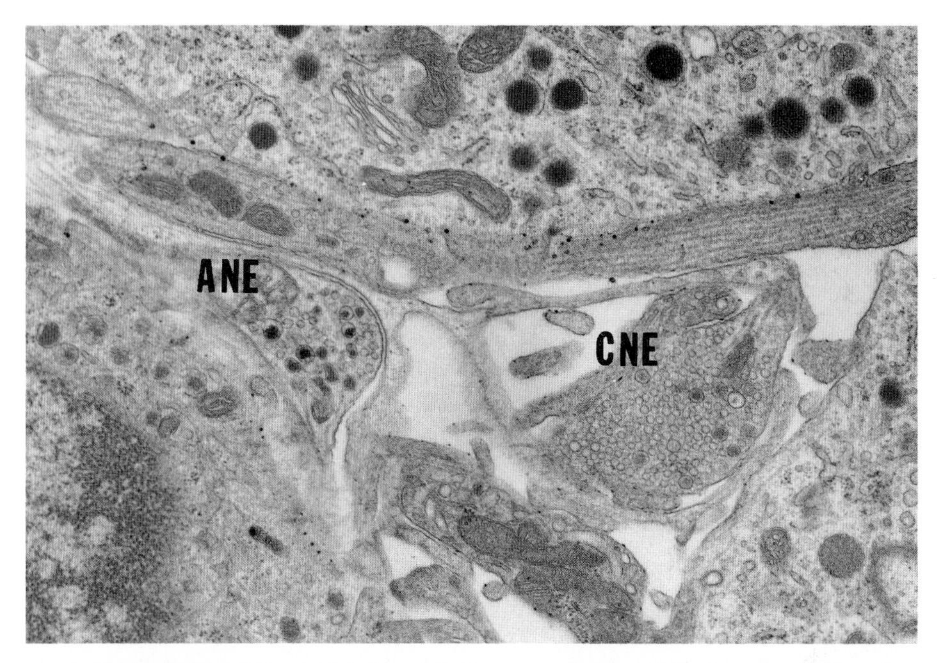

Fig. 7.4. "Adrenergic" (*ANE*) and "cholinergic" (*CNE*) nerve terminals in the islet of a rhesus monkey (Epple and Brinn 1975); ×25 000

tidergic and will be referred to as such in the remainder of this discussion. Islet nerves with such large, peptidergic granules may be widespread among "lower" vertebrates, since they have been identified meanwhile in several teleosts (Fig. 7.5 A), an amphibian, *Necturus maculosus* (Fig. 7.5 B), additional species of snakes (Buchan et al. 1984) and the alligator, *Alligator mississippiensis* (Buchan et al. 1982). Possibly the neuropeptide in most of these nerve endings in poikilotherms is VIP, which has been found by immunocytochemical staining in pancreatic nerves of man, rat, pig, dog, cat, mink, duck, chicken, alligator, snakes, and fish (Larsson et al. 1978; Sundler et al. 1978; Bishop et al. 1980; Van Noorden and Patent 1980; Buchan 1984). However, the presence of PHI, a peptide very similar to VIP, in intestinal neurons requires a reinvestigation of these data. In the perfused rat pancreas, PHI has strong insulotropic effects (cf. Szecòwka et al. 1983). Similarly, the presence of NPY immunoreactive neurons in the islets of the anglerfish (*L. americanus*) suggests the presence of further regulatory peptides in islet nerves (Noe et al. 1986). In the cat and pig, VIP positive fibers supply intrapancreatic ganglion cells, but not the islets (L.-I. Larsson et al. 1978). Instead, the islets of the cat contain many nerve terminals that are positive for an antiserum that reacted with both cholecystokinin (CCK) and gastrin (cf. Cantor and Rehfeld 1984). This observation was confirmed by Rehfeld et al. (1980), who further showed that the C-terminal tetrapeptide of CCK and gastrin is a potent releaser of islet hormones. In the dog islets, Fujita et al. (1982) describe four different types of neurosecretory terminals. It is interesting to note that the peptidergic terminals of the islet nerves in homeotherms usually do not show the large granules of "lower" vertebrates.

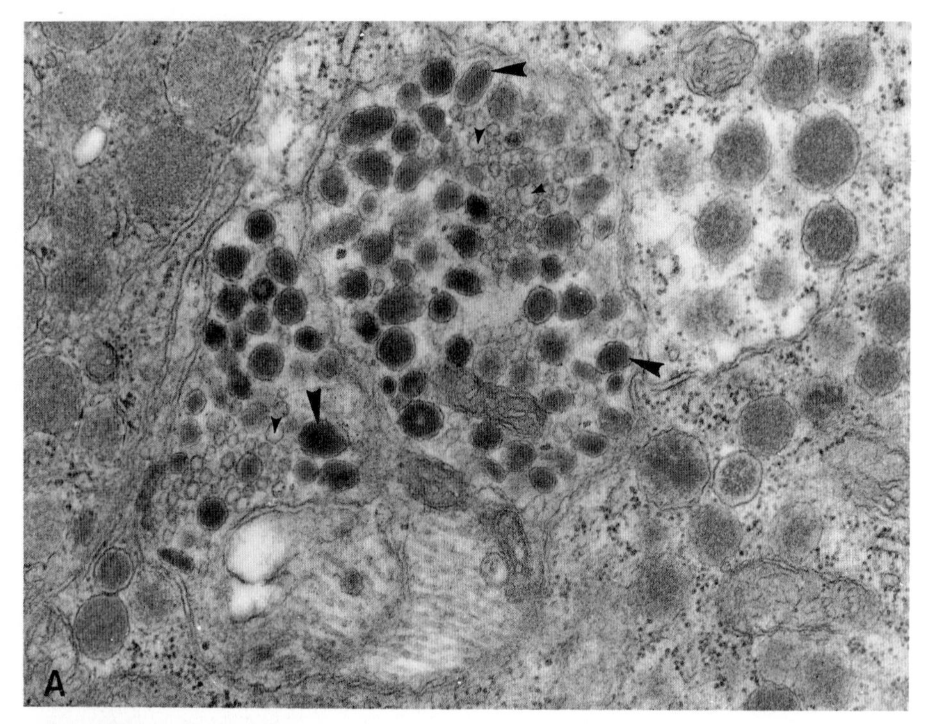

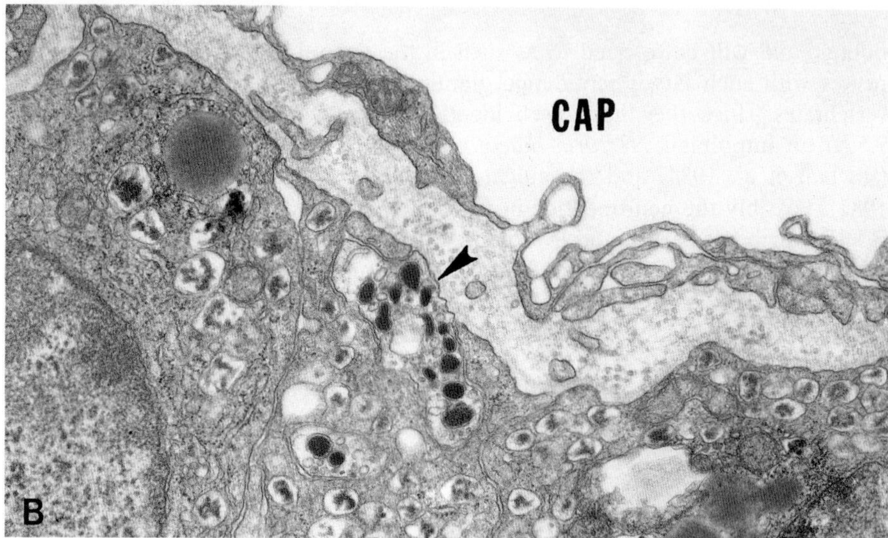

Fig. 7.5. Peptidergic nerve terminals in two poikilotherms. *A* Two terminals in the islet of an American eel (*Anguilla rostrata*). The larger dark ganules (*large arrowheads*) probably contain VIP, while the contents of the smaller clear vesicles (*small arrowheads*) are unknown. The larger granules surrounding the nerve endings are contained in endocrine cells; ×39900. *B* Nerve terminal (*arrowhead*) in the islet of an urodele amphibian, *Necturus maculosus.* Note that the unusually large granules reach the size of the granules in the adjacent B-cells. In this case, it is not clear if, (1) the granules represent neurosecretion, to be released into the subendothelial space (adjacent light area) of the capillary (CAP); (2) their contents are transmitter substances; or (3) serve both purposes; ×25000

The coexistence of classical neurotransmitters and messenger peptides in central and peripheral neurons is an accepted fact, and it will be necessary to subdivide both the sympathetic and parasympathetic nervous systems, probably according to the peptides present (Lundberg et al. 1982; Lundberg and Hökfelt 1983). Thus, one wonders how the inventory of islet nerves with various combinations of transmitters and peptides will appear in another decade. A foretaste may be provided by a report on primate neuromuscular junctions, in which Chan-Palay et al. (1982) found coexistence of enzymes synthesizing acetylcholine, catecholamines, taurine, and GABA.

The relationship of the nerve endings with islet cells varies phylogenetically, with the greatest specialization occurring in the teleosts. In this group it is not unusual to find prominent synaptic complexes with presynaptic membrane thickenings and other features common to central nervous system synapses (Brinn 1975). Patent et al. (1978) described two populations of nerve endings in the islets of a teleost, *Gillichthys mirabilis:* (1) vagally derived parasympathetic endings with large (130 nm) electron-dense granules and (2) sympathetic fibers from the coeliac ganglion which contained 100-nm electron-dense granules. Synaptic specializations are generally not found in islets of higher vertebrates, the only other type of direct neuroendocrine contact thus far observed is the gap junction between nerves and islet cells in the rat (Orci et al. 1973). In most species, nerves simply terminate in the intracellular space, so that one wonders if this is not a way to supply simultaneously different types of neighboring islet cells with an "option" to respond or not to respond to the nervous signal, depending on the state of their receptors. Peculiar associations between nerves and islet cells have been found in the dog, where Schwann cells send delicate processes around groups of endocrine cells, probably forming microdomains (P. H. Smith 1975). In the mouse, islet cells are in juxtaposition to axon bundles and nerve cell bodies, with or without intercalated Schwann cell processes (Fujita et al. 1979; Fujita 1983). The functional significance of these neuro-insular complexes remains to be elucidated.

It appears that, depending on the species, all four major types of islet cells may be supplied by neurons. Interestingly, there are not only quantitative species differences in islet innervation. Even one and the same type of neuron may supply different types of islet cells in different mammals (cf. P. H. Smith and Davis 1983).

Considering the ease with which the APUD paraneuron system can create or remove complete endocrine glands (Epple and Brinn 1980), we should not be too surprised about the strong phylogenetic variations of the islet innervation. On the other hand, the large number of potential factors controlling the activity of a pancreatic B-cell (Fig. 7.6), makes one wonder if this cell may sometimes receive more signals than necessary. Perhaps variations in the islet innervation reflect to some extent removal of duplication of services. The phylogenetic variations of the degree of the islet innervation are usually paralleled by the nervous control of the adenohypophysis (Epple and Brinn 1975). As in the case of the islet organ, the nervous control of the adenohypophysis appears to be strongest in the teleosts, and weakest in the cyclostomes.

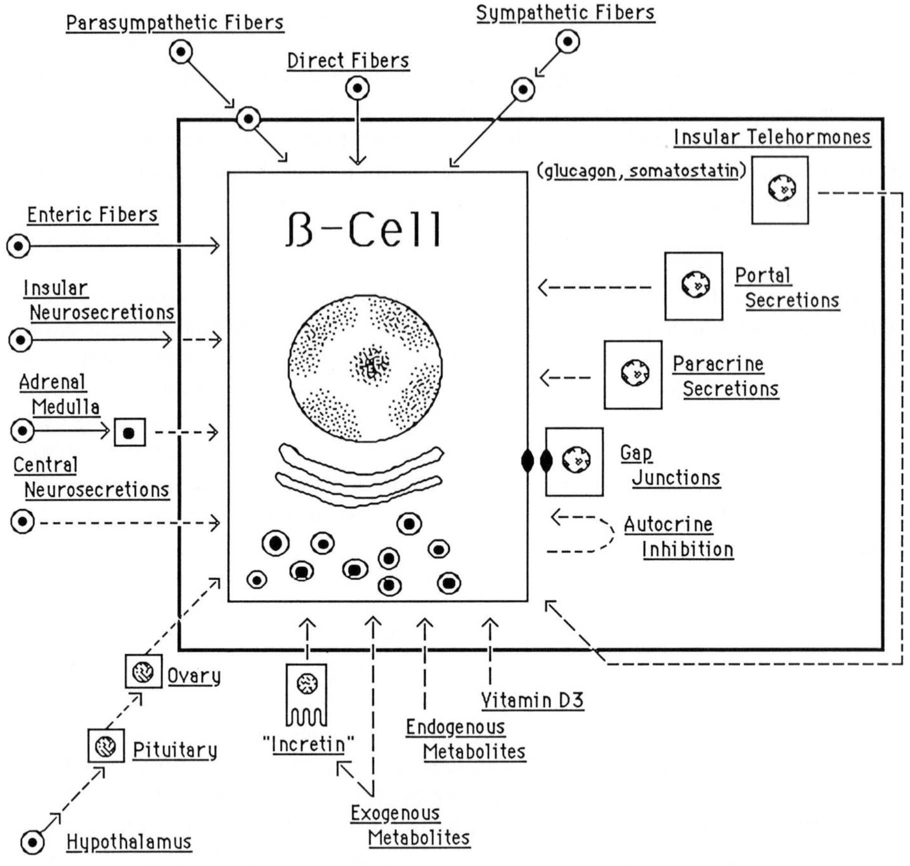

Fig. 7.6. Factors known or postulated to affect B-cell activity. For simplicity, this selection does not include blood flow, norepinephrine from intra-adrenal sources, additional pituitary and peripheral hormones of suspected insulotropic impact, glucagon and somatostatin from gastrointestinal sources, and factors modulating receptor sites. For further discussion, see Chapters 8 and 9. *Outer rectangle*: border of islet; *round cell bodies* neuronal perikarya; *uninterrupted lines*: axons; *rectangular cell bodies*: endocrine cells; *cell with microvilli*: "open" incretin cell of gut mucosa; *interrupted lines*: humoral pathways

7.2 Control of Islet Hormone Secretion

It is difficult to identify the impact of the islet innervation on hormone release in vivo since it acts in concert with other factors, the major ones being hormones and metabolites. The relative contribution of these factors seems to vary considerably with the overall situation (e.g., fed or fasting state), and certainly among species. In all likelihood, such differences explain the varying effects of pancreas denervation, which are particularly inconclusive after sympathectomy (for literature see Watanabe 1983). Berthoud (1984), who has recently tried to estimate the contributions of the nervous system, hormones and metabolites on the total insulin response during a meal in the rat, found that neurally mediated insulin response amounts to at least 26%, while the contributions of hormones and glucose were 30% and 20%, respectively. About 23% of the total insulin response could not be segregated.

From the preceding (Sect. 7.1) it is clear that the nervous control of the islet secretions can no longer be interpreted as a system of cholinergic and adrenergic counterbalances. Rather, a variety of co-released substances from nerve terminals may act in vivo, the spectrum of which (a) may be as large as that of the APUD secretions (see Chaps. 8.1 und 9.1), and (b) may vary greatly among species. Hence, in vitro-studies with individual neurotransmitters (or putative neurotransmitters), as summarized in the following sections, must be interpreted with great caution. Furthermore, as discussed with respect to the catecholamines in Sect. 8.1, the issue is often beclouded because the design of many studies does not allow to draw conclusions beyond the presence of receptor types. It is obviously easy to ignore the more complex questions, namely (a) to determine if an effect mimics either hormonal or synaptic mechanisms; or (b) if it is biologically meaningful at all. However, it is now clear that sympathetic neural and adrenomedullary activation can be dissociated (J. B. Young et al. 1984); hopefully, this possibility will be considered in future studies on islet innervation.

7.2.1 Insulin

According to the widespread "classical" picture acetylcholine released from parasympathetic nerve terminals is the most potent nervous stimulus of insulin secretion in mammals (Porte and Woods 1983; P. H. Smith and Davis 1983; Curry 1984; Berthoud 1984). This effect is antagonized by norepinephrine, released from sympathetic terminals, the impact of which depends on the type of receptor and the duration of the stimulus (see also Chap. 8.1.1). A short-term stimulus acts on alpha-adrenergic receptors to inhibit insulin release, whereas a prolonged stimulus activates beta receptors and enhances insulin release (cf. P. H. Smith and Davis 1983). However, this picture is probably too simple since the net effects of acetylcholine and norepinephrine on insulin secretion seem to be modified by rather complex interactions with peptidergic co-secretions (Raghu et al. 1984) as well as metabolic and hormonal inputs (Campfield and Smith 1983). Very little seems to be known about the nervous control of release of insulin in nonmammalian vertebrates. Without proof of a direct B-cell innervation in the chicken (see, however, Watanabe 1983), it is interesting that perfusion of the pancreas of this species with acetylcholine and 14 mM glucose causes a greater rise in insulin secretion than 42 mM glucose alone (Honey and Weir 1980). On the other hand, D. L. King and Hazelwood (1976) could not demonstrate vagal control over release of insulin in chickens. We are not aware of any publication on cholinergic modulation of insulin release in poikilotherms; the possible role of adrenergic mechanisms will be discussed in Chap. 8.1.

7.2.2 Glucagon

As in the case of insulin, glucagon secretion in mammals appears to be frequently controlled by interplay of sympathetic and parasympathetic stimuli, the impact of which is modulated, inter alia, by hormones and metabolites (Palmer and Porte 1983; Steffans and Strubbe 1983). In experiments designed according to the "classical" pattern, splanchnic stimulation and beta-adrenergic agonists cause glucagon secretion, particularly when plasma-glucose levels are below normal (Holst et al. 1981b; Unger 1983); also, vagal stimulation and cholinergic mechanisms enhance glucagon release (cf. P. H. Smith and Davis 1983). The glucose-dependent beta-adrenergic effect is considered to be a mechanism whereby glucose can be quickly mobilized in stress situations (Unger 1983). Thus, while the A-cell has basically the same neural input as the B-cell, the impact of the stimuli is very different. This has been demonstrated very effectively in the rat, in which stimulation of the splanchnic nerve increased secretion of glucagon and decreased that of insulin (Bobbioni et al. 1983). Using the alpha-blocker, phentolamine, Hisatomi et al. (1985) found inhibition of glucagon release from the perfused, isolated (!) pancreas of the rat. Since it is unlikely that cut distal axons respond to this treatment, their preparation may have contained "ectopic" sympathetic ganglion cells. Alternatively, the drug could have blocked the effect of insular catecholaminergic co-secretions (see Chap. 9.1.3). A role of synaptic co-secretions in the control of glucagon release has been shown in several studies. For example, vagus-stimulated glucagon secretion from the perfused pig pancreas is blocked by hexamethonium and not atropine, suggesting that a peptide (VIP?) is the causative agent (Holst et al. 1981a). VIP augments the carbachol-stimulated glucagon release in perfused mouse pancreas (Ahrèn and Lundquist 1982), and CCK_4 is a potent stimulant of glucagon release in the porcine pancreas (Rehfeld et al. 1980). No comparable data on the impact of neuropeptides on the glucagon release in other groups of vertebrates seem to exist; the only study on the impact of acetylcholine known to us was conducted in the perfused chicken pancreas in which this compound also stimulated glucagon release (Honey and Weir 1980).

7.2.3 Pancreatic Polypeptide

Vagal activity appears to be a major stimulus of PP secretion (Floyd and Vinik 1981; Hazelwood 1981; R. E. Miller 1981; Edwards et al. 1980; Schwartz 1980) although, at least in the dog, the vagus may suppress PP secretion that occurs in response to intravenous amino acids or glucose (A. Inui et al. 1986). Electrical stimulation of the porcine vagus increases PP release (Schwartz et al. 1978), a response that is mostly blocked by atropine and completely inhibited by hexamethonium. The blockage by hexamethonium and the potent PP secretagogue effect of CCK_4 (Rehfeld et al. 1980) suggest that neurotransmitters other than acetylcholine are involved in PP stimulation. In fetal sheep, baseline values are decreased by atropine (Shulkes and Hardy 1982). The feeding-associated increase in PP, particularly the first phase, which is attributable in part to vagal input to the pancreas (see references in Floyd and Vinik 1981), is blocked by vagotomy and parasympathetic blockade. In man, the second phase is present after truncal vagotomy but is inhibited by atropine, suggesting an extravagal cholinergic mechanism (Glaser et al. 1983). Beta-adrenergic mechanisms have also been implicated in post-prandial PP secretion (Linnestad et al. 1983). In chickens, left vagotomy blunts the PP secretory response to feeding and atropine causes a sharp decline in plasma PP after feeding (Kimmel et al. 1978). No data are available for poikilotherms.

7.2.4 Somatostatin

Like the other islet cells, the D-cells of many species are innervated. However, the precise nature and significance of this innervation are largely unknown. Experimental data from mammals suggest that splanchnic nerve stimulation and acetylcholine decrease somatostatin release from the islets, which is exactly the opposite of the effect of this neurotransmitter on insulin, glucagon, and PP release. Beta-adrenergic stimulation increases somatostatin release, while alpha-adrenergic stimulation decreases it. VIP has been found to stimulate somatostatin release. Thus, it appears that adrenergic stimuli and VIP, contrary to acetylcholine, affect the secretions of A-, B- and D-cells in a like manner (for literature see R. E. Miller 1981; Smith and Madson 1981; Woods et al. 1981; Gerich 1983a; Porte and Woods 1983; Smith and Davis 1983). It is difficult to evaluate the physiological significance of these data, which are derived from mammalian systems, often from perfusion or in vitro studies with very high doses. Furthermore, there can be little doubt that comparative studies will uncover considerable species or taxon differences in the nature or role of D-cell innervation. Thus, while the nervous control of pancreatic somatostatin release in mammals awaits further clarification, no experimental data from other groups of vertebrates seem to exist.

7.3 Conclusions

Despite the frequent allusion to the pancreatic islets as heavily innervated structures, an unbiassed evaluation of the pertinent literature reveals enormous interspecific differences, even among mammals. There is no question that an understanding of the islet innervation will require many more in-depth studies in well-selected species, and that the classical one synapse-one transmitter concept must be replaced by an APUD-type synaptic cocktail model (see e.g., Fig. 9.1).

Chapter 8

Hormonal Control of Pancreatic Islet Functions

A hormonal control of the pancreatic islets has been suspected for many decades, and a number of mechanisms have been implicated. In particular, the association of diabetes mellitus with pituitary- and adrenal-related aberrations (acromegaly, Cushing's syndrome), the alleviation of diabetic symptoms by removal of pituitary and adrenal glands, and the induction of diabetic symptoms or islet alterations by injections of pituitary and adrenal hormones, raises the question of specific insulotropic hormonal effects (see e.g., Bratusch-Marrain 1983; Ganda and Soeldner 1983; Pek and Spangler 1983; Schade and Eaton 1983; Volk and Wellmann 1985e). However, no specific adenohypophysial or pituitary-dependent insulotropic hormones have been identified with certainty. Rather, the effects of these hormones (with the possible exception of the ovarian and thyroid secretions) seem to have been due to indirect, mainly metabolic, interactions. On the other hand, direct insulotropic effects of adrenomedullary hormones have been identified, and new insulotropic candidates have been ascribed to two additional endocrine systems: (1) hypothalamus; (2) gastrointestinal mucosa. The hypothalamic insulotropic secretions require further studies; the gastrointestinal "incretins" are poorly characterized even in mammals. Since we discussed the hypothalamic insulotropic factors in Chap. 7, we will deal in the following only with the adrenomedullary secretions and the incretins. However, it is by no means certain that the list of insulotropic hormones is closed. Intra-insular receptors for vitamin D_3 and estrogen have been described and may prove to be of physiological significance (S. A. Clark et al. 1980; Winborn et al. 1983). In the chicken pancreas, vitamin D-dependent calcium-binding protein is exclusively localized in B-cells (Roth et al. 1982). Recently, it has been suggested that 1,25-dihydroxyvitamin D_3 plays an essential role in normal insulin secretion (Kadowaki and Norman 1985a, b).

8.1 Adrenomedullary-Insular Interactions

Although the frequent association between pheochromocytomas and diabetes mellitus has been noted for many years, the interactions between plasma catecholamines and islet secretions have remained an area of confusion, with a flood of often contradictory in vivo and in vitro data in stark contrast with a few findings of proven biological relevance (for literature see Charles 1981; R. E. Miller 1981; Woods et al. 1981; Gross and Mialhe 1982; Meglasson and Hazelwood 1983; Palmer and Porte 1983; Cryer et al. 1984; Vranic et al. 1984; Landsberg and Young

1985; Tilzey et al. 1985b). A major reason for this state of affairs has been the use of doses of exogenous catecholamines that, in many cases, may have been too high to correspond to plasma titers and too low to mimic synaptic concentrations; and a lavish use of synthetic agonists and antagonists, although it is largely unknown how their impact can be equated with variations of the endogenous catecholamine fractions. With the advent of new, highly sensitive, and specific radioenzymatic assays (Da Prada and Zürcher 1976; Peuler and Johnson 1977) for the simultaneous determination of dopamine, norepinephrine and epinephrine, it has become clear that intravascular bolus injections of $1-2\ \mu g\ kg^{-1}$ of these compounds create plasma titers within the physiological range while doses above $5\ \mu g\ kg^{-1}$ are excessive (see e.g., Dashow and Epple 1983; Epple and Nibbio 1985). However, the vast majority of investigators applied essentially higher doses. The same problem arises with earlier studies using continuous infusion, and the dilemma is still perpetuated in many in vitro investigations. Furthermore, it is worthwhile to note that the widespread use of insulin-induced hypoglycemia as a tool for studies on adrenomedullary counter-actions may be of limited physiological relevance. In the human, adrenalectomy has no impact in the return of the blood sugar to normal levels following insulin injections (Järhult et al. 1981).

8.1.1 Problems in the Use of Exogenous Catecholamines

Recent findings suggest that three additional factors may be of importance for both design and interpretation of studies with plasma catecholamines: (1) catecholaminotropic effects of circulating catecholamines; (2) the effects of co-released messenger peptides; (3) a kick-like rather than tonic mode of action of catecholamines, at least on some target cells.

The catecholaminotropic effect of catecholamines was first described in the lamprey (*Petromyzon marinus*) in which epinephrine causes a dose-related release of dopamine and norepinephrine (Dashow and Epple 1983), and has been confirmed for the eel, *Anguilla rostrata* (Epple and Nibbio 1985) and the rat (Epple et al. in prep.). The issue is rather complex because in the eel, but not in the lamprey, dopamine and norepinephrine cause, respectively the release of the respective other two catecholamines; in the rat dopamine causes the release of epinephrine and probably norepinephrine, also. Thus, in vivo application of one catecholamine may stimulate the release of one or two other catecholamines, and the resulting hormonal constellation must inevitably affect the integration of alpha- and beta-adrenergic, and dopaminergic mechanisms in a manner different from that of a single catecholamine applied in vitro. At this point, we can only speculate on the extent of supposedly overlapping receptor effects of individual catecholamines in vivo that were actually due to other, endogenously released catecholamines, and how differing ratios of circulating catecholamines, induced by varying quantities of epinephrine, modulate adrenergic mechanisms. At least on theoretical grounds, one could postulate that in some cases a beta-adrenergic glucogenic effect of epinephrine is counterbalanced by an alpha-adrenergic inhibitory effect of endogenous norepinephrine; perhaps this type of counterbalance by endogenous norepinephrine explains why only high doses of epinephrine cause an alpha-adrenergic inhibition of insulin secretion in the human (cf. Halter et al. 1984), a species in which epinephrine seems to be a physiological stimulator of norepinephrine release (Musgrave et al. 1985). If one adds to species-related catecholaminotropic effects species-related varying ratios of alpha- and beta-adrenergic receptors on target cells (see e.g., Cherksey et al. 1981), one cannot be surprised to find considerable interspecific differences in insulotropic catecholamine effects (cf. Loubatieres-Mariani et al. 1977, 1980; Clutter et al. 1980; Gray et al. 1980; Meglasson and Hazelwood 1983, 1984; Tilzey et al. 1985 b). Clearly, the impact of plasma catecholamine *interactions* must be considered in the evaluation of adrenergic mechanisms of the islet organ, particularly during stress or application of large exogenous doses of catecholamines in vivo.

The functional implications of co-released messenger substances during the secretion of APUD cells have been discussed in Sect. 8.1. In the adrenal medulla the smaller secretions (i.e., the catecholamines) are believed to represent *the* hormones, while the co-released peptides serve auxiliary functions. Whether the situation is really so simple should become clear within a few years. At this point, we are faced with a considerable number of adrenomedullary co-peptides, such as enkephalins, neuropeptide Y, VIP, substance P, neurotensin and somatostatin (cf. Polak and Bloom 1982; Adrian et al. 1983; Leboulenger et al. 1983 a, b; Viveros et al. 1983; Wu et al. 1983; Boarder and McArdle 1984; Kondo and Yui 1984; Varndell et al. 1984; Jarry et al. 1985). Conceivably, one or more of these peptides may modulate the interactions between plasma catecholamines and islet hormones at one or more sites (El-Tayeb et al. 1985), and one wonders if or how these adrenal-derived substances might interact with identical or related co-peptides of the islet organ (see Table 9.1). Based on findings in dogs, Werther et al. (1984) suggest that opioids, co-released during stress, may increase the insulin effect at peripheral sites by enhancing glucose entry into muscle and other tissues, while simultaneously increasing glucose output by antagonizing insulin in the liver. Clearly, the possible impact of co-released peptides from chromaffin cells adds another caveat for the interpretation of adrenomedullary-islet interactions, especially when based on in vitro experiments.

The question of a kick-like rather than tonic mode of catecholamine action arose with observations in the eel (Epple and Nibbio 1985). In this species, small doses of epinephrine, which scarcely affected the "normal" plasma titer of this substance ($<100\ \mathrm{pg\ ml^{-1}}$), caused a significant hyperglycemia, whereas slightly higher doses caused hyperglycemia plus a strong increase in plasma dopamine and norepinephrine. However, there was no hyperglycemia in stressed eels, even when the epinephrine titer exceeded $1000\ \mathrm{pg\ ml^{-1}}$; yet, regardless of the baseline titers, injection of epinephrine caused a strong increase of dopamine and norepinephrine (within 3 min), as well as of glucose (within 1 h), just as in unstressed animals. The most likely explanation of this seeming paradox is that the catecholaminotropic and glycemic effects of circulating epinephrine depend on spurt-like increases, and not on its plasma titer per se. That this "kick" effect may also be operating in the glycemic control of humans is suggested by the data of Rizza et al. (1979) and Sherwin and Sacca (1984). In both publications, the respective Fig. 1 shows an initial increase in plasma glucose in response to epinephrine infusion that soon reaches a plateau, in parallel with the epinephrine titer. However, after an increase of the infusion rate, there is a second drastic increase of glucose production. Interestingly, in neither study was there a significant change in the plasma titers of insulin and glucagon in response to the increase in the epinephrine delivery. Since an effect of epinephrine on the B-cells of mammals is clearly established (see below), it remains to be seen if the B-cells respond indeed to a mode of stimulation different from epinephrine-sensitive structures involved in the glycemic and catecholaminotropic effects. Finally, it appears possible that discrepancies in the impact of catecholamines on insular blood flow (cf. Rooth et al. 1985) are due to differences in the delivery (injection vs. extended infusion) of these substances.

8.1.2 Comparative Aspects

In mammals, the hyperglycemic effects of epinephrine have often been linked to its impact on secretions of insulin and/or glucagon. However, more recently it became clear that epinephrine acts via a number of mechanisms, both insular and extra-insular: (1) direct stimulation of hepatic glycogenolysis; (2) direct (?) stimulation of hepatic gluconeogenesis; (3) mobilization of glucogenic and ketogenic substrate from muscle and adipose tissue; (4) direct suppression of glucose uptake caused by the "mass effect" of hyperglycemia; (5) suppression of glucose clearance; (6) modulation of insulin release; (7) possibly, stimulation of glucagon release (for literature see Bahnsen et al. 1984; Cherrington et al. 1984; Cryer et al. 1984; Halter et al. 1984; Knudtzon 1984; Meglasson and Hazelwood 1984; Sherwin and Sacca 1984; Cryer 1985). In many mammals, plasma catecholamines seem to stimulate insulin release via a beta-adrenergic mechanism (see also Chap. 7.2.1), and to inhibit it by an $alpha_2$ mechanism, the (ob/ob) mouse being an exception since in this strain the $alpha_1$-receptor agonist phenylephrine inhibits insulin secretion (El-Denshari et al. 1981). However, this picture may be too simplistic, since Ahrèn et al. (1984) recently presented data suggesting that in the rat $alpha_1$-, $alpha_2$- and beta-adrenoceptors are involved in the regulation of basal insulin release. In the human, the suppressive alpha-adrenergic effect of epinephrine on insulin secretion (Rizza et al. 1980a, b) seems to play a role only in stress situations, since in this species the acute inhibition of basal insulin levels (Clutter et al. 1980) requires epinephrine titers of about 400 pg ml^{-1}, i.e., in the lower stress range (Halter et al. 1984). On the other hand, infusion of epinephrine in the dog causes an immediate sharp increase of insulin output (in the absence of a glycemic effect), suggesting that, under the conditions of these experiments, the beta-receptors of the canine B-cells were more activated (or sensitive?) than the alpha-receptors (cf. Gray et al. 1980; Loubatieres-Mariani et al. 1980). The release of glucagon via beta-adrenergic stimulation is well established (cf. Halter et al. 1984; Schuit and Pipeleers 1986), but it remains to be seen if this is a physiological phenomenon (Rizza et al. 1979; Clutter et al. 1980; Gray et al. 1980; Cryer 1985). The data on the possible adrenergic modulation of PP-release are particularly confusing (cf. Meglasson and Hazelwood 1983). However, in humans epinephrine infusions are without effect on PP plasma titers (Sive et al. 1980), and the hypoglycemia-induced rise in PP titers occurs also after adrenalectomy

(Järhult et al. 1981). The adrenergic control of the islet secretions in birds has been discussed by Gross and Mialhe (1982) and Meglasson and Hazelwood (1984). A decrease of plasma levels of insulin in the duck following infusions of both epinephrine and norepinephrine has been reported by Tyler and Kajinuma (1972), an observation that has been corroborated by findings of Gross and Mialhe (1982). Since, on the other hand, an inhibitory effect of the beta-agonist isoproterenol on glucagon release (Tyler and Kajinuma 1972) was probably due to a slight hyperglycemia, Gross and Mialhe (1982) concluded from a study with the beta-adrenergic blocker propanolol in the duck (1) that a normal beta-adrenergic stimulation is necessary for the B-cell to maintain basal insulin levels and its sensitivity to glycemic variations; and (2) that this is not the case for the A-cell, which remains sensitive to glycemic changes despite the presence of beta-adrenergic blockade. In the chicken, no clear physiological effect of catecholamines could be shown in vivo or in vitro. If there is any impact of circulating catecholamines on insulin release at all in this species it may be stimulatory and mediated via beta receptors (cf. Meglasson and Hazelwood 1984). On the other hand, in vitro studies with isoproterenol in the chicken suggest that secretion of PP may be stimulated via a beta-adrenergic mechanism (Meglasson and Hazelwood 1983). Very little is known on the direct impact of plasma catecholamines on the islet organ of poikilotherms. In many studies, epinephrine raised the blood sugar level, while on the other hand, the effects of norepinephrine varied greatly (for literature see Plisetskaya 1975). In vitro, high concentrations of epinephrine inhibited the insulin release in long-term cultures of urodelan (*Amphiuma means*) pancreas (Gater and Balls 1977). In the in situ perfused pancreas of the eel (*Anguilla anguilla*) high concentrations of epinephrine (10 and 50 ng ml^{-1}) reduced total insulin secretion in response to 10 mM lysine (Ince 1980); in intact, cannulated animals of the same species, single injections of epinephrine (25 and 50 μg kg^{-1}) and norepinephrine (50 and 100 μg kg^{-1}) caused a biphasic response insulin release, with an initial depression lasting 30–60 min. Repeated injections of 100 μg kg^{-1} of norepinephrine suppressed the subsequent elevation of plasma insulin, which otherwise peaked 180 min after the initial injection of epinephrine. On the other hand, an insulin response within a few minutes to an arginine stimulus (25 mg kg^{-1}) was totally suppressed by the simultaneous injection of 50 μg kg^{-1} epinephrine (Ince and Thorpe 1977). Injections of epinephrine (1–2 mg kg^{-1}) into scorpion fish (*Scorpaena porcus*) yielded very similar biphasic responses of plasma insulin though the results were modified by the pre-injection titer on insulin (Plisetskaya et al. 1976). Using an in vitro (superfusion) system, Tilzey et al. (1985a, b) found that physiological concentrations of epinephrine (10^{-10} M) inhibited insulin release from Brockmann bodies of the rainbow trout (*Salmo gairdneri*) while excessive doses (10^{-6} M) caused a biphasic release. In the holocephalian, *Hydrolagus colliei*, injections of 0.5 mg kg^{-1} of norepinephrine, but not of epinephrine, caused a continuous decrease in blood-sugar level, which conceivably may have been due to stimulation of insulin release. No such effect of norepinephrine was observed in a shark, *Squalus acanthias* (Patent 1970). No data on a direct impact of catecholamines on plasma insulin in ammocoetes seem to be available; in adult lampreys (*Lampetra fluviatilis*), huge doses of epinephrine affect the insulin titer though the impact must depend either on the season or the endogenous insulin titer present (see Chap. 10.4.2). On the other hand, insulin provoked a mammalian-like increase in plasma epinephrine, presumably in response to the ensuing hypoglycemia (Plisetskaya and Prozorovskaya 1971). In vitro, 1 mM of epinephrine-inhibited insulin release from islets of the Atlantic hagfish, *Myxine glutinosa* (Emdin 1982b). As pointed out above, it has become clear in recent studies with radioenzymatic methods that doses of exogenous catecholamines of more than 5 $\mu g\ kg^{-1}$ are excessive; it may be added here that the highest stress titers of catecholamines in intact, unanesthetized lampreys and eels so far identified radioenzymatically are below 5 $ng\ ml^{-1}$ (Dashow et al. 1982; Epple and Nibbio 1985) though injections of 2 $\mu g\ kg^{-1}$ or more of exogenous catecholamines may cause essentially higher levels of norepinephrine. In one extreme case, the norepinephrine titer of a stressed eel rose after 16 $\mu g\ kg^{-1}$ of epinephrine from about 2.5 to 112 $ng\ ml^{-1}$). Thus, a realistic assessment of our current state of knowledge suggests that, in nonmammalian vertebrates, we have no convincing evidence of an impact of circulating catecholamines on the release of islet hormones from intact animals under physiological conditions.

This state of affairs is particularly regrettable, since findings of Clutter et al. (1980) in the human should provoke a comparison with the evolutionary stages of catecholamine functions. From this work, it appears that (1) the chronotropic cardiac effects of epinephrine occur at lower doses than all other effects of this hormone; (2) lipolytic effects (blood glycerol) appear at lower doses than glycemic and ketogenic effects, (3) suppression of insulin release requires essentially higher levels; and (4) stimulation of glucagon secretion must require epinephrine titers in the upper stress or pharmacological range. Does this hierarchy of responses reflect, at least to some extent, evolution of functions of plasma epinephrine? Epinephrine has cardiovascular and respiratory, but apparently no metabolic functions in cyclostomes (Dashow and Epple 1983, 1985; Macey et al. 1984; Plisetskaya et al. 1984b). Only in the lower gnathostomes (studies in the eel), do metabolic effects appear in addition to the other functions (Dashow et al. 1983; Epple and Nibbio 1985). Similarly, one wonders if the findings of Clutter et al. (1980) can be related to the ontogeny of the epinephrine functions, since catecholamines are involved in fetal cardiovascular and respiratory functions (Phillippe 1983) before they become critical for the initiation of the glucose control by the newborn (Sperling et al. 1984). Are, perhaps, the insulotropic effects of plasma epinephrine phylogenetically and/or ontogenetically very late developments foreshadowed by the sensitivity of the islet cells of lower vertebrates to excessive doses of this hormone?

8.2 The Entero-Insular Axis

Teleologically, it would make sense to supply the islet organ with messages from various regions of the alimentary tube so that optimal integration of digestive and metabolic activities is assured (see also Chap. 14). By definition, an entero-insular axis can only occur when the islet organ has formed during phylogeny, i.e., at the ammocoete state (see Chap. 2.3). However, the interaction between different types of gastrointestinal paraneurons may well have evolved at an earlier stage, perhaps even among the coelenterates. Here, it seems indeed difficult to separate clearly neuronal and paraneuronal function (Fig. 1.1). We have emphasized (Epple et al. 1980) that islet hormones can reach the gastrointestinal tube only after passage through the liver and the general circulation; it is also clear that, vice versa, gastrointestinal hormones can reach the islets only by the same telehormonal route, since there are no direct vascular connections between gut and the islet organ. Consequently, most insulotropic messenger substances (if present) must pass through a dual portal system in the ammocoete stage, and a single hepatic portal system in adult lampreys and gnathostomes (Fig. 2.6).

We do not yet know when the "enteroinsular axis" (Unger and Eisentraut 1969) appeared phylogenetically. It has been speculated that it may already function at the cyclostome level, since the intestine of the hagfish contains "open" cells with GIP immunoreactivity (Falkmer et al. 1980), and since identical cells may occur in the gastrointestinal mucosa of all vertebrates (Falkmer et al. 1984). Unfortunately, there is no proof that GIP has a physiological role in the regulation of islet activity (cf. Creutzfeldt and Ebert 1985) and as in the case of the catecholamines, we are haunted by a flood of data obtained with unphysiological doses. The original name of GIP, i.e., "*G*astric *I*nhibitory *P*olypeptide", was based on the observation that this substance inhibits the gastric secretion, which is possibly effected through mediation by release of gastric somatostatin (cf. Flaten 1983). Only more recently, the acronym has been reinterpreted as *G*lucose-dependent *I*nsulinotropic *P*olypeptide, based on the observation that exogenous GIP augments glucose-stimulated insulin release. Now, however, it appears that, at least in the human, the insulinotropic effect of GIP only occurs at pharmacological doses (cf. Sarson et al. 1984). Thus, we are left (a) with the likelihood that the mammalian upper small intestine sends out message(s) enhancing insulin response to oral glucose loads; and (b) with the uncertainty as to the nature of the substance(s) involved. Besides GIP, other candidates for this "incretin", a term originally coined for the insulinotropic activity in intestinal extracts (cf. LaBarre and Stille 1930), include gastrin, secretin, VIP, cholecystokinin-pancreozymin, enteroglucagon, the new "brain-gut" peptide PHI (cf. Szecòwka et al. 1983) and perhaps even glucagon-like peptide I (Schmidt et al. 1985). It appears possible that a cumulative effect of several gastrointestinal hormones, or a so-far unidentified substance may be involved. Furthermore, the possible role of neural interactions remains to be clarified (cf. Sarson et al. 1984; Creutzfeldt and Ebert 1985).

The current state of knowledge concerning the impact of gastrointestinal hormones on glucagon secretion by islets has been reviewed by Pek and Spangler (1983). As in the case of insulin, there are many data, but no biological conclusions appear possible. There seems to be an entero-insular axis for PP release from the islet organ (Tsuda et al. 1983), which may play an important role in the second increase of plasma PP during the biphasic responses of this hormone to food. Both the timing and different nature of the prevailing stimuli (protein and fat) indicate that the PP-tropic enteric mechanisms must be different from those involved in insulin release (Schwartz 1980). Several known gut hormones elicit PP release, but none of them appears to have a physiological function (Floyd and Vinik 1981). Gingerich and Kramer (1983) isolated two peptides from mammalian duodenum, both with a significant PP secretagogue activity. Their primary structure is unknown. In man, exogenous somatostatin (S-14 and S-28) suppresses basal PP secretion as well as the response of this hormone to a protein-rich meal (Marco et al. 1983). Although it is not known that the effect is physiological, the extra-insular distribution of the F-cells in many species (see Chap. 4.2) would make a gastrointestinal rather than insular somatostatin effect more likely.

In conclusion, there is no doubt that, at least in mammals, an entero-insular axis exists. It is likely, in the case of insulin and PP, that this axis involves as yet unidentified messengers from the gastrointestinal mucosa. The situation is more uncertain with respect to glucagon. In all three cases, the role of possible neural links remains to be established.

Chapter 9

Functional Strategies of the Islet Organ

Many endocrine glands do not simply function by release of hormones into the circulation. Instead, they deploy a host of "functional strategies", which range from mechanisms at the molecular level to topographic associations (Epple 1982). Probably no other endocrine gland surpasses the islet organ in the number and diversity of such strategies. As a consequence, probably no other endocrine organ presents investigators with more difficulties. The following discussion should make the point.

9.1 Multiple Secretions

A basic fact, which is all too often overlooked in experimental design, is the heterogeneity of the granule contents of the APUD/paraneuron cells (see Chaps. 7.2 and 8.1.1). With the sampling of possible substances in an "APUD granule" illustrated in Fig. 9.1, it must be recalled that probably all of these are released simultaneously with *the* hormone. All islet cells contain variations of this type of granule; furthermore, the four major islet cells differ not only in their hormone production, but also in the elaboration of the additional granule substances such as co-released peptides and small messenger molecules. The significance of these variations is in most cases unknown, and the problem is increased by inter- and ultraspecific as well as age-related, or even seasonal differences (Lange 1973; Feldman 1979; El-Salhy and Grimelius 1981). Furthermore, islet cells need not at all be identical with gastrointestinal endocrines that produce the same hormone(s). For example, by ultrastructural criteria, F-cells of the islet type seem to occur in the pyloric mucosa of the dog; however, in this location they also show gastrin immunoreactivity. Other PP-immunoreactive cells in the fundic stomach of the same species correspond to a fraction of glucagon-immunoreactive cells (Solcia et al. 1980). The difficulty increases when both pancreatic and gastrointestinal endocrine tissues release related and immunologically similar substances, but in differing ratios. This is, for example, the well-known fact with somatostatin (see Chap. 13) and with glucagon (cf. Baldissera and Holst 1984); we have also already outlined a similar problem with the X-cell of the holocephalians (see Chap. 4.2). Furthermore, it appears that the ratio of glucagon to related molecules within the A-granules may change during the development. In summary, even very similar ultrastructure and a shared hormone are insufficient to establish the identity of two APUD cells. An analysis of their full spectra of messenger substances, together with embryological and functional criteria, probably must be added to

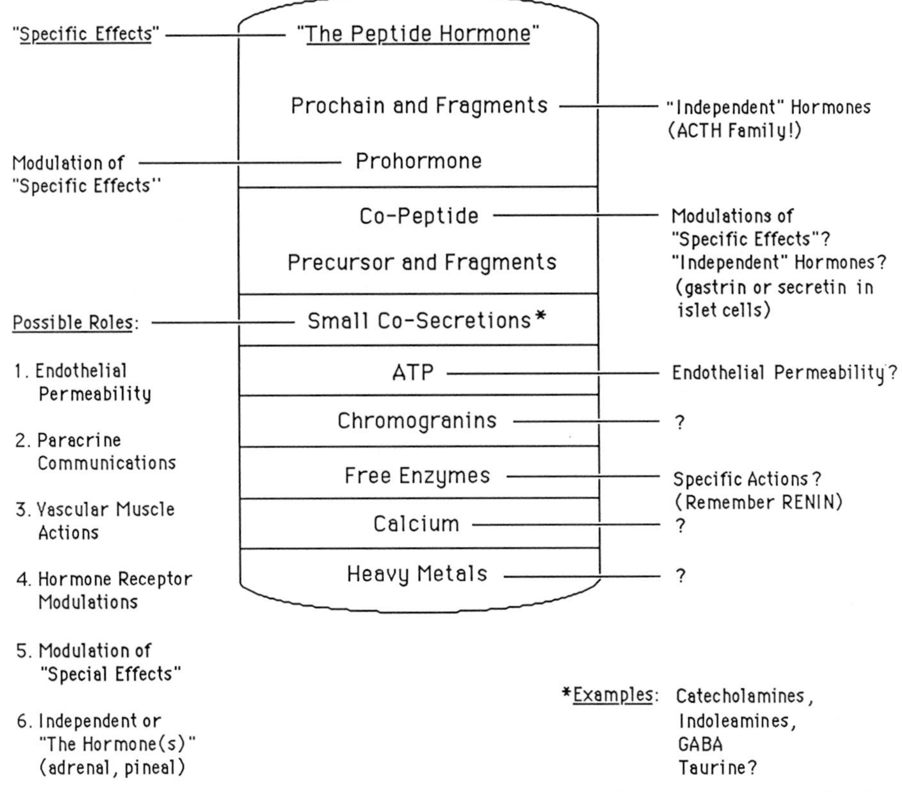

Fig. 9.1. Possible substances that may be released simultaneously during the exocytosis of an "APUD"granule, and some of their known and suggested functions

ascertain their complete identity. For practical purposes, this notion may be ignored in many cases. However, when it comes to the evaluation of cell-to-cell interactions, or to tumor identification and etiology (cf. Creutzfeldt 1985; Krejs 1985; Leshin 1985), it may be useful not to call apples and pears by the same name.

In the following, we concentrate on four groups of messenger substances found in islet granules, and we also consider the co-released glycoprotein (CGA/SP-I), the function of which is still unknown. Possibly, other messenger substances (e.g., melatonin, histamine, or taurine) will yet be identified in the islet granules of some species; it may be that the high but variable calcium content of the B-cell granules (Wollheim and Sharp 1981; Gagliardino 1983) modulates insulin secretion (Steinberg et al. 1984) when released at the cell surface. Similarly, an extracellular effect of non-membrane-bound granule enzymes such as dopamine beta-hydroxylase remains a possibility. Finally, it must be kept in mind that release of messengers of probably nongranular origin, such as prostaglandins (cf. Charles 1981; Robertson 1981; Halushka and Colwell 1983; Luyckx and Lefèbvre 1983), could profoundly modulate the effects of the other islet secretions.

9.1.1 Major Islet Hormones

Recent advances in the methodology of molecular cloning, DNA sequencing, peptide purification and microsequencing have dramatically increased our understanding of synthesis of messenger peptides. For example, it is now possible to predict from the sequence analysis of cDNA the structure of preprohormones, and to confirm the prediction by translation of mRNA in cell-free systems. Also, because post-translational processing has been studied in considerable detail, it is clear that the post-translational formation of different descendents of prohormones is a common event, which results in heterogeneous forms of a single messenger peptide. Considerable progress in this field was due to use of mammalian islet tumors and principal islets of teleosts. The high-molecular weight precursors of insulin, glucagon and somatostatin have been characterized (see Chaps. 10.1.1, 11.1 and 13.1); several post-translational products with distinct biological activities have been identified in the glucagon and somatostatin families (see Chaps. 11.1 and 13.1). Usually the final steps of post-translational processing are mediated by enzymes that are associated with the membranes of the Golgi apparatus and of the secretory granules (see e.g., Orci 1985). However, Koranyi et al. (1981) have shown that generation of glucagon from a precusror can also take place in circulating blood. Possibly all four major islet cell types are capable of producing and releasing simultaneously two or more forms of the same hormone.

Considerable differences between insulin of brain and pancreas in the guinea pig have been pointed out in Chap. 2.2.1. More than one form of pancreas insulin has been identified in certain teleosts (see Chap. 10). In the rat, two insulin genes express two insulins, which differ in only three residues (L. F. Smith 1966). Under normal conditions, about 10–20% of the circulating immunoreactivity of human insulin is due to its precursor, proinsulin, but this ratio may change during stimulated insulin release to 20–50% (Charles 1981; Kitabchi 1983). A high ratio of proinsulin to insulin may be diagnostic of insulinoma (Rubenstein et al. 1977a, b). Because glucagon, the GLP's and related intermediates stem from a common precursor (see Chap. 11.1), they may be released in varying proportions from pancreatic A-cells and other elements of the digestive tract (cf. Moody and Thim 1983; Baldissera and Holst 1984). It appears that there is a family of somatostatin genes (Shen and Rutter 1984). Thus, it is not surprising that teleost islets produce several forms of somatostatin (cf. Plisetskaya et al. 1986c), and that at least two forms of somatostatin are released from the islets and other tissues of mammals and some other vertebrates (see Chap. 13).

Variation in the biological potencies and spectra of related hormonal peptides provide the basis of a functional strategy of some endocrine cells. In the case of the adult B-cell, co-released proinsulin probably is of little physiological significance, since it represents a small fraction of the secretion and, in addition, has only about 5–10% of the biological activity of insulin (Kitabchi 1977, 1983). However, interaction between insulin and related growth factors at the receptor sites of various tissues (see Chap. 10) causes one to wonder if, perhaps, during the prenatal life similar interactions between insulin and proinsulin exist in some tissues (cf. Nissley et al. 1977; G. L. King and Kahn 1981). Another interesting question is whether the two rather different forms of guinea pig insulin (see Chap.

2.2.1) have different action spectra. Coincident with the several forms of circulating glucagon and glucagon-like immunoreactivity, it has been shown that they affect different targets, although there is overlap in binding to varying degree (Holst 1977). Indeed, in reactive hypoglycemia excess of "gut-type" glucagons seems to block the hepatic receptors so that pancreas glucagon cannot induce postprandial glycogenolysis (Rehfeld et al. 1973). On the other hand, it remains to be seen to what extent the functions of the GLPs (see Chap. 11.1) overlap with those of pancreas glucagon. The two forms of circulating somatostatin (S-14 and S-28) in mammals and birds vary considerably in their potency at target organs (cf. Conlon 1983; Strosser et al. 1984) with, in general, the larger form exerting profounder or longer-lasting effects (Marco et al. 1983).

9.1.2 Co-Released Peptides

The presence of more than one hormone in a single islet cell was originally postulated upon the discovery of amphiphil islet cells in poikilotherm vertebrates (Epple 1967). However, confirmation of this phenomenon was only possible with the advent of refined immunocytochemical techniques, and after the isolation of PP, somatostatin and other "brain-gut" peptides. In recent years, the number of reports on co-existing messenger peptides in the islet organ has grown to such pro-

Table 9.1. Co-peptides of the islet organ

Substance	Cell type	Species	Reference
Gastrin	D-cell	Man	Erlandsen et al. (1976)
	Independent?	Lizard	El-Salhy and Grimelius (1981)
GIP	A-cell	Rat	P.H. Smith et al. (1977)
	A-cell	Man, pig, dog, cat, guinea pig	Alumets et al. (1978b)
	Independent?	Lizard	El-Salhy and Grimelius (1981)
CCK	A-cell	Human, rat	Grube et al. (1978a)
LHRH	B-cell	Rat	Seppälä et al. (1979)
TRH	A-, B-cells	Rat	Kawano et al. (1983)
CRF	A-cell	Catfish, toad, lizard chicken, rat, mouse, cat, monkey, man	Petrusz et al. (1983)
Endorphin	A-cell	Rat	Grube et al. (1978b)
	D-cell	Rat, guinea pig, man	Watkins et al. (1980)
	F-cell	Lizard	El-Salhy and Grimelius (1981)
	?	Human	Bruni et al. (1979)
ACTH	?	Man, dog, rat	Larsson (1977b)
	A-cell	Rat	Graf (1981)
Prolactin	B-cell	Rat	Meuris et al. (1983)
Neurotensin	?	Mouse	Berelowitz and Frohman (1983)
PP	A-cell	Frog	Kaung and Elde (1980) El-Salhy et al. (1982)
PP	A-cell	Fetal man and rat	Ali-Rachedi et al. (1987)

This compilation is based on immunoreactivities. Hence, the precise nature of some of these peptides remains to be confirmed (Epple and Brinn 1986).

portions that it is difficult to retrieve all pertinent data. There is no question that some reports were overoptimistic, and that it will take quite a while until all controversies are settled (see e.g., Buffa et al. 1979; Grube and Aebert 1981; Yalow and Eng 1981). In Table 9.1, which is an attempt to organize the available information, it is likely that some data have been overlooked. Nevertheless, the table may serve to identify some frequent associations, although it must be kept in mind that this compilation probably reflects a bias due to the availability of certain antisera, and other factors as well. Both typical "inhibitory" and "stimulatory" messenger peptides coexist within the same cell type, and probably also in some (or most?) cases within the same granule. Interestingly, it has been suggested that insulin and TRH are released from rat islets upon different stimuli although both peptides seem to be co-localized in B-cell granules (Lamberton et al. 1985). Whatever the reasons for the secretory peptide associations, it appears currently impossible to exclude a priori of any kind of combination of messenger peptides in any APUD/paraneuron cell, and probably also in neurons. Two important questions arise with the preferred location of some primary tumors, such as, for example, the gastrinomas and VIPomas of the pancreas (cf. Ch'ng et al. 1984; Bosman et al. 1985). Are these, perhaps, related to (1) the renewed expression of normally transient embryonic features of islet cells in the "wrong" environment of the adult pancreas? or (2) the differentiation of "true" gastrin or VIP cells in the "wrong" place? In this connection, it must be recalled that there is a transient population of gastrin cells in the fetal pancreas that disappears after birth (L.-I. Larsson et al. 1976a).

9.1.3 Small Messenger Molecules

Monoamines. Dopamine (DA), norepinephrine (NE) and/or serotonin (5-hydroxytryptamine, 5-HT) have been identified in the islets of ratfish, amphibians, birds and mammals (cf. Cegrell 1968; Lange 1973; Lebovitz and Feldman 1973; Falkmer et al. 1984), although they have been reported to be absent from reptilian islets (Trandaburu 1976). There are great species variations in the quantity and type of monoamines. In addition, monoamines may be detectable only at certain stages of the development (see e.g., Teitelman et al. 1981a, b). Variations in the islet uptake of monoamines and their precursors are also considerable, and so is the islet content of monoamine oxidase, which degrades them (Feldman and Chapman 1975; Mahoney and Feldman 1977). As pointed out by Lange (1973), because the conditions that affect the monoamine content of islet cells are poorly understood, demonstration of a cell's capability to handle monoamines or their precursors may be more important than a negative histochemical reaction. The precise roles of the islet monoamines are largely unknown (Robertson 1981). One must bear in mind that their titers in the interstitial fluid of the islets should be influenced also by fractions from other organs, such as the adrenal medulla, paraganglia, CNS and peripheral nerves ("synaptic leakage"). The important and essentially open question concerns the *intragranular* monoamines. Do they act inside the cell, outside the cell, or in both locations? Since they are released together with the other contents of the islet cell granules, one must suspect that they have functions outside the cells of origin. DA has been reported to have both stimulatory and inhibitory effects on insulin secretion, but the contradictory results may be due in part to impact on blood flow. However, the general consensus seems to be that DA inhibits secretion of both insulin and glucagon (cf. Woods et al. 1981). A further complication in the study of DA is the possibility that it is converted to NE in islet cells (George and Bailey 1978). NE has mixed effects on the B-cell, although an inhibitory influence may prevail; it also stimulates release of glucagon. Since NE acts on both alpha- and beta-adrenergic receptors, which in turn are subdivided into further classes, an endless series of pharmacological experiments with agonists and antagonists can be envisioned. However, what is needed is an elegant method that reveals the site of action of islet cell-released NE. Because there seems to be general agreement that 5-TH inhibits release of both insulin and glucagon, it has been suggested that it is involved in the development of diabetes mellitus (cf. Woods et al. 1981; Lindström and Sehlin 1983). Islet-released 5-HT may also cause the release of NE from the sympathetic innervation of nerves, thus reinforcing its direct inhibitory effect on insulin release (Quickel et al. 1971). A new aspect of the mechanism of monoamine action on the islet cells has been reported by Lindstrom and Sehlin (1983), who find in in vitro studies that the precursor of 5-TH, 5-hydroxytryptophan, induces effects that are opposite to those of 5-TH. When added into the incubation medium, 5-hydroxytryptophan causes insulin release and Ca^{2+} uptake! One wonders if there is not in some cases a similar situation with respect to the catecholamines, i. e., when DOPA or DA are given (see also George and Bailey 1978). Epinephrine (E) seems to be present also in islet cells (Hansen and Hedeskov 1977); furthermore, it is clear that circulating E may interfere with the effects of NE at the level of the adrenergic receptors of islet cells.

Gamma-Aminobutyric Acid (GABA). This amino acid occurs in the central and peripheral nervous system (Jessen et al. 1979; Racagni et al. 1982), enteric neurons (cf. Saffrey et al. 1983) and in the pancreatic islets of mammals (for literature see Vincent et al. 1983; Reussens-Billen et al. 1984) and teleosts (Gerber and Hare 1979). It has been found to coexist with somatostatin (Oertel et al. 1982) and motilin (Chan-Palay et al. 1981; Nilaver et al. 1982) in neurons of the thalamus and medulla, respectively. In addition, GABAergic fibers occur in the median eminence and the neuro-intermediate lobe of the pituitary gland (Vincent et al. 1982). GABA, which is generally believed to be a major inhibitory neurotransmitter (Krogsgaard-Larsen et al. 1981), also seems to be involved as a portal blood messenger in the regulation of the adenohypophysis (Oertel et al. 1982; Ondo et al. 1982; Racagni et al. 1982). Until recently, the cellular localization to a distinct type of islet cell has not been firmly established, though its presence in an insulinoma (Okada et al. 1976) suggests that it occurs in B-cells. In a comprehensive study on normal and diabetic rats, GABA was found to be restricted to the B-cells (Vincent et al. 1983). In addition, both the GABA-synthesizing enzyme glutamate decarboxylase and the GABA-metabolizing enzyme, GABA-transaminase, were localized to the B-cells. From these observations the authors suggest that, contrary to the situation in neurons, in the B-cell glutamate is utilized via the GABA shunt, and that GABA, therefore, serves a metabolic role. However, it seems that

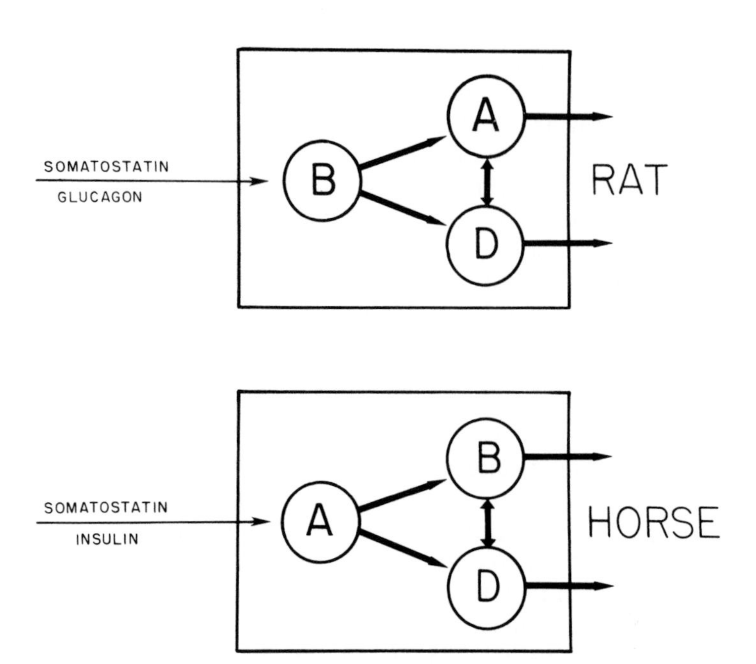

Fig. 9.2. Hormonal concentrations (indicated by *thickness of arrows*) in the capillary blood of the islets of two different mammals. Note that Unger's (1983) model for the rat (*top*) is supported by findings in the dog (Kawai et al. 1982b) although the intra-insular distribution of islet cells in this species seems to be different (Ferner 1952; Munger et al. 1965). One wonders if and how Unger's functional principle applies to the horse (*bottom*), whose islet structure does not seem to allow A-cell control by high local insulin titers (Epple and Brinn 1986)

the precise location of GABA inside the B-cell is not known. On the other hand, Reussens-Billen et al. (1984) localized high-affinity binding sites for GABA in the Schwann cells, intralobular ducts, and confirming data of Taniguchi et al. (1980a, b), in certain peripheral islet cells in the rat pancreas. Furthermore, the peri-insular innervation showed considerable labeling, a finding that should not be surprising if one considers the wide distribution of GABA in enteric neurons (cf. Saffrey et al. 1983). The observations of Reussens-Billen et al. (1984) raise a question as to the origin of the GABA which normally binds to the peripheral islet cells (probably D-cells), and neurons. The new interpretation of the intra-islet circulation in the rat (see Fig. 9.2) would make it possible that the B-cells act via co-released GABA on presumed D-cells, which would be consistent with the secretion of GABA from pieces of rabbit pancreas in response to high glucose concentrations (Gerber and Hare 1980). On the other hand, the high levels of circulating GABA are not reduced by induction of streptozotocin diabetes (Gerber and Hare 1979, 1980). Hence, circulating GABA from extrapancreatic sources could conceivably modulate the function of the presumed D-cells. Whatever the origin of GABA that normally binds to the peripheral islet cells, there is good evidence that GABA depresses somatostatin release from both the hypothalamus (Gamse et al. 1980) and the pancreatic islets (Robbins et al. 1981; Taniguchi et al. 1982). Unfortunately, the data on the effects of GABA on release of insulin and glucagon are utterly confusing (cf. Vincent et al. 1983; Reussens-Billen et al. 1984). At this point, we are left with strong arguments, both in favor of and against, a messenger function of GABA in the islet organ of the rat.

Purinergic Compounds. There appears to be no proof yet that ATP, when released from APUD granules, acts via extracellular pathways. However, modulation of both A- and B-cell activity by exogenous nucleosides and nucleotides has been reported (cf. Charles 1981; Weir 1983); it is relevant thereto to note that there is good evidence for adenosine neurotransmission in the CNS (cf. Dunwiddie 1985).

9.1.4 Chromogranin A (CGA/SP-I)

The secretory granules of many APUD cells contain a glycoprotein with a molecular weight of about 70000. In the granules of the adrenal medulla, where this substance was first identified as chromogranin A (CGA), it comprises about 40% of the soluble proteins (Winkler and Westhead 1980). Recently, it became clear that CGA is very similar or identical with secretory protein I (SP-I), which occurs in the granules of the parathyroid gland (Cohn et al. 1982, 1984). Furthermore, CGA/SP-I may be identical with a substance in several endocrine glands that reacts with a monoclonal antibody to an unknown component of a human pheochromocytoma (Lloyd and Wilson 1983). The distribution of CGA/SP-I seems to be restricted to certain neurons and the paraneurons, since it was not detected in the thyroid follicle cells and a variety of nonendocrine tissues. The reports on its absence in some APUD/paraneuron cells vary. It has been localized to islet cells in three different studies (O'Connor et al. 1983; Lloyd and Wilson 1983; Cohn et al. 1984), but there is no agreement as to its precise cellular distribution. The function of CGA/SP-I is as yet unknown, but its presence in serum is noteworthy (O'Connor 1983). It is possible that this glycoprotein, similar to neuron-specific enolase (Sheppard et al. 1982), will prove to be an important molecular marker of normal and abnormal APUD/paraneuron cells, and a circulating indicator of APUDomas.

9.2 Multiple Routes of Communication

Differing routes of communication are another set of strategies of the islet organ. We have already seen in Chap. 6 that the intrapancreatic distribution of the islet tissue has provoked considerable amount of speculation as to a possible functional principle involved. However, the possible intra-insular interactions of the islet cells have caused an even greater volume of controversial publications and speculations (for literature see Hopcroft et al. 1985; Samols et al. 1986). Basically, there are two schools of thought: (1) the intra-insular distribution of the individual islet cells has no functional significance; (2) the islets are highly organized microcomputers with functions dependent on precise intra-insular arrangements. Before we present our personal interpretations, we will summarize the known and possible actions of the islet cells that depend on, or use, topographic factors.

Autocriny. This term refers to the effect of secretions on the cells that release them. It has been suggested that autocriny is an important mechanism in inhibitory cellular self-regulation. Failure of autocrine regulation has been implicated in the development of cancer (Sporn and Todaro 1980). In the islet organ, the possibility of autocrine control has been shown for A-, B- and D-cells, the secretions of which can be inhibited by their respective hormones, i.e., glucagon (Kawai and Rouiller 1981), insulin (Iversen and Miles 1971; Rappaport et al. 1972; Beischer et al. 1978; Liljenquist et al. 1978) and somatostatin (or at least its analogues: Ipp et al. 1979). However, Shima et al. (1977) were unable to inhibit insulin release with exogenous insulin in the human. It appears beforehand unknown if autocrine mechanisms act within the islet organ in vivo, although somatostatin receptors on D-cells make it likely for this cell type (Patel et al. 1982). Considering the multitude of other substances co-released with the islet hormones, it is conceivable that also one or more of the former are involved.

Gap Junctions. After the demonstration of gap junctions in the islets of Langerhans by Orci et al. (1973, 1975), these structures have been investigated in several electrophysiological, dye- and metabolite-transfer investigations (for literature see Meda et al. 1983; Orci 1983). The combination of freeze-fracture studies with the application of the fluorescent dye Lucifer yellow revealed that in the islets of adult rats small territories of coupled B-cells may exist, and that the pattern of such territories can be modulated via changes in the number of gap junctions. However, earlier hopes that an increase in gap junctions would be a specific indicator of insulin secretion could not be confirmed, since dye-coupling increased after both stimulation and inhibition of insulin release (Meda et al. 1983). Thus, the significance of the gap junctions between the B-cells remains unknown, though it appears that they create small functional syncytia in which members may communicate via electrotonic, ionic and other signals (Kohen et al. 1979; Meda et al. 1981). Since there are also gap junctions between neurons and islet cells (Orci et al. 1973), one wonders if electrotonic coupling via gap junctions provides a means by which a single neuron can affect simultaneously a group of B-cells (see Eddlestone and Rojas 1980; Samols et al. 1983). However, there are gap junctions also between B- and other islet cells (Orci et al. 1975; Michaels and Sheridan 1981; Meda et al. 1982, 1983), and between islet and exocrine cells (Alumets et al. 1978a; Bendyan 1982; Agulleiro et al. 1985). As suggested by Meda et al. (1983), the type of intercellular communication in these cases may be different from that between B-cells. Thus, while it appears that different types of islet cells can use gap junctions to communicate with each other, and possibly also with neurons and exocrine cells, the functional significance of these communications remains to be established. No information seems to exist on the noninnervated (!) islets of the spiny mouse, or on any nonmammalian species.

Cytotransmission. We have recently proposed this term (Epple 1982) for a mechanism discovered by L.-I. Larsson et al. (1979). It refers to a situation in which a paraneuron transfers secretions to the surface of another cell by release from the bulbous tip of a long basal extension (L.-I. Larsson 1984). This mechanism, resembling neurotransmission more than paracrine secretion, has been observed in the mammalian stomach. Here, the extensions of "open" somatostatin cells make contact with gastrin and parietal cells (L.-I. Larsson et al. 1979). Perhaps this represents a mechanism by which local messages from the antral lumen reinforce inhibitory messages conveyed by circulating somatostatin from other sites. No such arrangement has yet been demonstrated in the islet organ, although the D-cells of many species tend to form long basal extensions.

Paracrine Secretions. When properly applied, this term refers to local message transfer via the interstitial fluid (Feyrter 1953), as distinguished from transport via the vascular circulation. Paracrine secretion has been suspected to be a major channel of communication among islet cells (for literature see Fujimoto et al. 1983; Samols et al. 1983, 1986; Taborsky 1983; Unger 1983; Maruyama et al. 1984); the heterogeneous composition of islet cell granules provokes inevitably the thought that not only the hormones, but also other messenger substances (e.g., TRH: Morley et al. 1979; CRF: Moltz and Fawcett 1985) may be involved. However, no hard data seem to exist. Furthermore, there is no proof that tight junctions create in vivo intra-islet fluid compartments which would favor such interactions. Instead, In't Veld et al. (1984) recently presented strong evidence against the existence of tight junctions in the islets of rat and humans under normal conditions. They suggest that the tight junctions in the isolated islets are temporary adaptive mechanisms that protect the intra-islet environment against perturbations. If this is so, then one must assume that either no intra-islet microdomains exist at all, or that cell contacts other than tight junctions seal off such territories. Perhaps small canaliculi between adjacent islet cells serve as channels of communication (Fujita et al. 1981c). Unger's (1983) recent suggestion that somatostatin acts on the B-cells via systemic blood would make a previously postulated sealed somatostatin-poor compartment (Kawai et al. 1982b) unnecessary. However, without reliable information on islet angioarchitecture in many mammalian species, Unger's (1983) model remains only as a tempting hypothesis (see Fig. 9.2).

At least in some nonmammalian taxa, the morphology does not favor certain types of paracrine interactions among islet cells. Thus, the topography of the light and dark islets of birds (see Chap. 4.2) excludes a major paracrine effect of insulin on A-cells though, on the other hand, the presence of D-cells in both types of islets could provide a nonvascular route for influence of somatostatin on both A- and B-cells (see Chap. 14). In turtles, at least in *Chrysemys picta*, the D-cells occur largely outside the islets (Gapp and Polak 1983; Agulleiro et al. 1985). In larval lampreys, paracrine interactions between islet cells must be absent because the follicles of Langerhans consist of B-cells only. In adult lampreys, B- and somatostatin cells occur largely in separate follicles (see Chap. 4.1). As pointed out in Chap. 6, paracrine islet secretions seem to stimulate the formation of exocrine "zymogen mantles" around the islets of the chicken.

Portal Hormone Actions. In contrast to paracrine secretion, this type of communication involves transport of high concentrations of hormones to nearby targets via blood vessels. Portal communication can occur in three different ways: (1) downstream communication within a single network of capillaries or "leaky" venules, as postulated for insulo-acinar systems; (2) communication involving intermediate venules, as in the hypothalamo-hypophysial system, or via efferent arterioles, and a second capillary network; (3) communication involving an intermediate vein, in which the secretions are partly diluted by blood from other sources, before they reach the capillary network of their target, as in the insulo-hepatic portal system. It seems that only the first and third option can apply to the islets since neither a second intrapancreatic postvenous network, nor efferent islet arterioles have been reported (see also Chap. 6). Intra-insular communication via a single capillary network has been postulated for many mammals, including the human (for literature see Samols et al. 1983, 1986). Bonner-Weir and Orci (1982) demonstrated in the rat that arterioles first penetrate the B-cell regions, before they give rise to capillaries that then supply the peripheral A- and D-cells. In accordance with Samols et al. (1983), Unger (1983) suggests that the functional consequences of such a system might be as follows: "(1) the B-cells would be the first islet cells to receive glycemic signals; (2) the insulin concentrations perfusing the A-cells would be the highest in the body; (3) the B-cells would be exposed to systemic glucagon concentrations. Such an anatomical arrangement would skew the insulin-glucagon feedback system in favor of insulin's inhibiting effect on A-cell secretion and reduce the importance of glucagon's stimulatory effect on the B-cells, thereby placing insulin in the anatomical position to serve the role of release-inhibiting factor for glucagon." This model (Fig. 9.2) would also explain the effect of low concentrations of systemic somatostatin on B-cells, since the latter would not be exposed to the higher titers of somatostatin, which is released further downstream. However, in the cat, sphincter-like arrangements at the sites where the capillaries leave the islets (Syed Ali 1984) raise the question of a possible backward transport of somatostatin and glucagon into B-cell regions. The extent to which the structural basis of Unger's new functional principle can be applied to species other than certain rodents remains to be ascertained. In the islets of the horse, the blood passes first through centrally located A-cells before it reaches the peripheral B- and D-cells (Fujita 1973). Obviously, the situation in this species must be very different from that in the rat (Fig. 9.2). In birds, the already-mentioned intrapancreatic distribution of "light" and "dark" islets makes it impossible for insulin to reach the vast majority of A-cells in other than low systemic concentrations. In the Brockmann bodies of teleosts, Lange (1984) could not identify a consistent pattern of islet cells along the capillaries, such as would be needed for the type of hormonal interactions postulated for mammals (see, however, Syed Ali 1986). Of course, there are no A-cells in the islet follicles of lamprey larvae at all. Not surprisingly then, neither somatostatin nor glucagon affected the glucose- or lysine-stimulated insulin release from eel (*Anguilla anguilla*) and lamprey (*Lampetra fluviatilis*) islets in vitro (Thorpe and Duve 1985).

The evolution and functional principle of the insulo-hepatic portal system have been outlined in Chap. 2.3. Since the impact of the individual islet hormones on the liver will be discussed in the following chapters, it may suffice here to recall that (1) the liver is the first organ to be reached by the islet hormones, before they are diluted in the systemic circulation, and (2) the liver controls the amounts of these hormones that finally enter the systemic circulation. Thus, the liver is both target and checkpoint of islet secretions. Although information on amounts of islet hormones that are extracted in the liver varies greatly, and it appears that in mammals the hepatic extraction of insulin exceeds that of glucagon (Rubenstein et al. 1972; Ishida et al. 1980). In the perfused rat liver, hepatic uptake of somatostatin-14 is greater than that of SS-28 (Sacks and Terry 1981). There seem to be almost no information on possible interactions between the liver and co-secretions of major islet hormones; however, opiates can suppress hepatic glucose production (cf. El-Tayeb et al. 1985).

Telehormone Actions. The term telehormone (L. G. Leibson 1981) refers to the "classical" type of hormone that acts after dilution in the general circulation on distant target(s). All four major islet hormones function as telehormones in addition to their known or postulated paracrine and/or portal actions. Details of their "teleactions" will be discussed in the following chapters.

9.3 Local Interactions of Islet Cell Secretions: Terra Incognita

The uncertainty concerning the interactions among the types of islet cells involves a multitude of possible but, at least in vivo, largely unproven options (see e.g., Samols et al. 1986). By sheer logic, one would speculate that, e.g., antagonism, that is observed whenever (stimulating) TRH and (inhibitory) somatostatin act on common targets elsewhere (see e.g., Reichlin 1983a), also plays a role in intra-islet regulations. On the other hand, the large number of co-released "regulatory" peptides in the A-cells of the rat suggests a much more complex situation (see Table 9.1). Figure 9.3 shows only one simple set of theoretical options for communications sent out by one islet cell type, the B-cell, and it assumes that communications are undirectional. By adding some channels for the A- and D-cells to "talk back", the number of imaginable intra-islet interaction systems can be raised to scores. If this were the main problem, one would feel confident about approaching it in the rat. However, one wonders if the rat islet is not one of nature's more tasty red herrings. Why should the rat islet be a priori more typical than the horse islet, where the distribution of A- and B-cells is exactly the opposite? Clearly by invoking variations in intra-islet communications, interstitial circulation, receptor variations and nervous input, one could construct a theoretical model in which ultimately the routes of inter-islet cell communication in both rat and horse are the same. At this point, it may be more fruitful to ignore the mammalian islet morphology for a while and to establish, by biologically meaningful in vivo-studies, the interactions of the islet secretions: is there, among mammals, indeed a general pattern? Furthermore, it may be worthwhile to search for an animal model that allows to combine in vivo the simultaneous application and measurement of circulating substances, in both interstitial fluid and islet blood, with recordings of the electrical islet cell activity. Unfortunately, such a model may not exist among mammals. However, certain teleosts with large Brockmann

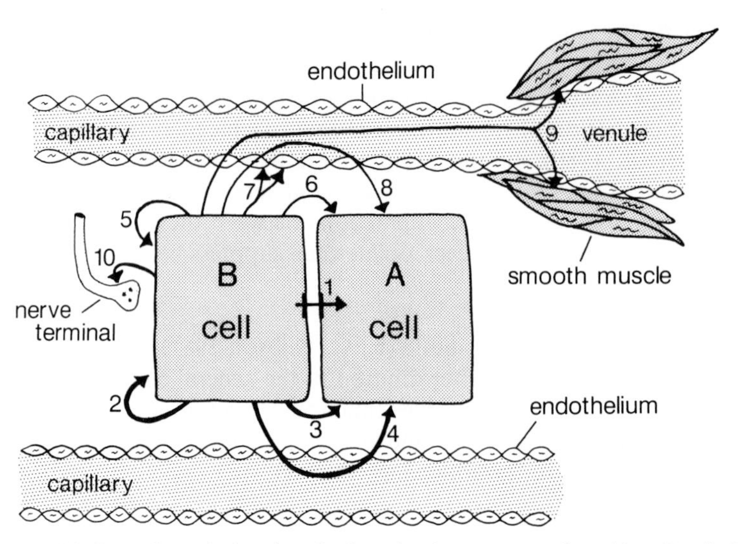

Fig. 9.3. Some theoretical options for intra-insular messages released by a B-cell. *Heavy lines* in *lower part* of the figure: insulin effects; *fine lines* in *upper part* of the figure: effects of co-secretions. *1* Gap junctions; *2* and *5* autocrine regulation; *3* and *6* paracrine messages; *4* and *8* portal messages; *7* endothelial permeability; *9* control of vascular muscle; *10* presynaptic modulation

bodies, Squamata with intra-splenic islets or birds (splenic lobe), may be suitable for the simultaneous focal perfusion of islet tissue for the collection of interstitial fluid (for technical details, see Jarry et al. 1985) and cannulation of the islet vasculature. In this way, it could be possible to obtain *basic hard data* whose relevance to the mammalian situation could then be tested. An added bonus to the system in teleosts would be the strong islet innervation (see Chap. 7), which would be removed if so indicated by the experimental protocol.

Of course, co-release of messenger substances may not just serve in stimulatory or inhibitory communications among islet cells. It remains to be seen whether co-released APUD-substances also affect endothelial permeability and/or vascular muscle (cf. Andrew 1982; Epple 1982). Even hormonal effects different from those of the "major" islet hormones cannot be ruled out at present, since as pointed out in Sect. 9.2, the liver could be the site where the co-released islet secretions modulate the effects of the "major" islet hormones. Mutatis mutandis, this would resemble the situation in peripheral nerves, where peptides may serve to sustain the effects initiated by catecholamines (Lundberg and Hökfelt 1983). Also, co-released substances could modify the metabolite uptake of their own cells. Such a possibility was suggested by Uvnäs-Wallensten (1981; discussion) for insulin co-released at peripheral nerve terminals, where the peptide may enhance local glucose uptake.

Finally, one cannot exclude the possibility that granular exocytosis of islet cells may also serve to remove excess substances from the cell. Perhaps this is the case with the uptake and release of catecholamines or their precursors during experimental overloading of cells that normally do not show monamine fluorescence (see e.g., L.-I. Larsson 1984).

From the preceding discussion, it is obvious that our insights even into the most basic aspects of islet cell interactions are very limited. However, two points emerge: (1) the study of intra-islet interactions should greatly benefit from the development of a new, better-suited model for in vivo studies; (2) interpretation of pertinent data that ignores the possible effects of co-released islet secretions may lead to wrong conclusions.

Chapter 10

Insulin

10.1 Molecular Structure and Related Compounds

10.1.1 The Insulin Molecule

Like other messenger peptides, insulin is synthesized via a large single-chain precursor (cf. Docherty and Steiner 1982; Steiner et al. 1984). The initial steps of the formation of the final two-chain peptide involve synthesis of preproinsulin (Fig. 10.1), removal of the signal sequence and transfer of the resulting proinsulin to the Golgi apparatus. Proinsulin consists of three regions (A-, C- and B-chains), and cleavage of the large, connecting C-peptide during the final steps of biosynthesis results in the formation of the two-chain peptide. The latter contains a pair of inter-chain disulfide bonds, and it has a molecular weight to about 6000. The C-chain, which is phylogenetically highly variable, seems to have no biological function. In mammals, a 29-amino acid fragment of the C-peptide is secreted, whereas in an elasmobranch (*Torpedo marmorata*) Conlon and Thim (1986) isolated a small, truncated C-peptide. The latter consists of 17 amino acid residues; mechanism(s) of its formation remain(s) to be clarified. On the other hand, the structure of insulin has been highly conserved during evolution. Recently, there has been great progress in the analysis of the cellular compartments involved in insulin biosynthesis (Orci 1985); details on the molecular structure of insulin, its precursors and of the insulin gene may be found in Dayhoff (1978), Albert (1982), Blundell and Wood (1982), Sorokin et al. (1982) and Steiner et al. (1984). Insulin-immunoreactivity has been identified in unicellular organisms and many invertebrates (see Chap. 2.1). The primary structure of insulin from representatives of all vertebrate classes except the amphibians has been described (Fig. 10.2). Mammalian insulin consists of 51 amino acids, 21 in the A-chain and 30 in the B-chain. It has been suggested that the sequence 22–26 of the B-chain is the site needed for the biological activities of insulin, and that the B-chain is the most conserved part of the insulin molecule during evolution (cf. El-Salhy et al. 1984). The insulin molecule can generally form dimers in solution through hydrogen bonding and association of hydrophobic regions primarily between the B-chains. Although some exceptions will be noted, most insulins also form hexameric crystals in association with zinc ions binding to the B 10 histidine residues; this may be the form in which the hormone is stored in the secretion granules of B-cells (Hodgkin and Mercola 1972). Excepting the hystricomorphs (guinea pig and relations), the primary structure of mammalian pancreas insulins is rather uniform. Rats and mice have two insulins, rat insulins I and II being identical to mouse insulins I and II, respectively (Bünzli et al. 1972). The two peptides repre-

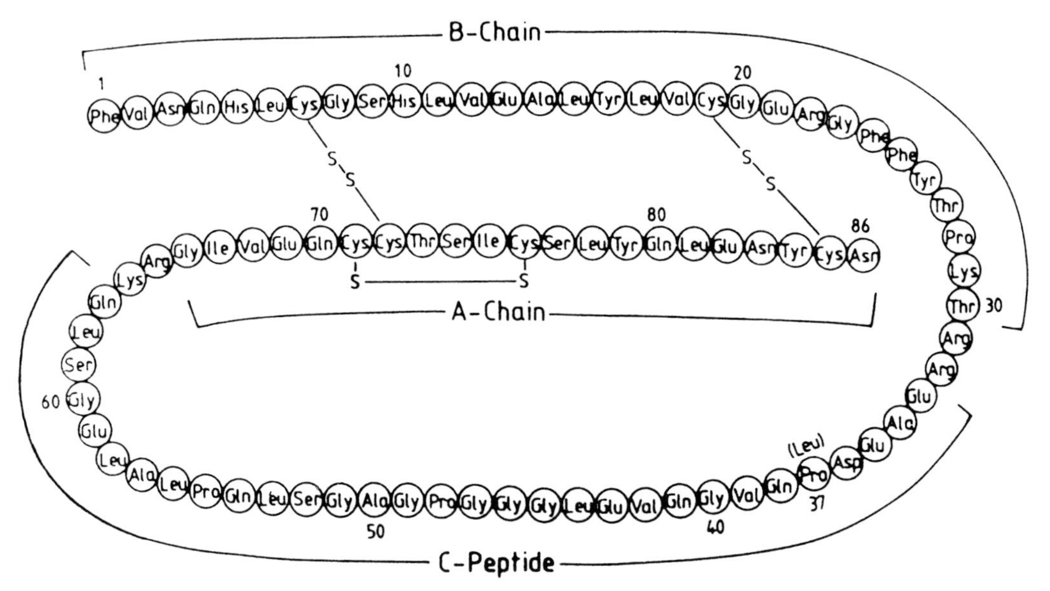

Fig. 10.1. Proposed amino acid sequence of proinsulin of the rhesus monkey (*Macaca mulatta*). Note that it is identical with human proinsulin except for position 37 of the C-peptide which in the human is occupied by leucine (Naithani et al. 1984)

sent nonallelic genes and differ in positions B9 and B29 (L. F. Smith 1966). A similar duplicity occurs in teleosts (see below). The problem of two different insulins in the guinea pig has been discussed in Chap. 2.2.1.

Although multifactorial causes of human diabetes mellitus have long been the subject of debate, it has only recently become evident that an additional contributing factor may be mutations in the insulin gene. Tager et al. (1979) described a diabetic patient with hyperinsulinemia but without typical resistance to exogenous insulin. Biochemical analysis of a pancreatic sample revealed an abnormal insulin with a leucine for phenylalanine substitution at position B25 (Shoelson et al. 1983), a part of the B-chain which is otherwise invariant from hagfish to mammals. Furthermore, this insulin showed decreased binding to lymphocytes and rat adipocyte membranes as well as decreased stimulation of 2-deoxyglucose transport and ^{14}C-glucose metabolism in rat adipocytes. This patient also had some normal circulating insulin in a ratio of 40:60 with the mutant form, leading the authors to speculate whether (1) codominance of parental genes or (2) the expression of nonallelic genes, as in rats and mice, was the contributing genetic factor. Two additional mutant insulins have also been reported in diabetic humans, but the exact amino acid substitutions could not be determined (Shoelson et al. 1983), Niall (1982) speculates that gene duplication may be a strong factor in hormone evolution, leaving the original hormone to carry out its normal functions while the duplicate gene is then free to mutate, perhaps leading to a product that can interact with different receptors and produce different metabolic sequelae. Compared with other vertebrates, mammals have an exaggerated dependence on insulin in the regulation of carbohydrate metabolism.

Insulin A-Chain Amino Acid Sequences

										1										2		
	1	2	3	4	5	6	7	8	9	0	1	2	3	4	5	6	7	8	9	0	1	2
Human[1]	G	I	V	E	Q	C	C	T	S	I	C	S	L	Y	Q	L	E	N	Y	C	N	
Bovine	-	-	-	-	-	-	-	A	-	V	-	-	-	-	-	-	-	-	-	-	-	
Sei whale	-	-	-	-	-	-	-	A	-	T	-	-	-	-	-	-	-	-	-	-	-	
Sheep,goat	-	-	-	-	-	-	-	A	G	V	-	-	-	-	-	-	-	-	-	-	-	
Elephant	-	-	-	-	-	-	-	-	G	V	-	-	-	-	-	-	-	-	-	-	-	
Rat,mouse I,II	-	-	-	D	-	-	-	-	-	-	-	-	-	-	-	-	-	-	-	-	-	
Coypu	-	-	-	D	-	-	-	-	N	-	-	-	R	N	-	-	M	S	-	-	-	D
Guinea pig	-	-	-	D	-	-	-	-	G	T	-	T	R	H	-	-	Q	S	-	-	-	
Horse	-	-	-	-	-	-	-	-	G	-	-	-	-	-	-	-	-	-	-	-	-	
Chicken,turkey	-	-	-	-	-	-	-	H	N	T	-	-	-	-	-	-	-	-	-	-	-	
Alligator	-	-	-	-	-	-	-	H	N	T	-	-	-	-	-	-	-	-	-	-	-	
Rattlesnake	-	-	-	-	-	-	-	E	N	T	-	-	-	-	-	-	-	-	-	-	-	
Codfish	-	-	-	D	-	-	-	H	R	P	-	D	I	F	D	-	Q	-	-	-	-	
Coho salmon	-	-	-	-	-	-	-	H	K	P	-	N	I	F	D	-	Q	-	-	-	-	
Chum salmon	-	-	-	-	-	-	-	H	R	P	-	N	I	F	D	-	Q	-	-	-	-	
Bonito	-	-	-	-	-	-	-	H	K	P	-	D	I	F	Q	-	-	-	-	-	-	
Carp	-	-	-	-	-	-	-	H	K	P	-	-	I	F	E	-	Q	-	-	-	-	
Anglerfish	-	-	-	-	-	-	-	H	R	P	-	N	I	F	D	-	Q	-	-	-	-	
Toadfish I	-	-	-	-	-	-	-	H	R	P	-	N	I	F	D	-	Q	-	-	-	-	
Toadfish II	-	-	-	-	-	-	-	H	R	P	-	D	K	F	D	-	Q	S	-	-	-	
Dogfish	-	-	-	-	H	-	-	H	N	T	-	-	-	-	D	-	Q	G	-	-	-	Q
Hagfish	-	-	-	-	-	-	-	H	K	R	-	-	I	-	N	-	Q	-	-	-	-	-

Insulin B-Chain Amino Acid Sequences

												1										2										3	
	-	-	1	2	3	4	5	6	7	8	9	0	1	2	3	4	5	6	7	8	9	0	1	2	3	4	5	6	7	8	9	0	1
Human,elephant			F	V	N	Q	H	L	C	G	S	H	L	V	E	A	L	Y	L	V	C	G	E	R	G	F	F	Y	T	P	K	T	
Bovine[2]			-	-	-	-	-	-	-	-	-	-	-	-	-	-	-	-	-	-	-	-	-	-	-	-	-	-	-	-	-	A	
Rabbit			-	-	-	-	-	-	-	-	-	-	-	-	-	-	-	-	-	-	-	-	-	-	-	-	-	-	-	-	-	S	
Rat,mouse I			-	-	K	-	-	-	-	-	-	-	-	-	-	-	-	-	-	-	-	-	-	-	-	-	-	-	-	-	-	S	
Rat,mouse II			-	-	K	-	-	-	-	-	-	-	-	-	-	-	-	-	-	-	-	-	-	-	-	-	-	-	-	-	M	S	
Coypu			Y	-	S	-	R	-	-	-	-	Q	-	-	D	T	-	-	S	-	-	R	H	-	-	-	Y	R	P	N	D		
Guinea pig			-	-	S	R	-	-	-	-	-	N	-	-	-	T	-	-	S	-	-	Q	D	D	-	-	-	-	I	-	-	D	
Chicken,turkey			A	A	-	-	-	-	-	-	-	-	-	-	-	-	-	-	-	-	-	-	-	-	-	-	-	-	S	-	-	A	
Alligator			A	A	-	-	R	-	-	-	-	-	-	-	D	-	-	-	-	-	-	-	-	-	-	-	-	-	S	-	-	G	
Rattlesnake			A	P	-	-	R	-	-	-	-	-	-	-	-	-	-	-	-	I	-	-	-	-	-	-	Y	-	S	-	R	S	
Codfish		M	A	P	P	-	-	-	-	-	-	-	-	-	D	-	-	-	-	-	-	-	D	-	-	-	-	-	N	-	-		
Coho salmon			A	A	A	-	-	-	-	-	-	-	-	-	D	-	-	-	-	-	-	-	-	K	-	-	-	-	N	-	-		
Chum salmon			A	A	A	-	-	-	-	-	-	-	-	-	D	-	-	-	-	-	-	-	-	K	-	-	-	-	-	-	-		
Bonito			A	A	-	P	-	-	-	-	-	-	-	-	D	-	-	-	-	-	-	-	-	-	-	-	-	-	Q	-	-		
Carp	N	A	G	A	P	-	-	-	-	-	-	-	-	-	-	-	-	-	-	-	-	-	P	T	-	-	-	-	N	-	-		
Anglerfish		V	A	P	A	-	-	-	-	-	-	-	-	-	D	-	-	-	-	-	-	-	D	-	-	-	-	-	N	-	-		
Toadfish I		M	A	P	P	-	-	-	-	-	-	-	-	-	D	-	-	-	-	-	-	-	D	-	-	-	-	-	N	-	-		
Toadfish II		M	A	P	P	-	-	-	-	-	-	-	-	-	D	-	-	-	-	-	-	-	D	-	-	-	-	-	N	S			
Dogfish			L	P	S	-	-	-	-	-	-	-	-	-	-	-	-	-	F	-	-	-	P	K	-	-	Y	-	L	-	-	+	G
Hagfish			R	T	T	G	-	-	-	-	K	D	-	-	N	-	-	-	I	A	-	-	V	-	-	-	-	-	D	-	T	K	M

+ = D or G

[1]Identical with pig, rabbit, dog, and sperm whale.

[2]Identical with pig, horse, dog, sheep, goat, sperm whale, and sei whale.

Fig. 10.2. Primary structure of the insulins of various vertebrates. (Compiled from Humbel et al. 1972; Dayhoff 1978; Bajaj et al. 1983; Lance et al. 1984; Plisetskaya et al. 1985)

When a mutant insulin is less active in this respect than the original, and when it competes with the original hormone for receptor sites, then the duplication may be an unfavorable one and subject to removal from the evolutionary scheme by selective pressures.

The pancreatic insulins of guinea pig and swine differ by 17 amino acids (L. F. Smith 1966). The unusual structure of the guinea pig pancreas insulin is paralleled by modifications of the guinea pig insulin receptor (Horuk et al. 1979), and both an elevated pancreas content (A. E. Zimmerman and Yip 1974) and plasma titer of this hormone (A. E. Zimmerman et al. 1974). Both the guinea pig and its relative the coypu lack the B 10 histidine residue and, therefore, do not bind zinc. However, the unique insulin of the teleost *Cottus scorpius* does not readily form crystals containing zinc-insulin hexamers, despite the presence of the B 10 histidine residue (Cutfield et al. 1986). The guinea pig differs from the coypu by having a B 22 substitution of arginine by aspartic acid, which may explain the inability of guinea pig insulin to dimerize (A. E. Zimmerman and Yip 1974). Blundell et al. (1972) point out that zinc-binding insulins elicit a strong immunologic response in three hystricomorphs, the guinea pig, coypu, and capybara, and in two new-world monkeys (*Saimiri* and *Cebus*), all of which are South American species. The presence of unusual insulin in the two species of South American primates (both Cebidae) contrasts with the "regular" type of insulin in Old World monkeys and apes (Mann and Crofford 1970; Naithani et al. 1984). It seems that this existing observation has never been pursued by structural analyses and study of the insulin of the third group of monkeys, the Callitrichidae, which is restricted to the New World.

Data on insulin structure in nonmammalian vertebrates are rather sketchy (Fig. 10.2). Chicken and turkey insulins are identical (L. F. Smith 1966) and differ from porcine insulin by four residues (7 – 10) in the A-chain and three (1, 2, 27) in the B-chain. Alligator insulin has more similarities to avian insulin than to the insulin of the two snakes (*Crotalus atrox* and *Zaocys dhumnades*) studied so far. Its A-chain is identical to that of the chicken A-chain, while the B-chain has three conservative substitutions (Lance et al. 1984). The similarity between alligator and avian insulins is in perfect agreement with the phylogenetic relationship of these animals, but contrasts with the great differences in their islet structure and functions (see Chaps. 4.2 and 10.4). Insulin of the rattlesnake (*Crotalus atrox*) differs from its porcine homolog by substitutions in the highly variable region of A 8, A 9, and A 10, and also nine in the B-chain (Kimmel et al. 1976). The substitutions at B 16, B 18, and B 30 are unique, whereas those at B 5 and B 25 have previously been reported only in the coypu. The structures of the insulins of several teleosts (see Fig. 10.2) have been identified (for literature see Plisetskaya et al. 1985, 1986a; Cutfield et al. 1986). The tuna and toadfish resemble rodents in having two distinct insulins (Humbel et al. 1972). The complete sequence of tuna-I insulin is unknown. When compared with mammalian insulins the peptides from cod, tuna, anglerfish and toadfish show a frame-shift in the B-chain so that the amino acids are sequenced from B 0 to B 29. Toadfish-II-insulin differs further by having only 28 amino acids in its B-chain. The primary structure of the insulin of *Cottus scorpuis* is unique among teleosts in that there is no N-terminal extension of the B-chain (Fig. 10.2).

The insulin of a shark (*Squalus acanthias*) has an A-chain consisting of 22 amino acids, a phenomenon so far known for only one other species, the coypu; its B-chain seems to have an extension at the carboxyl terminus by two residues. Despite various substitutions when compared with other gnathostome insulins, the biological potency of that of *S. acanthias* is similar. Since it shows homology with certain residues of IGF-2, Bajaj et al. (1983) wonder if this indicates a greater growth-promoting activity than shown by other fish insulins (see also Sect. 10.1.2). Another elasmobranch, *Torpedo marmorata*, has an insulin very similar to that of the dogfish (Conlon and Thim 1986). However, it contains 21 amino acid residues in the A-chain and 30 in the B-chain (Fig. 10.2). Peterson et al. (1975) first reported the amino acid sequence of insulin of the hagfish (*Myxine glutinosa*), noting that it contains 52 residues, the additional residue being at the B-chain carboxy-terminus. Like the rattlesnake, the hagfish has an unique substitution at B 18 (valine to alanine). With a total of 23 position differences with porcine insulin, Emdin and Falkmer (1977) consider only 12 of them to be "radical". Hagfish insulin forms dimers in solution, but lacks the zinc-binding B 10 histidine and, therefore, like hystricomorph insulins, it does not form hexamers. Further details on structure and function of hagfish insulin have been summarized recently by Emdin et al. (1985). Analysis of the insulin of the lamprey (*Lampetra fluviatilis*) suggests a difference of nine residues from hagfish insulin (Emdin and Falkmer 1977).

The relatively strong conservation of the insulin molecule is paralleled in the conservation of its receptor. Thus, the subunit structure of the hepatic insulin receptor in *Myxine* is similar to that of the mammalian insulin receptor. Furthermore, the former is similar to the latter with respect to its affinities to different vertebrate insulins (cf. Leibush 1983; Plisetskaya 1985); and it has been suggested that the insulin receptor has been functionally better conserved during evolution than the insulin molecule itself (Muggeo et al. 1979). On the other hand, there seem to be structural differences between the insulin receptors in the brain and peripheral target tissues (Heidenreich et al. 1983).

In conclusion, much of the basic structure of insulin appears to have been conserved during vertebrate evolution, with approximately half of the amino acid residues being considered as invariant. The connecting (C-) peptide of proinsulin is highly variable (for literature see Kitabchi 1983; Steiner et al. 1984), but this residue sequence is probably of little more physiological significance than providing proper alignment of the subsequent A- and B-chains of insulin. The conservation of certain regions of the insulin molecule no doubt reflects a consistency of physical properties and a common biosynthetic pathway, but certain minor amino acid substitutions can drastically alter function. It is, therefore, conceivable that a number of substitutions noted from hagfish to birds and mammals altered the functional spectra of the hormone across class or even species lines. However, as we will see below, even the main function(s) of insulin in most nonmammalian vertebrates elude us.

10.1.2 Related Compounds

Several hormones of the gut and pancreas appear to have common genetic origins and are, therefore, easily classified into families. Insulin has structural relationships with several groups of peptides of diverse origin that include (1) the somatomedins and insulin-like growth factors (IGFs), (2) relaxin, and (3) nerve growth factor (NGF).

Blundell and Humbel (1980), Zapf et al. (1981), Blundell and Wood (1982), Froesch et al. (1983), and Froesch and Zapf (1985) have reviewed the structural similarities of the somatomedins and IGFs with insulin. The somatomedins are defined by four criteria: (1) serum levels vary positively with growth hormone levels, (2) stimulation of sulfate incorporation into cartilage, (3) fibroblast mitogenic activity, and (4) some insulin-like activities in muscle and adipose tissue. Peptides fitting these criteria include IGF-I, somatomedins A and C and multiplication stimulating activity (MSA) (Blundell and Humbel 1980). The weak responsiveness of IGF-II to growth hormone should set this substance apart from typical somatomedins. The IGFs and insulin stem from a common precursor molecule whose encoding gene underwent duplication about 600 million years ago. The similarity between the IGF-I receptor and the insulin receptor suggests that there was a concomitant duplication of the gene(s) responsible for the synthesis of the precursor receptor (cf. Froesch and Zapf 1985). A subsequent gene duplication resulting in separate IGF-I and IGF-II peptides seems to have taken place just before the appearance of reptiles. Daughaday et al. (1985) found both IGFs in the serum of several species of mammals, the chicken and a turtle (*Pseudemys scripta*), while the sera of a toad (*Bufo marinus*) and a teleost (*Salmo gairdneri*) contained only IGF-I. The two IGFs, which were purified from a serum fraction originally characterized as having a nonsuppressible insulin-like activity (NSILA), are identical with the factors that stimulate sulfate incorporation into cartilage. Each IGF consists of a single polypeptide chain containing three disulfide bridges, bearing striking structural resemblances to proinsulin. They contain 70 and 67 amino acids, respectively, with 62% of the amino acid positions being identical. In the human, about 45% of the amino acids in positions corresponding to the A- and B-chains of insulin are identical with amino acids in insulin (Fig. 10.3). However, there is no resemblance between the C-peptides of insulin and the corresponding regions of the IGFs (Engberg et al. 1984). IGF-I is identical to the somatomedins A, and C, whereas IGF-II has no somatomedin partner; instead, its homolog is MSA from the rat. The liver seems to be a major, but not exclusive, site of IGF production. Synthesis of IGFs is widespread in embryonic tissues (for literature see Bassas et al. 1985; Underwood et al. 1986). Synthesis and secretion of IGF-I, as the major growth factor in vivo, are controlled by growth hormone, insulin, and nutrition. IGF-II is much less dependent on growth hormone control. In contrast to insulin, the secretion of both IGFs seems to be a constant, slow process; they circulate in blood bound to two highly specific carrier proteins. The latter are also secreted by the liver but are not identical with prepro-regions of the nascent IGFs. As a consequence of their protein binding, the IGFs have much longer half-lives than insulin, i.e., 4 h in the rat and 16 h in the human. The specific spectra of the physiological effects of the IGFs are not

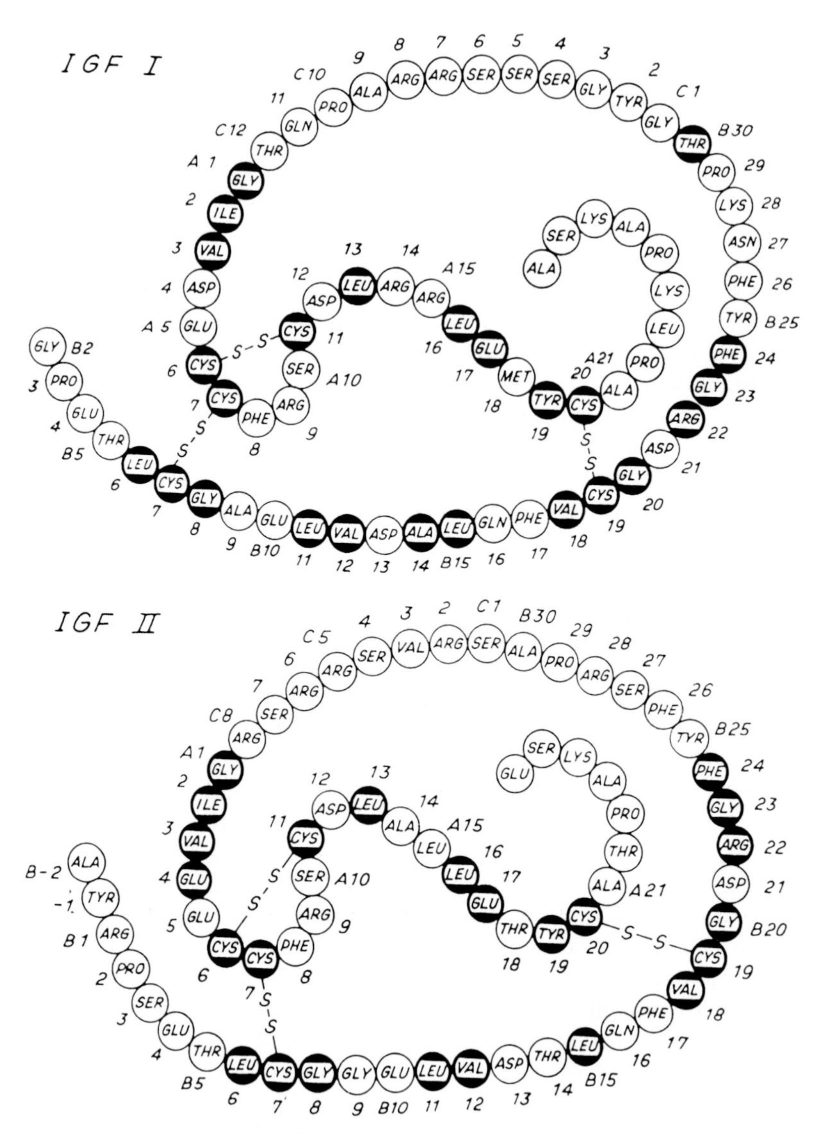

Fig. 10.3. Primary structures of IGF-I and IGF-II. Alignment of the amino acids has been chosen to give maximal homology with human proinsulin. *Black circles* indicate residues identical in proinsulin and the IGFs (Zapf et al. 1981)

yet completely known. In vitro studies have shown that both are important for cell differentiations, IGF-I strongly affecting growth of mesenchymal derivatives and IGF-II affecting tissue repair. In the ovary, IGFs may exert an autocrine role in the regulation of follicular growth and development (Adashi et al. 1985; Hammond et al. 1985). However, the distribution of IGF receptors indicates a much wider spectrum of target tissues (see below). Low levels of IGF-I in pygmies, (human) Laron dwarfs and toy poodles suggest that this hormone is imporant in growth control (cf. Froesch and Zapf 1985). IGF-I has also been proposed to be an antler-stimulating hormone in the red deer (Suttie et al. 1985). In mammalian and chick embryo cells, two types of IGF receptors have been identified, and the affinity of both receptors for the two IGFs has been shown to be overlapping, but not identical. Interactions between insulin and the IGFs at the receptor level are illustrated in Fig. 10.4; further details are discussed in a review by Czech et al. (1985). In the chick embryo, IGF receptors are present on day 2, i.e., 1 – 2 days before the insulin receptors. At this point, neurulation is completed, and the corresponding stages would be approximately day 11 in the rat and week 4 in human embryogenesis. From this observation, as well as others on chicken and mammals (including human), it appears that IGF receptors occur in many (perhaps all?) embryonic/fetal tissues, including the brain, and that the pattern of IGF receptors is organ-specific. Furthermore, it now appears that the teratogenic effects of excessive doses of insulin in the chicken embryo in the classical experiments of Landauer (1945) were mediated by IGF receptors (cf. Bassas et al. 1985; De Pablo et al. 1985a). Clearly, the interactions between receptors for IGFs and insulin (see Sect. 10.3) in prenatal development promise to be an area of most fruitful research for years to come, and there is increasing evidence that insulin and the IGFs have complementary anabolic and growth-promoting actions in many tissues, including the placenta (Froesch et al. 1983; Stuart et al. 1984; Bhaumick and Bala 1985).

Relaxin, a product of the ovary and several other, mainly reproductive, tissues (cf. Weiss 1984; Eldridge and Fields 1985), is structurally similar to insulin. It consists of two polypeptide chains linked by disulfide bridges, and is synthesized as a single chain precursor. Schwabe et al. (1982) have determined the amino acid sequences of rat and porcine relaxin and a partial sequence of the shark molecule,

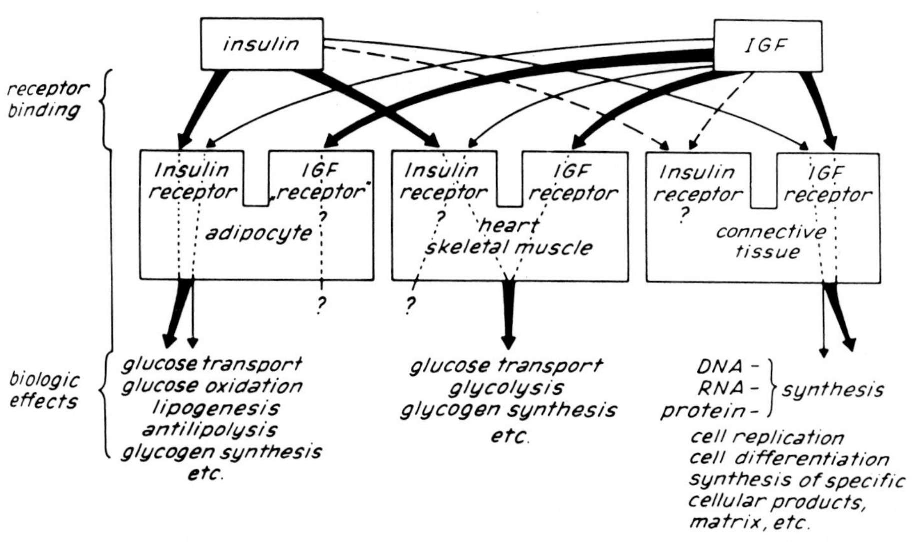

Fig. 10.4. Complementary actions of insulin and IGFs. Insulin is a prerequisite without which growth processes cannot be initiated. The IGFs may have taken over from the common ancestral molecule the regulation of the slow growth processes and the stimulation of the differentiation of some mesenchymal cells (Froesch and Zapf 1985)

pointing out that insulin-relaxin differences, position for position, are usually about 75%. Furthermore, relaxin does not compete with insulin for the latter's receptors. Blundell and Humbel (1980) are uncertain as to whether insulin and relaxin are products of convergent or divergent evolution; Schwabe et al. (1982) conclude that insulin and relaxin are not the result of gene duplication. The physiological role(s) of relaxin are not yet clear. It has been implicated in several functions related to pregnancy, parturition, and lactation, such as increasing vascularization and glycogen content of the uterus, inhibition of uterine contractions, relaxation of the pelvic ligaments and dilation of the cervix (for literature see Eldridge and Fields 1985).

Nerve growth factor (NGF) occurs in many organs, including the submandibular gland and the prostate of rodents, and in smaller quantities in snake venoms. Its richest source seems to be bovine seminal fluid (cf. Shikata et al. 1984; Siigur et al. 1985). A homology between insulin and NGF was suggested by Frazier et al. (1972). However, unlike insulin, NGF has not conserved a hydrophobic core, so that any substitutions in that region cause conformational changes in what would be equivalent to the B-chain of insulin (Blundell and Humbel 1980). NGF also does not compete with insulin for its receptors. The NGF plays an important role in the development and maintenance of the peripheral nervous system, but additional roles appear likely (cf. Ng and Wong 1985). In particular, it may act as a retrograde messenger between target organs and their innervating neurons (Heumann and Schwab 1985).

Although the evidence suggests that insulin and the structurally similar NGF and relaxin are not divergent molecules, there must be some biological significance in the conservation of the common three-dimensional architecture of these molecules.

10.2 Antigenicity, Assays, and Endogenous Titers

Insulin is a weak antigen (Arquilla et al. 1972). However, because of marked differences between pancreatic insulin of the guinea pig and other mammals, it has been the species of choice in which to raise antibodies against mammalian insulins. The guinea pig's strong immune response and its ready availability also make it useful for deriving antibodies to other vertebrate insulins. The ideal radioimmunoassay or immunocytochemical procedure would employ an antibody directed against insulin of the experimental species as well as using insulin of that species as both standard and tracer. Because unavailability of the native hormone often renders the homologous test system impossible, porcine-insulin antiserum combined with porcine insulin as standard and tracer are frequently used to estimate insulin levels in other species. This assay system has worked well in birds (Naber and Hazelwood 1977), reptiles (Rhoten 1974a, b), and amphibians (Schlaghecke and Blüm 1981). In an early attempt to measure insulin in fish by radioimmunoassay, Tashima and Cahill (1968) used guinea pig anti-pork insulin antiserum with bonito insulin standards to measure toadfish insulin secretory responses. However, they experienced weak cross-reactivity between the endogenous toadfish insulin and the antiserum. Patent and Foá (1971) refined this approach by immunizing guinea pigs with cod insulin and using that antibody with cod insulin as the standard and tracer to estimate insulin levels in goldfish and toadfish. Plisetskaya and Leibush (1974) estimated insulin in scorpionfish with a porcine insulin antiserum combined with scorpionfish insulin as the standard and tracer. The scorpionfish insulin apparently cross-reacted very well (>85%) with the pork insulin antiserum. Thorpe and Ince (1976) used a modification of the Patent and Foá (1971) procedure to measure insulin in several teleosts, finding the greatest sensitivity in the codfish. Other workers have used guinea pig anti-anglerfish insulin antiserum (Fletcher et al. 1978; Wagner and McKeown 1982) or guinea pig anticarp insulin antiserum (Huth and Rapoport 1982) to measure insulin in teleost systems. On the other hand, the cross reactivity of the insulins from three different species of *Oncorhynchus* with antisera to mammalian insulins is very low (cf. Plisetskaya et al. 1985). Recently, Tilzey et al. (1985a) reported a "homologous" radioimmunoassay for teleost insulin that was most sensitive for the determination of concentrations of 50–500 μg/ml of insulin from *Salmo gairdneri* in vitro; these investigators applied antibodies against insulin from *Katsuwonus pelamis* raised in rabbits.

Using an adaptation of a method for mammalian insulin, Gutierrez et al. (1986) found greatly varying titers in 11 teleost species, ranging from almost 12 ng/ml in *Barbus comiza* to about 1 ng/ml in *Ictalurus punctatus*. As in the case of glucagon (see Chap. 11.2), the insulin titer in the dogfish (0.65±0.09 ng/ml) was lower than in any teleost. As one might expect from the structural differences between teleost and mammalian insulins the homologous assay gives values that are usually several times higher than those obtained from heterologous systems, often being in the range of 4–6 ng/ml (cf. Plisetskaya et al. 1976, 1986a; Thorpe 1976; Thorpe and Ince 1976; Maximovich et al. 1978; Bondareva et al. 1980; Furuichi et al. 1980; Albert 1982). With a fully homologous RIA system, Plisetskaya et al. (1985) found stage- and sex-dependent differences in the plasma titers of salmon (*Oncorhynchus kisutch*) insulin: in juveniles (Smolt): 3.4±0.3 ng/ml; postspawning males and females: 2.7±0.6 and 1.2±0.26 ng/ml, respectively. Because of poor cross-reactivity between hagfish insulin and anti-porcine insulin antiserum (0.1–0.5%), Emdin and Steiner (1980) used guinea pig anti-hagfish insulin antiserum along with hagfish insulin as standard and tracer in a sensitive radioimmunoassay system, detecting levels as low as 1 ng/ml. In summary, there may be sufficient cross-reactivity between insulin of a given species and anti-porcine insulin antiserum to allow reasonably good estimates of insulin levels in some species, e.g., scorpionfish. However, in most lower vertebrates there must be a closer phylogenetic match between the antigenic insulin and the one being measured in order to carry out accurate radioimmunoassays. Another complicating factor in an insulin radioimmunoassay is the degree to which proinsulin binds to the insulin antibody. In mammalian assays, proinsulin is variously reported to account for from 10–50 percent of the measured insulin activity (see Chap. 9.1.1), whereas proinsulin of the catfish (*Ictalurus punctatus*) is reported to compete only 10 percent as effectively as native insulin for binding sites on a specific antibody (Albert 1982). The degree of interference by proinsulin in other vertebrate insulin radioimmunoassays appears to be unknown.

For many years, insulin, as well as several other hormones, has been assayed by activity units based on older bioassay methods. The activity units have been relatively consistent in mammals, but when lower vertebrate insulins are subjected to bioassays to determine the activity units per milligram of hormone, the results may be inconsistent (Patent and Foà 1971; Plisetskaya et al. 1976; Thorpe and Ince 1976). Recently, however, there has been a trend to standardize all of the islet hormone assays, including insulin, by expressing their quantities in mass units. In mammalian determinations the conversion from activity units (μU/ml) to mass units (ng/ml) is as follows: ng/ml = μU/ml × 1/25 (Charles 1981).

As indicated above, comparisons of insulin measurements in different species are often complicated by the lack of homologous insulin for radioimmunoassay purposes. The result of this is that many assays have employed antiporcine insulin antiserum and porcine insulin as the standard and tracer. Others have used the porcine insulin antiserum with combinations of homologous or near-homologous insulins as standards and pork insulin as the tracer. Consequently, in many instances it is only possible to estimate insulin levels. Since attempts to analyze

Table 10.1. Plasma insulin levels in "lower" vertebrates

Species	Age/condition	Insulin (ng/ml)	Reference
Chicken	Adult	0.6 – 1.9	Raheja (1973), Benzo and Green (1974), Colca and Hazelwood (1976)
Penguin	24-h fast	1.6	Chieri et al. (1972)
Penguin	12-day fast	0.4	Chieri et al. (1972)
Frog	Autumn	2.4	Schlaghecke and Blüm (1981)
Frog	Spring	1.2	Schlaghecke and Blüm (1981)
Goldfish	Fasted 5 days	1.88 [a]	Patent and Foá (1971)
Codfish	Fed	3.53 [a]	Patent and Foá (1971)
Goldfish	Adult	6.35	Thorpe and Ince (1976)
Rainbow trout	Adult	3.00	Thorpe and Ince (1976)
Rainbow trout	7-day fast	1.60	Thorpe and Ince (1976)
Rainbow trout	Fed 7 days	6.80	Thorpe and Ince (1976)
Rainbow trout	Juvenile	1.4 – 4.5	Wagner and McKeown (1982)
Dab	Adult	2.75	Thorpe and Ince (1976)
Eel	Adult	1.12	Thorpe and Ince (1976)
Plaice	Adult	0.58	Thorpe and Ince (1976)
Pike	Adult	0.26	Thorpe and Ince (1976)
Coho salmon	Juvenile	3.4	Plisetskaya et al. (1985)
Coho salmon	Adult males	2.7	Plisetskaya et al. (1985)
Coho salmon	Adult females	1.2	Plisetskaya et al. (1985)
River bream	Adult	0.27	Murat et al. (1981)
River bream	Juvenile	0.46	Murat et al. (1981)
Sturgeon	Immature	0.42	Murat et al. (1981)
Sturgeon	Mature	0.74	Murat et al. (1981)
Sturgeon	Prespawning	0.13	Murat et al. (1981)
River lamprey	Prespawning (Oct)	0.21	Murat et al. (1981)
River lamprey	Prespawning (Feb)	1.2	Murat et al. (1981)
Sea bass	Captive	5 – 11	Gutierrez et al. (1984)

[a] Calculated from codfish insulin standard at 19.2 U/ml.

various assay data based on the use of homologous and nonhomologous insulin standards are inconclusive, no distinctions as to assay standards have been made in Table 10.1. Wherever possible, activity units have been converted to mass units. In general, the data compiled in this table indicate that the plasma titers of insulin in "lower" vertebrates are higher than those in mammals; this suggestion is strongly supported by the likelihood that the heterologous assay, applied in most cases, underestimates the actual values. The biological significance of these differences requires further study. Leibush (1983) suggests that the higher titer may compensate for the lower affinity of the receptors of species to their own insulin. On the other hand, it appears possible that mammals have developed a more efficient system for control of the insulin titers. Since the onset of insulin release following a stimulus is rather rapid in poikilotherms (see below), their elimination mechanism may be slower than in mammals. However, studies by Ince (1982, 1983) in the eel and trout show that at least exogenous insulin can be disposed of rather quickly. With respect to aquatic vertebrates, it must be kept in mind that in addition to several routes known for mammals (cf. Charles 1981), the gills may be an organ for elimination of polypeptide hormones such as insulin; and last but not least, that the dynamics of circulating insulin are certain to be strongly affected by the body temperature (cf. Ince 1983).

10.3 The Physiology of Insulin Before Birth

As in phylogeny, insulin also precedes the appearance of B-cells in ontogeny. Immunoreactive and bioactive insulin is present in chicken eggs before fertilization, and in chicken embryos on day 2, i.e., before the development of the endocrine pancreas, the beginning of metabolic liver functions and the appearance of insulin receptors (Trenkle and Hopkins 1971; DePablo et al. 1982; Bassas et al. 1985). Possibly the pre-pancreatic insulin is a local growth and/or differentiation factor acting via IGF- or other receptors. Nevertheless, in the mammals studied (human, calf, sheep, rat, rabbit), insulin appears during early gestation in both pancreas and blood, suggesting also a role of pancreas insulin in prenatal development (cf. D. Agostino et al. 1985). The situation in the guinea pig (see Chap. 2.2.1) makes one wonder if, at this point, several forms of insulin are generally present. Most of the earlier information on the possible functions of insulin in prenatal vertebrates has been reviewed by Baxter-Grillo et al. (1981). However, it is noteworthy that prenatal insulin receptors have binding characteristics similar to adult receptors, but by contrast to the latter, cannot be down-regulated by high ambient insulin concentrations (cf. Peyron et al. 1985). It has long been known that newborn children of diabetic mothers are often oversized. This phenomenon has been explained by an overstimulation of their B-cells by increased fetal blood sugar levels, due to the maternal hyperglycemia (see below). In such cases, fetal hyperinsulinemia may be associated with increased adiposity and enlargement of various organs. Recent studies have confirmed that these growth effects are indeed due to fetal insulin. Such effects have been induced experimentally in rhesus monkeys even in the absence of hyperglycemia. Furthermore, since no evidence of an increased somatomedin stimulation by insulin has been found, it appears that insulin itself has a direct growth-promoting effect in the fetus (Susa et al. 1984a, b). On the other hand, Cooke and Nicoll (1984) conclude from studies on transplanted fetal rats paws that the effects of insulin in this case are more growth-supporting than growth-promoting. Taken together, the most recent evidence suggests that during pre- and early postnatal development insulin acts as a selective growth, differentiation, and/or maturation factor on some organs, and that these functions may appear before the metabolic effects of insulin. This conclusion is supported by the early appearance of insulin receptors in the embryonic nervous system and other organs of the chicken (cf. Bassas et al. 1985); insulin effects on macromolecular synthesis and neuronal maturation in the fetal nervous system (cf. Puro and Agardh 1984); increased availability of circulating insulin to the newborn brain (Frank et al. 1985); the presence of insulin receptors in endoderm derivatives such as fetal bronchi, gut and liver (cf. Sodoyez-Goffaux et al. 1985), and mesenchyme derivatives such as brown adipose tissue (Peyron et al. 1985); direct insulin effects on mesoderm/mesenchyme derivatives including the osteoblast, stimulating via the latter cell type collagen formation in fetal rat bone (Kream et al. 1985); growth retardation when insulin antibodies are given to early chicken embryos (DePablo et al. 1985b). Furthermore, insulin binding seems to reach a maximum in late fetal and early postnatal life (cf. Peyron et al. 1985). From this compilation, it appears doubtful that the impact of embryonic/fetal insulin can be related to any specific organ system; and it is furthermore obvious that the prenatal interactions of insulin (possibly also of proinsulin: Nissley et al. 1977), various growth factors (see Sect. 10.1), and growth hormone must be extremely complex.

Although insulin production in the developing avian pancreas has been studied by a number of investigators (cf. Foltzer et al. 1982), little is known about its functional competence. Benzo and Green (1974) detected insulin in the plasma and pancreas of 5-day chick embryos, slightly later than found by immunocytochemical methods, and correlated the subsequent increases in insulin levels with those of the liver glycogen. When given at day 5, exogenous insulin already causes hypoglycemia (Zwilling 1948). L. Leibson et al. (1976a) showed that glucose injections into 12- to 16-day-old chicken embryos provoke a marked increase in plasma insulin; they mention in the same paper that, according to Mazina and Vizek (1969), insulin injections into chicken embryos cause a marked drop of blood FFA. The latter effect is different from that seen in postnatal chickens (see Sect. 10.4.8). In another paper, L. Leibson et al. (1976b) discuss the sensitivity of the developing cardiac and skeletal muscle of the chicken to exogenous insulin. Taken together with the data of Foà's group, their findings suggest that the embryonic cardiac muscle responds to insulin much earlier (day 6) than the skeletal muscle used (day 15). In the embryonic chicken heart, insulin stimulates uptake of both glucose and amino acid (cf. Guidotti et al. 1969). In a study on splenic lobes of the chicken pancreas in vitro, Foltzer et al. (1982) found that the responsiveness of the B-cells to a glucose stimulus declined after hatching. Virtually nothing is known about the role of insulin in the development of poikilotherms, although findings of Frye (1962) indicate that salamander larva can live without islets. On the other hand, embryonic pancreas of the bullfrog contained sizable amounts of insulin immunoreactivity (Hulsebus and Farrar 1985), suggesting a function at this stage. Meiotic maturation of *Xenopus laevis* oocytes can be induced by exogenous insulin (El-Etr et al. 1979). Considering the current state of knowledge, it is clear that the role of insulin in prenatal vertebrates, and its interactions with the related growth factors, will be a most fertile field for future investigations.

10.4 The Physiology of Insulin in Postnatal Vertebrates

Insulin is the only islet hormone to occur in all vertebrates including the ammocoete; and in all vertebrates, exogenous insulin causes a decrease in blood sugar. Beyond these two statements, it becomes difficult to make generalizations. It appears to be axiomatic to interpret insulin as *the* anabolic hormone, but we know almost nothing about the function of insulin in the early deuterostomians in which it occurs in the "open" mucosa cells of the gut (see Chaps. 2.1 and 14). In vertebrates, insulin has been implicated in the regulation of carbohydrate, lipid, and protein metabolism; growth; mineral metabolism and osmoregulation; and release of other hormones. Unfortunately, the attempt to identify the physiological roles of insulin is fraught with a large number of data of uncertain, often most doubtful biological relevance. The nemesis of endocrine research, i.e., the use of excessive but "effective" doses of hormones and metabolites, extends to almost all taxa studied. Hence, we will begin with a look at the realms of insulin control in the taxon for which the most reliable information is available: mammals. Subsequently, we will try to trace the evolution of the insulin functions, beginning with the cyclostomes.

10.4.1 Mammalia

In the *Introduction* we pointed out that serendipity was a major factor in the discovery and early successful therapeutic application of insulin. However, serendipity was twice also a major factor in the discovery of experimental diabetes. Von Mehring and Minkowski (1889) as well as Dunn et al. (1943) happened to perform their experiments with extremely well-suited species: the dog and the rabbit. If Von Mehring and Minkowski had used a herbivorous mammal, the result of total pancreatectomy would have been much less drastic; and if they had worked with rats, there would probably have been no diabetes at all because it is very difficult to remove enough (>90%) of the rather diffuse rat pancreas to achieve metabolic derangements (cf. Scow 1957). On the other hand, if Dunn and coworkers had studied the effect of alloxan in the guinea pig, they would not have obtained diabetes mellitus because the B-cells of this species are highly alloxan-resistent (for literature see F. G. Young 1963; Sirek 1969; Cooperstein and Watkins 1981). These examples illustrate some of the problems encountered in the study of experimental diabetes mellitus, and the need to perform thorough histological analysis of the pancreas when a new species is used. In addition, they are a good introduction to the question of an insulin sensitivity or insulin dependence which are frequently invoked to explain sequelae of islet- or pancreatectomy. Nevertheless, we can take the immediate effects of total pancreatectomy in the dog as an example of "classical" insulin-deficiency diabetes mellitus: hyperglycemia, glucosuria, hyperlipemia, hypercholesterolemia, ketonemia, ketonuria, polyphagia, polydipsia, polyuria; depletion of glycogen stores in liver and muscle, depletion of lipid stores in the adipose tissue, and wasting of muscle substance. These defects reflect the unique importance of insulin in mammals, in which this hormone probably controls more metabolic processes and tissues than in any other group of vertebrates. The opposite side of the coin is obvious: acute insulin deficiency and/or destruction of the B-cells is incompatible with mammalian survival, and it is an interesting question why insulin could achieve such a dangerously powerful role in mammals. Certainly, there must have been considerable advantages to be gained for the chances taken, especially since the mammalian B-cells appear to be most sensitive to viruses (Rayfield and Yoon 1981; Craighead 1985), toxins (as exemplified by the easy induction alloxan and/or streptozotocin diabetes in many species; cf. Cooperstein and Watkins 1981; Chang and Diani 1985), and endocrine/metabolic dysregulations (Soret and Dulin 1981). Unfortunately, the current state of knowledge is insufficient for the interpretation of the role of mammalian insulin in an evolutionary perspective.

To evaluate the multifaceted actions of insulin in mammals, a look at the situation in the human may be most instructive. Excellent and comprehensive reviews on this topic have recently been written by Charles (1981), Felig (1983), P. Cryer (1985), and Unger and Foster (1985). The following sections are largely based on their accounts.

As in most mammals, glucose is the primary B-cell stimulator in the human, both in vivo and in vitro (Cerasi 1975; Lacy 1977; Hedeskov 1980; Ashcroft 1981; Charles 1981; Täljedal 1981; P. Cryer 1985). Its effects on insulin titers are greater when administered orally or intraduodenally than by any parenteral route, emphasizing the presence of endogenous secretagogue and/or nervous pathways which are largely unknown (see Chaps. 7 and 8). Glucose, as well as other agents, when acutely increased in concentration, cause a biphasic release of insulin, and much effort has been devoted to the explanation of this phenomenon (cf. Berthoud 1984). It is very rare, however, that under normal conditions the islets experience such a sharp increase in glucose concentration (Charles 1981). Therefore, the physiological significance of a biphasic pattern of insulin release in the day-to-day or minute-to-minute activity of the islet organ is questionable, although it can be elicited also in other mammalian and nonmammalian species (see below). In the intact rat, e.g., oral glucose causes two preabsorptive peaks of insulin release which are followed by a third, glucose-induced postabsorptive peak (Louis-Sylvestre 1978; see also Chap. 7). Hence, at least the early in vitro response of the B-cells cannot be equated with the in vivo events that are neurally and hormonally induced.

The impact of amino acids on insulin release is somewhat controversial (cf. Ashcroft 1981; Fukagawa et al. 1986). Charles (1981) emphasizes that some confusion exists over the effects of amino acids on insulin secretion because of the frequent use of pharmacological doses. However, the branched-chain amino acids valine, leucine, and isoleucine, particularly when infused as mixtures, are more potent insulin stimulants than the straight-chain ones. Like glucose, amino acids elecit a stronger insulin release when administered intraduodenally (Charles 1981).

In the human, fats also stimulate insulin release when ingested rather than injected, although their impact is weak when compared with glucose or amino acids. In the dog, increased plasma FFA seem to stimulate insulin release directly (Charles 1981). In this context, it has been occasionally stated that the ruminants are exceptional among mammals by the fact that volatile fatty acids are the major insulin secretagogues (Lacy 1977; Charles 1981). However, it should be emphasized that, paradoxically, the effects of these fatty acids are greatest after intravenous injection, less effective when injected into the abomasum, and ineffective when injected into the rumen. When a mixture of acetate, the major fatty acid of the rumen, butyrate and proprionate is infused into the rumen for 2 h, there is only a small transient peak of insulin release (Bassett 1972). Therefore, caution is due in interpreting the effects of volatile fatty acids on insulin secretion in ruminants, a group with many metabolic differences from simple-stomached mammals (cf. Brockman and Laarveld 1985; Vernon et al. 1985). Ketone bodies stimulate insulin release in some mammals (human, rat, and dog), while they have an inconsistent affect on sheep and none on rabbit pancreas in vitro (Robinson and Williamson 1980). The metabolic functions of insulin at its three major target organs (cf. Charles 1981; Felig 1983) listed in Table 10.2, and the major metabolic pathways under insulin control are summarized in Figures 10.5 – 10.7. However, it must be understood that the actions of insulin are dependent on, or strongly modified by, other hormones. In particular, the hepatic effects of insulin depend on the ratio between circulating insulin and glucagon (see Sect. 11.5).

Carbohydrate Metabolism. In the "basal", postabsorptive state (Fig. 10.5) the liver appears to be the sole source of blood sugar. About 75 percent of the glucose

Table 10.2. Major metabolic functions of insulin in mammals (Felig 1983)

	Liver	Adipose tissue	Muscle
Anticatabolic effects	Decreased glycogenolysis Decreased gluconeognesis Decreased ketogenesis	Decreased lipolysis	Decreased protein catabolism Decreased amino acid output Decreased amino acid oxidation
Anabolic effects	Increased glycogen synthesis Increased fatty acid synthesis	Increased glycerol synthesis Increased fatty acid synthesis	Increased amino acid uptake Increased protein synthesis Increased glycogen synthesis

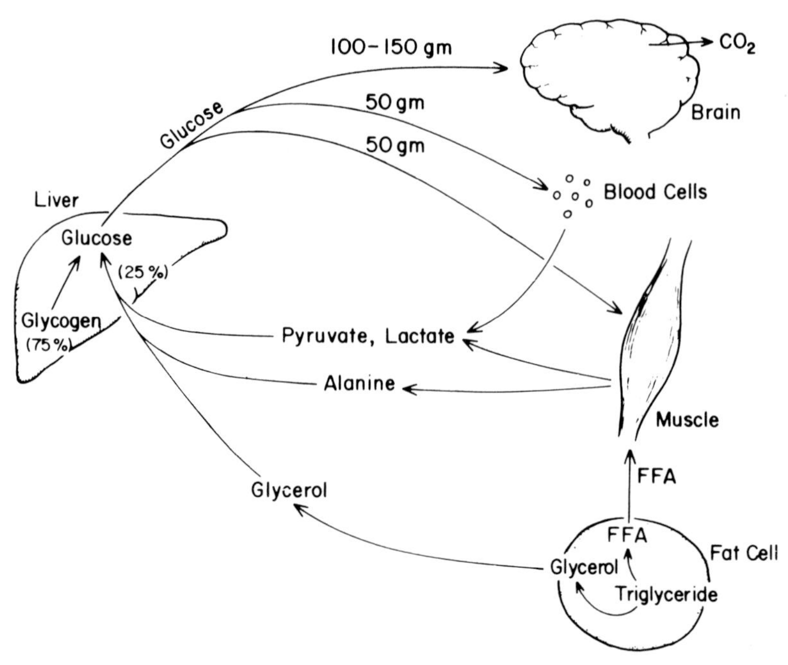

Fig. 10.5. Glucose production and utilization in the human after 10–14 h of fasting. In this "basal" state, only a small proportion of glucose uptake (mainly by muscle) is insulin-dependent (Felig 1983)

production occurs via glycogenolysis, the remainder being supplied via gluconeogenesis from pyruvate, lactate, glycerol, and amino acids (mainly alanine). The brain is the major site of glucose uptake, while, among others, the blood cells and "resting" skeletal muscle also use glucose, though to a lesser degree. However, the glucose utilization of both brain and blood cells is not insulin-dependent; consequently, more than 75 percent or more of the glucose utilization at this state does not require the direct action of insulin. Nevertheless, an appropriate insulin/glucagon ratio is necessary to maintain a blood sugar level above approximately 70 mg%, since otherwise the glucose needs of the brain and nervous system (about 6 g/h) cannot be met. During this time, insulin declines to plasma levels of about 0.6 ng/ml, which is sufficient to reduce the inhibitory impact of this hormone on hepatic glycogenolysis and gluconeogenesis. A different scenario evolves when the glycogen reserves are exhausted, which occurs after 1–2 days of fasting. Now increased gluconeogenesis occurs, and insulin levels decrease further so that the ratio in favor of the catabolic antagonists (cortisol and glucagon) allows a stronger release of gluconeogenic substrates from the periphery (mainly skeletal muscle) and their subsequent conversion to glucose in the liver. During the postprandial, "fed" state, the B-cell responds to the increased levels of plasma glucose and other metabolites with an enhanced insulin release, which results in plasma concentrations of 30–80 μU/ml. The important anabolic role of insulin becomes obvious from an analysis of the events during the first 3 h following an oral glucose load of 100 g. During this time, the liver extracts about

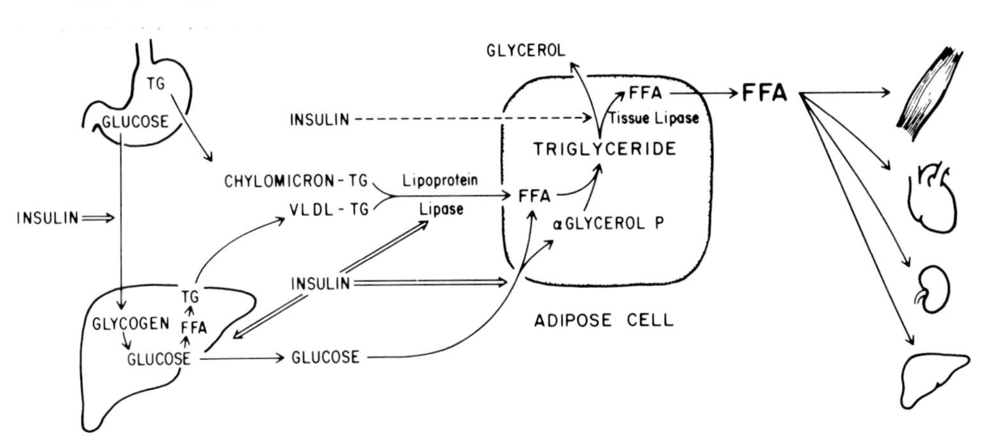

Fig. 10.6. Impact of insulin on the lipid homeostasis of the human. Stimulatory effects of insulin on liver (FFA synthesis from glucose and their esterification to form triglycerides), and on the adipocyte (lipoprotein lipase activity and glucose uptake) are indicated by *double-lined arrows.* The inhibitory effect of insulin on lipase within the adipocyte is indicated by the *broken arrow* (Felig 1983)

60 percent of the total glucose entering the blood, and of this already 10 percent during the first passage through the portal system. Simultaneously, hepatic glucose production is reduced to 20–25 percent of the preingestion level, while the newly extracted glucose is used for synthesis of hepatic glycogen and triglycerides. Of the remainder of the glucose load, about 25 percent is utilized by the brain and some other insulin-independent tissues (mainly blood cells), and about 15 percent by muscle and adipose tissue. Thus, the three insulin-dependent major tissues, i.e., liver, adipose tissue, and muscle together take up about 75 percent of the total glucose load. The sensitivity of these three tissues varies: suppression of hepatic glucose production occurs at insulin levels less than half of those required for stimulation of "peripheral" glucose utilization. Studies with modified insulin molecules suggest that hepatic and "peripheral" actions of insulin are linked to different sites of the molecule (Tompkins et al. 1981). Furthermore, the cellular locations of the insulin action differ. Because the liver cell membrane is freely permeable to glucose, the effect of insulin in these cells is intracellular. The glucoregulatory hepatic actions of insulin occur via three key enzymes: glycogen synthetase and phosphorylase, and glucokinase. Changes of the blood sugar level elicit an immediate response by changing insulin titers, which, in turn, have a rapid effect on glycogen synthetase and phosphorylase. The interplay of insulin-stimulated increase of glycogen synthetase and decrease of phosphorylase (and the opposite change when plasma insulin decreases) is apparently a major event in the insulin-mediated rapid glucostatic controls. On the other hand, an increase of plasma insulin for several hours or longer enhances the levels of hepatic glucokinase, i.e., the enzyme that "traps" glucose inside the

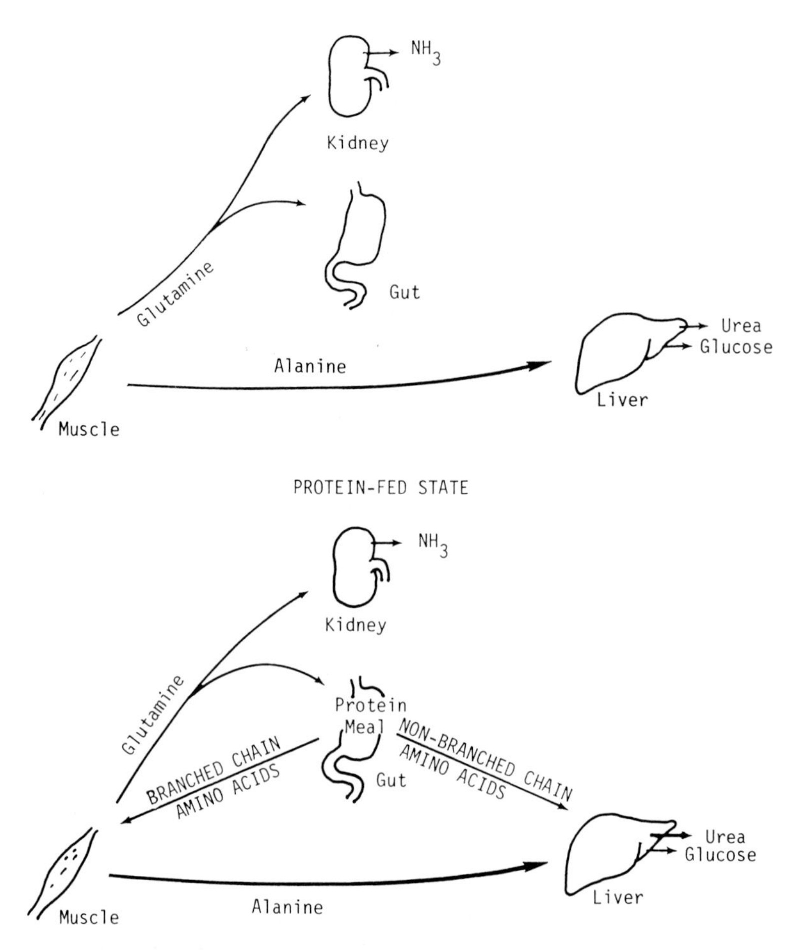

Fig. 10.7. Inter-organ exchange of amino acids in the human during fasting (*top*) and after a protein meal (*bottom*). Muscle releases alanine and glutamine in the fed as well as fasted state. After protein feeding, nitrogen repletion in muscle occurs primarily as a result of net uptake of the branched-chain amino acids, valine, leucine, and iso-leucine. The increase in plasma insulin that occurs in normal subjects after protein feeding stimulates the uptake of amino acids by muscle tissue. In diabetics the uptake by muscle of the branched-chain amino acids is diminished, resulting in postprandial elevations of these amino acids (Felig 1983)

liver cells via phosphorylation. The increase in glucokinase is induced by formation of glycolytic intermediates, which involves GMP.

In contrast to the liver, membranes of both adipose tissue and muscle are not freely permeable to glucose. In these tissues, the major role of insulin involves glucose transport, although in addition, it also increases glycogen synthetase and phosphofructokinase. In the human adipose tissue, insulin-stimulated glucose uptake provides mainly the alpha-glycerophosphate for the esterification of the FFA to triglycerides. In contrast to some mammalian species such as the rat, synthesis of FFA from glucose plays only a minor role in human adipocytes (see below).

In skeletal muscle, insulin-mediated glucose uptake serves mainly the production of glycogen. Exercise enhances the insulin sensitivity of the muscle considerably, although the precise mechanisms appear unknown.

Lipid Metabolism. The direct effects of insulin in human lipid metabolism involve mainly two target tissues: liver and adipose tissue. However, also ketone uptake and oxidation of ketones by muscle are accelerated in the presence of insulin (Sherwin et al. 1976). Other organs are affected indirectly by insulin, FFA being a major fuel for skeletal muscle, heart, and kidney, and ketones being a major source of energy for the human brain during extended fasting. In the human, formation of FFA and subsequently of triglycerides occurs primarily in the liver, a process stimulated by several different actions of insulin. However, probably as in all mammals, no appreciable storage of triglycerides takes place in this organ under normal conditions. Instead, the newly synthesized triglycerides are transported in the form of very low density lipoproteins to the adipose tissue where they are stored. The uptake of FFA from both the endogenously produced very low density lipoprotein fraction and exogenously (food) derived chylomicron triglycerides into adipocytes is mediated by lipoprotein lipase, which is stimulated by insulin. Within the adipocyte, the FFA are re-esterified to triglycerides, and as already mentioned, the glycerol component is derived from insulin-stimulated glucose uptake and alpha-glycerophosphate formation. These anabolic effects of insulin are supported by its inhibitory action on the adipocyte lipase, with the antilipolytic action being stronger than the effect on glucose transport. A decline in plasma insulin (e.g., during fasting or exercise) leads to lipolysis and release of FFA from adipose tissue which then move to their sites of utilization, i.e., mainly muscle, heart, kidney, and liver. If the decrease in the insulin level continues, increasing quantities of FFA are released from the adipose tissue; the liver concomitantly increases their oxidation to acetyl CoA which is then converted to ketones (beta-hydroxybutyrate and acetoacetate). The increased production of ketones is fostered by the removal of a direct antiketogenic effect of insulin, which occurs via its impact on hepatic carnitine. Carnitine is necessary for the transfer of FFA across the mitochondrial membrane to the sites of the beta-oxidative enzymes; the level of carnitine is reduced by insulin. Thus, a decline in plasma insulin fosters hepatic ketogenesis by two mechanisms: (1) supply of precursors, i.e., FFA, and (2) increase of beta-oxidation. While a slow, moderate increase of plasma ketones during fasting will substitute for the deficient glucose supply to the brain, strong, sudden increases of plasma ketones during acute insulin deficiency may lower the pH of the blood to such a degree that the human brain can no longer function. The result is the well-known diabetic coma, which will lead to death if the situation is not corrected rapidly.

Protein Metabolism. In humans, the importance of insulin for protein metabolism is demonstrated by the wasting syndrome of severely diabetic (Type I) persons, which in the pre-insulin era inevitably led to death. In these patients, excessive breakdown of protein, especially in muscle, is accompanied by hyperaminoacidemia and a negative nitrogen balance. Application of insulin reverses this phenomenon by a combination of anabolic and anti-catabolic effects: (1) increased net uptake of amino acids, (2) decreased amino acid oxidation, (3) decreased protein catabolism, and (4) increased protein synthesis.

In the intact, postabsorptive human, muscle releases mainly alanine and glutamine. Alanine serves as a substrate of gluconeogenesis in the liver, a process controlled by the inhibitory action of insulin which, however, occurs under euglycemic conditions only when the plasma levels of this hormone are strongly elevated (16 ng/ml). In this case, the hepatic action of insulin seems to involve both interference with the uptake of alanine and alterations of the intrahepatic disposal of this amino acid. Since this situation is normally absent in the late postabsorptive state, net efflux of alanine from muscle and hepatic gluconeogenesis from this substrate increasingly supply the glucose needed for function of the brain. After a protein-rich meal, the efflux of glutamine and alanine from muscle tissue continues. However, it is offset by a strong net flow of amino acids (mainly branched-chain) in the opposite direction, which repletes nitrogen losses during fasting (Wool et al. 1972; Wahren et al. 1976). This process is controlled by insulin, the secretion of which is stimulated postprandially mainly by glucose and/or amino acids. In addition to an increased amino acid uptake, insulin fosters the anabolic state of the muscle by stimulating protein synthesis (Jefferson 1980) and inhibiting protein breakdown (Jefferson et al. 1974). The increase of plasma insulin in the "fed" state also has an anabolic overall effect on the liver, where gluconeogenesis is reduced and protein synthesis increased (Jefferson 1980). It is noteworthy that the inhibitory effect of insulin on hepatic gluconeogenesis requires much lower levels of this hormone when there is a concomitant increase of glycemia, i.e., the usual situation after a mixed meal.

Growth Promotion. Considering the overall anabolic role of insulin in the human, one would expect that insulin plays an important role also in regulation of postnatal growth. Indeed, it has long been known that the growth of poorly controlled diabetic children is retarded even when growth hormone is present in normal or supranormal plasma titers (Hayford et al. 1980; Tamborlane et al. 1981). Recent studies suggest that the growth-promoting effects of insulin in the postnatal state are not only due to its direct anabolic and anti-catabolic functions, but also to its stimulation of somatomedin production (cf. Felig 1983).

Electrolyte Metabolism. It has been suspected that an anti-natriuretic effect of insulin may be responsible for the edema that occurs both after insulin treatment of some diabetic patients, and after refeeding of malnourished persons. A direct effect of insulin on renal sodium reabsorption has been shown; and following a physiological increase of plasma insulin, urinary sodium drops in the absence of changes in the glomerular filtration rate or aldosterone excretion (DeFronzo 1981; Hammerman 1985). In addition to this natriuretic effect, insulin also seems to have a specific hypokalemic effect which is not related to its impact or glucose metabolism. However, the latter appears to be mediated via potassium uptake by muscle and liver (cf. Moore 1983; Zierler et al. 1985). An impact of insulin on potassium metabolism is also suggested by the occurrence of hyperkalemia in the absence of uremia or acidosis in some diabetic patients (DeFronzo et al. 1977). Finally, it appears possible that insulin also has a direct impact on the calcium and phosphate balance, since it can stimulate both phosphate uptake by liver and muscle, and a decrease in phosphaturia (Guntupalli et al. 1985). On the other hand, increased urinary calcium levels in diabetics can be corrected with improved metabolic control (Gertner et al. 1980).

10.4.2 Cyclostomata

Hagfishes. Since the previous reviews of Hardisty (1979), Epple et al. (1980), and Hardisty and Baker (1982), several important publications on the islet physiology of the hagfish have appeared. Many recent data have been reviewed by Plisetskaya (1985). There is a number of discordant findings that may be related to species differences between the Pacific *Eptatretus stouti* and the Atlantic *Myxine glutinosa* (see also Chap. 4.1), although it is likely that also varying experimental conditions are involved. Thus, mammalian antiinsulin had no effect in *M. glutinosa* (Falkmer and Wilson 1967) but increased plasma glucose, fatty acids, and amino nitrogen in *E. stouti* (Plisetskaya et al. 1983b). Isletectomy had no impact on the glycemia in *M. glutinosa* and did not affect the decrease of blood sugar after a glucose load (Schirner 1963a, b; Falkmer and Matty 1966). In contrast, in *E. stouti*, there was a hyperglycemic tendency after islectomy (Matty and Gorbman 1978). In *M. glutinosa*, glucose (1 – 5 mM) induced insulin release from the islet organ in vitro, an effect that was strongly enhanced by lowering the temperature (Emdin 1982a). However, incubation of islets of *E. stouti* showed no effect on insulin release when either 1.1 or 6.7 mM glucose was present in the medium (Stewart et al. 1978). In *M. glutinosa*, a crude hagfish insulin preparation caused a long-lasting hypoglycemia 2 – 3 days after the injection, whereas huge doses of bovine insulin were needed to obtain the same effect (Falkmer and Matty 1966). On the other hand, *E. stouti* was much more sensitive to the glycemic effect of the bovine preparation (Y. Inui and Gorbman 1977). Purified insulin from *M. glutinosa* had no effect on the titers of plasma constituents (glucose, amino nitrogen, triglycerides, and FFA) in *M. glutinosa* (Emdin 1982b) while mammalian insulin provoked a decrease in amino nitrogen in *E. stouti* (Y. Inui and Gorbman 1977). The latter observation is consistent with an insulin stimulation of glycine incorporation into muscle protein of this species (Y. Inui et al. 1978). In an earlier study, it was found that unfed *M. glutinosa* can maintain their glycemia over long periods of time (Falkmer and Matty 1966). More recently, Emdin (1982b) reported in the same species that blood glucose decreased drastically during 1 month of starvation, a finding that agrees with observations of Matty and Gorbman (1978), who find a similar decline of glycemia in *E. stouti* after 10 weeks of starvation. Clearly, this list of contradictions makes it difficult to identify features that could be termed "typical" or "basic" for hagfishes. In a comprehensive study on *M. glutinosa*, Emdin (1982b) found that serum insulin levels diminished strongly within 1 month of starvation (2.2 to 1.1 nM), simultaneously with the decline of blood glucose and a decrease of more than 90 percent of glycogen in the liver and skeletal muscle. At the same time, also serum amino nitrogen, triglycerides, and FFA decreased, in parallel with a smaller but significant decline in hepatic protein and triglycerides. In contrast, there was a modest increase in muscle protein and triglycerides. From these results, it appears that in the fasting hagfish skeletal muscle was the prime source of energy, a conclusion underlined by the large mass of this tissue (about 50 percent of the hagfish body weight: cf. Emdin 1982b). No insulin effects were seen on the liver, whereas the hormone induced an approximately twofold stimulation of the synthesis of glycogen, protein, and neutral lipids from [^{14}C]glucose in the skeletal muscle. Insulin also stimulated the synthesis of muscle glycogen, and of both muscle and liver protein from [^{14}C]leucine. In these studies, only about 10 percent of the radioactive dose was incorporated into the muscle and liver, and the metabolic effects of insulin contributed to only half of this fraction. Thus, the overall effect of the hagfish insulin was rather small, which is reflected in the absence of any quantitative insulin-induced changes in blood glucose, amino nitrogen, triglycerides, and FFA; or in muscle and liver glycogen and protein. In further experiments, single glucose loading increased the serum insulin level from 0.7 to 1.9 nM; a similar increase was also obtained with repeated doses of a mixture of glucose and amino acids. In an overall evaluation of these experiments, Emdin (1982b) concludes that the physiological role of insulin in regulation of skeletal muscle metabolism in hagfish is similar, but quantitatively smaller than in "higher" vertebrates. This conclusion agrees with the suggestion of Y. Inui et al. (1978) that stimulation of muscle protein synthesis by insulin is of primary importance in *E. stouti.* On the other hand, the results in both *M. glutinosa* and *E. stouti* suggest that the hagfish liver, which amounts to about 4 percent of the body weight and contains a much lower glycogen concentration than skeletal muscle, plays only a minor metabolic role (Y. Inui et al. 1978; Emdin 1982b). Hepatectomy in *E. stouti* is not followed by hypoglycemia or hyperaminoacidemia; a transitory appearance of postoperative stress hyperglycemia shows that other sources of blood sugar are readily available (Y. Inui and Gorbman 1978). Thus, just as in the lamprey (Larsen 1978), but in contrast to the Japanese eel (Y. Inui 1969; Y. Inui and Yokote 1975) and mammals, a small size, very limited fuel reserves, and a weak or absent effect of insulin on this organ (Y. Inui et al. 1978; Emdin 1982b) show that the hagfish liver is of little metabolic significance.

Another interesting issue arises with the slow rate of insulin biosynthesis and release in *M. glutinosa.* In perfusion experiments in vitro, the onset of glucose-induced insulin release was 17 min, which may be in part due in the time required for exocytotically released insulin to pass through the thick connective tissue layer surrounding the islet follicles (Emdin 1982a). However, about 40 h are needed for de novo synthesized insulin to be released into the medium in vitro (Emdin and Falkmer 1977). Furthermore, contrary to current information on other vertebrates, changing temperatures have opposite effects on insulin production and release. Biosynthesis is enhanced when the temperature is increased, while stimulated insulin release is roughly doubled when the temperature is lowered by 10 °C (Emdin and Falkmer 1977; Emdin 1982a). While this discrepancy cannot have an impact on the metabolism under basal conditions with a very low insulin release, one wonders how these mechanisms interact when a larger food load causes a prolonged stimulation of insulin release. On the other hand, there is evidence for intracellular degradation of insulin, a process that may be necessary when excess B-granules accumulate during long periods of starvation (Emdin and Falkmer 1977).

The probable absence of glucagon in the cyclostome islets (see Chap. 4.1) raises the question of an interaction between insulin and other hormones in the maintenance of glycemia and other metabolic functions. In islets of *E. stouti*, in which mixed follicles of B- and somatostatin cells are common (Epple and Brinn 1975), local "paracrine" interactions between somatostatin and insulin cells appear a distinct possibility. In vitro, somatostatin (S-14) suppressed insulin release from islets of *Eptatretus stouti* (Stewart et al. 1978). A possible functional relationship between thyroid hormones and insulin was suggested by a decrease in thyroid hormones following injections of anti-insulin (Plisetskaya et al. 1983b). Administration of thyroid hormones, especially T_3, decreased plasma glucose and elevated FFA, whereas the antithyroid compound 6-propylthiouracil raised plasma glucose and alpha-amino nitrogen, but lowered plasma FFA (Plisetskaya and Gorbman 1982; Plisetskaya et al. 1984b). Ignoring the effects on plasma FFA, these observations would be consistent with a mutual stimulatory relationship between insulin and thyroid hormones. However, in a pilot study there was no effect of T_3 on the immunoreactive insulin levels of *E. stouti* (Plisetskaya et al. 1984). There seems to be no evidence of an insulin-catecholamine interaction in the hagfishes under physiological conditions (see however, Chap. 8.1.2). In *M. glutinosa*, even very high doses of epinephrine have no effect on glycemia (Falkmer and Matty 1966). In *Eptatretus*, physiological doses of epinephrine do not effect plasma levels of glucose and FFA. The endocrine mechanisms involved in the stress hyperglycemia in hagfishes are unknown (Plisetskaya et al. 1984). In summary, there are many new and interesting data related to the insulin physiology of the hagfishes; however, until the biological significance of these data is ascertained, the function of the hagfish B-cells remains speculatory.

Ammocoetes. The widely accepted notion that insulin was phylogenetically the first islet hormone to "leave" the gut mucosa (Epple et al. 1980; Hardisty and Baker 1982) is supported by the presence of only B-cells in the ammocoete islets (see Chap. 4.1). However, the biological status of the ammocoete is a matter of debate (cf. Hardisty 1979, 1982; Mallatt 1985), and as pointed out in Chap. 2.3, the larvae of the northern lampreys (Petromyzontidae) may not be good representatives of the "basic" ammocoete. Unfortunately, there seem to be no studies on the islet physiology of the ammocoetes of the species of *Mordacia* and *Geotria*, and information on larval Petromyzontidae is limited (cf. L. Leibson and Plisetskaya 1969; Barrington 1972; Hardisty and Baker 1982). The blood sugar of larval *Lampetra planeri* appears to be well regulated (L. Leibson and Plisetskaya 1969; O'Boyle and Beamish 1977). Barrington (1942), in attempts to extirpate the islet follicles of larval *Petromyzon marinus* by cautery, found an increase in blood sugar. Since this work ignores the presence of scattered islet cells and follicles, many of which may never have been destroyed by Barrington's method (for literature see Hilliard et al. 1985), the hyperglycemia may have been due, at least to a large extent, to stress and/or other factors. An older, indirect attempt to demonstrate the involvement of B-cells in glycemic regulation was the study of the impact of heavy glucose loads. In the B-cells of larval lampreys, this leads to varying degrees of degranulation, or even of glycogen deposition, vacuolization, and necroses (Ermisch 1966; Morris and Islam 1969). Since in these studies the blood sugar levels must have exceeded normal glycemia considerably, the physiological significance of the B-cell changes remains uncertain. A more direct approach to the role of insulin in ammocoetes was attempted by injections of mammalian preparations of this hormone. The results, obtained with very high doses, were lasting hypoglycemia, and variable effects on tissue glycogen (Ermisch 1966; L. Leibson and Plisetskaya 1969; Morris and Islam 1969). Simultaneous injection of insulin alleviated the hyperglycemia induced by a glucose load (Morris and Islam 1969). As in adult lampreys, very high doses of epinephrine caused a hyperglycemia (L. Leibson and Plisetskaya 1969) of uncertain relationship to the islet organ (see Chap. 8).

From this compilation, it appears that knowledge of the physiology of insulin in the animals with the presumably most primitive islet organ is almost nonexistent. Endogenous insulin is present (Ermisch 1965; Van Noorden et al. 1972). Although it may have an impact on the carbohydrate metabolism, the evidence is indirect. Nothing seems to be known about the role of insulin in any other functional realm of the ammocoete.

Adult Lampreys. The conspicuous metamorphic changes of the organs of the lampreys, and the switch from microphagous larval to parasitic or predatory feeding style, make the adult lampreys animals very different from the ammocoetes (cf. Hardisty 1979, 1982; Potter 1980, 1986). Hence, it must be expected that the islet physiology differs considerably between the two stages, as emphasized by the differences in the islet cytology (see Chap. 4.1).

The literature on the islet physiology and related phenomena in adult lampreys has been covered largely in the reviews of Epple et al. (1980), Larsen (1980), Murat et al. (1981), Hardisty and Baker (1982) and Plisetskaya (1985). A striking feature of adult lampreys is their ability to maintain blood sugar level reasonably high despite the extended period of prespawning fasting during which it is clear that gluconeogenesis must provide the major fraction of glucose (Plisetskaya et al. 1976). The role of insulin in the regulation of the metabolism of adult lampreys is poorly understood, although in some cases the information appears more conclusive than in hagfishes. Ignoring various indirect approaches used in earlier studies (for literature see the above reviews), the following will concentrate on the results obtained with (a) isletectomy, (b) use of anti-insulin serum, (c) insulin injections, (d) studies of the responses of endogenous insulin to various manipulations, (e) study of the variations of endogenous insulin under natural conditions.

Hardisty and coworkers (Hardisty et al. 1975; Zelnik et al. 1977) studied the impact of isletectomy in *Lampetra fluviatilis*, a species with conspicuous "cranial" and "caudal" islet accumulations (Fig. 4.2). Although in these studies total removal of islet tissue was probably not achieved (the intermediate cords were not extirpated), the immunoreactive plasma insulin dropped to about half that of controls, and the blood sugar increased conspicuously. Partial isletectomy by removal of either the cranial or caudal islets resulted in no significant glycemic changes. There was no impact of cranial or total isletectomy on liver glycogen;

however, caudal isletectomy caused a strong decrease so that Hardisty et al. (1975) are inclined to relate this result to cytological differences between the cranial and caudal islets. In the "totally" pancreatectomized animals, there was delayed removal of blood sugar following a single glucose injection, which seemed to occur by urinary losses. The findings of Hardisty and coworkers are corroborated by studies of Plisetskaya and coworkers (Plisetskaya and Leibush 1972; Plisetskaya et al. 1976), who tried to induce insulin deficiency by injections of anti-insulin serum in the same species. These authors found hyperglycemia, increased plasma FFA, a decreased glycogen content in liver, heart, and tongue muscles, but no change in the large muscle layers of the body wall. In intact lampreys, a single glucose load provoked an insulin response that was significant after 8 h and still present after 24 h. Long-term injections of heavy glucose loads for 12 days also caused increased plasma insulin levels, while pharmacological amounts of epinephrine for 3 days suppressed insulin (Zelnik et al. 1977). Contrary to the observations of Hardisty et al. (1975), Plisetskaya et al. (1976) found no response of plasma insulin to glucose loads when measured after 210 min (except in one case when early migrants were studied), although arginine and lysine induced insulin release within 2 h. However, as in the study of Zelnik et al. (1977), huge doses of epinephrine caused a prolonged suppression of plasma insulin when its level was high (October), but curiously had the opposite effect when the insulin level was low (March), although hyperglycemia appeared in both cases. A large number of studies with exogenous insulin injections all had the expected effect of a decrease in blood sugar level (cf. Epple et al. 1980; Hardisty and Baker 1982), and Larsen (1976a) showed that lampreys can survive for days even when insulin injections lower the glycemia to nonmeasurable values. Another interesting result was the decrease of plasma FFA following insulin injections which was more rapid than that of the glycemia and also of shorter duration (cf. Plisetskaya 1980). On the other hand, the data on the insulin effect on liver glycogen vary greatly, a phenomenon also observed in gnathostome fishes (see below).

Prespawning *Lampetra fluviatilis* from the Baltic Sea display strong changes in the plasma titer of insulin (cf. Plisetskaya et al. 1976; Murat et al. 1981), which is high after cessation of food uptake in early fall, as they approach the Neva estuary from the Bay of Finland. Subsequently, the insulin level drops drastically to reach a nadir in February, and then rises again as spawning time approaches. Following spawning in May, there is a second, even more drastic drop of plasma insulin. These changes were delayed during an exceptionally cold and long winter. The physiological significance of these findings is emphasized by recent observations on the females of landlocked *Petromyzon marinus* in the Great Lakes in which plasma insulin undergoes almost identical variations between the onset of migration and death (Sower et al. 1985).

The target tissues of endogenous lamprey insulin require further studies. Mammalian insulin effects have been shown in organs with a high activity of glucose-6-phosphatase (cf. Plisetskaya et al. 1976; Murat et al. 1981). As in the hagfish, the metabolic importance of the liver must be very limited. It amounts to about 1 percent of the body weight of the lamprey, and its glycogen content is usually low (Plisetskaya and Kuzmina 1972). Removal of up to 95 percent of the liver tissue has little effect on the carbohydrate metabolism (blood sugar level;

glucose load; insulin-induced hypoglycemia) except for a smaller degree of stress hyperglycemia. However, vitellogenesis is practically absent (Larsen 1978). Another possible target organ of insulin is the kidney, which contains appreciable quantities of glycogen and a rather high glucose-6-phosphatase activity (Murat et al. 1979); yet it is not known if the lamprey kidney is sensitive to insulin. The brain and meninges contain considerable glycogen reserves which may be mobilized for the glucose supply of the brain, especially in emergencies (L. Leibson and Plisetskaya 1969; Rovainen et al. 1971; Plisetskaya 1975). Their depletion in response to anti-insulin serum could indicate that they depend on endogenous insulin (cf. Murat et al. 1981). Contrary to the situation in hagfishes, there is no evidence of an insulin effect on the largest store of energy reserves, the muscles of the body wall, which causes one also to wonder as to the source of increased plasma FFA after anti-insulin injections. Unfortunately, it seems that an effect of insulin on muscular fat has never been studied. The mechanisms of a peculiar growth-promoting effect of insulin in the absence of weight gain in *Lampetra fluviatilis* require an analysis (Larsen 1976a). In a sober evaluation of her pertinent experiments, Larsen (1976a, b) suggests that it may be unnecessary to invoke hormonal factors in the control of the carbohydrate metabolism in lampreys. Beforehand, the most striking peculiarities related to endogenous lamprey insulin seem to be (a) the hyperglycemia after islet removal, and (b) the lack of insulin sensitivity of the most important organ for fasting metabolism, i.e., the skeletal muscle of the body wall. The former finding suggests that there is a greater dependence of the carbohydrate metabolism on insulin in lampreys than in hagfishes and also in gnathostome fishes (elasmobranchs and eel; see below). The insulin insensitivity of skeletal muscle, if confirmed with lamprey insulin, may be unique among vertebrates.

Conclusions. Our understanding of the biological role of insulin in cyclostomes is most unsatisfactory, although we know a number of effects of exogenous insulin, and also know some effects of exogenous substances on endogenous insulin. Furthermore, there are conspicuous variations of endogenous insulin titers in adult lampreys. In comparison with gnathostomes, it appears that the role of the liver in the carbohydrate metabolism of the cyclostomes is rather negligible. The slow recovery from hyperglycemia at lower temperatures, the long-lasting effects of insulin-induced hypoglycemia and an apparently very slow insulin clearance, together with rather slow responses of endogenous insulin to various stimuli all indicate that in both hagfishes and lampreys the role of insulin must be linked to long-term effects. Therefore, instead of continuing to induce acute insulin effects, it may be profitable to design long-term studies at appropriate temperatures involving, for example, slowly delivering implants, isletectomy, monitoring of serum parameters via vascular cannulation and, of course, use of cyclostome insulins.

10.4.3 Chondrichthyes

Selachians. Although the islet organs of selachians represent structurally something like the basic model of the gnathostome pancreas (see Chap. 4.2), surprisingly little work has been done on their physiology. There are five studies on the effects of pancreatectomy, an operation easily achieved. However, mostly because of the design, the results are rather inconclusive except for those of Grant et al. (1969) on the little skate, *Raja erinacea.* In general, the findings suggest that pancreatectomy is not followed by a mammalian-type diabetes (cf. Epple and Lewis 1973; Patent 1973; Epple et al. 1980). The interpretation of the findings is complicated by the presence of "open" insulin cells in the selachian intestine (El-Salhy 1984), the "antidiabetic" impact of which is unknown. If the secretion of these cells is a pancreatic type of insulin or of equivalent biological action, it could conceivably blunt the hyperglycemic effect of pancreas removal and even cause the hypoglycemia observed by Grant et al. (1969). On the other hand, we cannot assume a priori that "open" gut cells respond to blood-borne messages created by pancreatectomy. Although there are great individual variations in the blood sugar, insulin injections consistently evoked hypoglycemia in all species studied. The latter, which increases with the dose of the hormone, may last for days (cf. Patent 1973; DeRoos and DeRoos 1979; DeRoos et al. 1985). In the dogfish, *Squalus acanthias*, a dogfish insulin preparation was found to have about the same effect as the mammalian hormone (Patent 1970). In the same species, large doses (50 IU/kg) of bovine insulin caused a long-lasting decline of the blood sugar during which the glycemia dropped from about 40 mg% to a nadir of 2.1 ± 0.6 mg% by day 3; however, no convulsions were observed. Return to normal levels occurred after about 14 days (cf. DeRoos and DeRoos 1979). On the other hand, Leibson and Plisetskaya (1968) report convulsions in skate, *Dasyatis pastinaca*, after injection of 30–50 IU/kg of insulin. In some studies, increases of glycogen in liver, skeletal and heart muscle were found following insulin injections (cf. L. Leibson and Plisetskaya 1968; Patent 1970), but great variations in the tissue content make it difficult to decide if the results after high doses of exogenous insulin reflect a physiological phenomenon; furthermore, the glycogen content in the liver is very low (2%, or considerably less). Fasting dogfish are capable of storing liver glycogen for short periods only (Patent 1970, 1973). Using doses of 50–250 IU/kg of mammalian insulin in *Squalus acanthias*, DeRoos et al. (1985) observed the expected decline of plasma glucose, and also of alanine for several days, while plasma lactate showed less clearcut variations. There was no clear impact on plasma beta-hydroxybutyrate, which is interesting considering that ketones may be the major fuel of elasmobranchs (Zamnit and Newsholme 1979). No other data on the effects of insulin in the selachians seem to be available.

Holocephalia. There is only a single study of the role of insulin in these fishes. Patent (1970) found a hypoglycemic effect of bovine, selachian, and holocephalian insulin similar to that in a shark (*Squalus acanthias*), but no effect on tissue glycogen.

From the preceding, it is clear that the available information is too sparse to allow general conclusions as to the physiological roles of insulin in the chondrichthyes.

10.4.4 Actinopterygii

Of the five major taxa of actinopterygians (Polypteridae included), two have been used in studies on the islet physiology. For one of the two, the Chondrostei, there are only data on one species, a Caspian sturgeon (*Acipenser güldenstädti*). However, since the measurements were performed in a mammalian system, the values are probably underestimates. Like other sturgeons, this species is a polycyclic spawner which repeatedly migrates from the sea to breeding grounds in the upper regions of rivers. In the Caspian Sea, sexually immature sturgeons have a plasma insulin titer of about 0.4 ng/ml. Sturgeons with maturing gonads are voracious feeders accumulating large quantities of carbohydrates and fats, with concomitant increases in the level of insulin to about 0.7 ng/ml. During their anadromous spawning migration, the high insulin titer drops, a process that seems to be related to their fasting state at this time, so that the lowest values (traces) are found during spawning. Thereafter, as the animals migrate back to the sea and resume feeding, the insulin titer rises strongly and finally reaches its original level of 0.7 ng/ml on their return to sea water (cf. Murat et al. 1981). The early changes in this sturgeon are almost identical with those in prespawning lampreys (see Sect. 10.4.2), which suggests that the role of insulin in both cases is very similar despite the wide phylogenetic gap between the taxa.

The only other group of actinopterygians used in studies on islet physiology are the teleosts in which the quantity of pertinent data is second only to those on mammals. Unfortunately, however, the state of knowledge is not commensurate. The findings are often at variance, which may partly be explained by species differences and variations of life style, stages in the life cycle, nutrition, environmental factors, etc. Moreover, too many experiments were repetitions of previous ones, showing that a large dose of mammalian insulin causes hypoglycemia in another species (cf. Epple 1969; Plisetskaya 1975; Epple et al. 1980; Ablett et al. 1981 a, b; Plisetskaya et al. 1984a, 1985). Isletectomy has been used only in a few studies, and except for the work on the eel, may have been subtotal (cf. Epple and Lewis 1977; Epple et al. 1980; Lewis and Epple 1984). The main problem with this procedure is that species with well-developed Brockmann bodies usually have a widely scattered, sometimes "diffuse" pancreas which contains smaller islets in surgically inaccessible regions. Since these islets tend to have a higher percentage of glucagon cells, postoperative hyperglycemia may simply reflect a changing ratio of insulin: glucagon, but not a complete insulin deficiency. We suspect that this may explain, at least in part, the consistent hyperglycemia in the studies on teleosts after removal of their Brockmann bodies (cf. Epple 1969; Khanna and Gill 1972; Chidambaram et al. 1973; Epple and Lewis 1973). On the other hand, complete islet removal in the eel requires total pancreatectomy and ligation of the portal vein, but leaves the bile duct intact. If this operation is carried out in sexually immature "yellow" eel, with empty gut and at moderate tem-

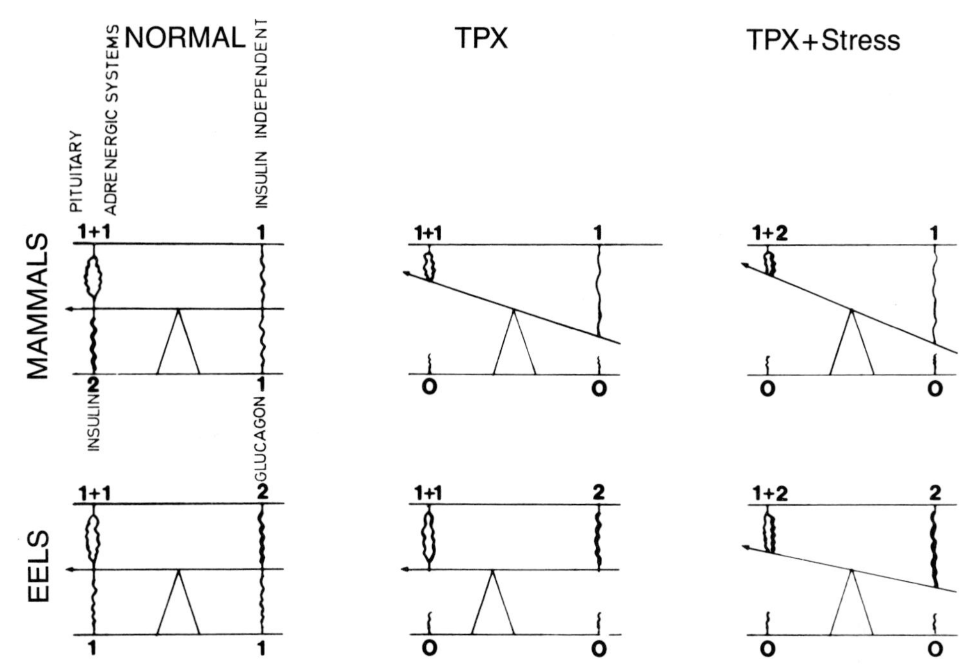

Fig. 10.8. Model approach to explain the difference in the control of glycemia between mammal and eels. The level of glycemia is indicated by the tip (*arrowhead*) of a doubly dampened beam of a balance. The dampening *springs* symbolize the controlling factors, with the *thickness of the springs* indicating their estimated relative contribution (1 or 2). The ratio of increasing to lowering factors causes different effects in the mammal and the eel, respectively, because of their different contribution to the glycemic balance. Since in the eel insulin and glucagon appear to be well balanced, pancreatectomy (TPX) will only result in a less dampened equilibrium. If stress is added, however, hyperglycemia will ensue (Epple and Lewis 1977)

perature (16 °C), viable experimental "models" are created, since both liver and anterior intestine continue to receive sufficient blood supply (Lewis et al. 1977). The study of pancreatectomized eel revealed that in these animals the glycemic regulation is in a state of undampened balance (Fig. 10.8), apparently due to the simultaneous removal of the opposing actions of insulin and glucagon. However, any kind of stress will cause severe hyperglycemia since the counteraction of insulin is absent. In particular, epinephrine may interfere since in the eel this hormone is (1) released instantly upon stress, and (2) is strongly hyperglycemic at physiological levels (Epple and Nibbio 1985). On the other hand, sustained postsurgical stress hyperglycemia may involve increased cortisol secretion (Chan and Woo 1978). Also, the metabolic state plays a role. In freshly captured eel, the operation affects far more metabolic parameters than in eels fasted 6 months (Table 10.3). Conceivably, higher temperatures could strongly modify the result of the operation, particularly when other neuroendocrine mechanisms are stimulated at differing degrees, but this question requires further study. Nevertheless, the picture that emerges from the studies on the pancreatectomized yellow eel is that (1) insulin deficiency does not necessarily cause hyperglycemia, and (2)

Table 10.3. Differences in the impact of pancreatectomy between freshly captured and fasted eel (*Anguilla rostrata*) (Lewis and Epple 1984)

	Statistically significant differences between PVL and TPX eels	
	Freshly captured eels[a]	Fasting eels[b]
Blood parameters		
Glucose	0	+ (10)
AAN	+ (10); − (20)	0
Urea	0	0
FFA	− (20)	− (10)
Cholesterol	− (10)	0
Hematocrit	− (10 and 20)	− (10 and 20)
Tissue parameters		
Liver glycogen	− (10)	0
Muscle glycogen	0	0
Liver FFA	+ (20)	0
Muscle FFA	0	0
Liver hydration	+ (10)	0
Muscle hydration	0	0
HSI	0	0
Abdominal fat deposits	+ (10 and 20)?	0?

[a] Source: Epple and Lewis (1977).
[b] Present study.
[c] 0 no effect, + increase, − decrease. The days after surgery when effects were present are given in parentheses.

the overall metabolic impact of islet hormones is reduced in the state of lowered (fasting) metabolism (Epple and Lewis 1977; Lewis and Epple 1984). The latter conclusion may apply to other teleosts as well, since Falkmer (1961) obtained a stronger response to isletectomy in fed than in fasted sculpin (*Cottus scorpius*). It is also relevant that Ottolenghi et al. (1982) observed in a catfish (*Ictalurus melas*) and others (Cowey et al. 1977a; Ablett et al. 1981b) in the rainbow trout (*Salmo gairdneri*) a number of differences between fed and fasted animals in responses to insulin injections. Anti-mammalian insulin serum has been applied in a few studies.

Falkmer and Wilson (1967) did not see a hyperglycemic effect in *Cottus scorpius*, though Plisetskaya and Bondareva (1972) obtained both hyperglycemia and decreased liver glycogen and two other species, *Scorpaena porcus* and *Spicara smaris.* Ince and Thorpe (1975) also obtained hyperglycemia in *Esox lucius.* Thus, in two out of three studies with anti-insulin, the effect was as expected. However, as pointed out by Thorpe (1976), confirmation of these data with species-specific antisera is needed. In captive carp, a peculiar type of diabetes develops spontaneously. This "Sekoke disease" involves, inter alia, hyperglycemia, ketoacidosis and muscle degeneration. B-cells are present, but the response to exogenous insulin is poor (Yokote 1970; Nakamura et al. 1971). Exogenous insulin caused in all teleosts studied a decrease of blood sugar. In rainbow trout, this effect was even seen when insulin was dissolved in the aquarium water since the hormone can obviously be taken up by a parenteral route from the environment (Munford and Greenwald 1974). Like certain other poikilotherms, teleosts are sometimes able to survive an insulin-induced "aglycemia" (cf. DeRoos and DeRoos 1979). The hypoglycemic effect of insulin is accompanied by a variable effect on liver glycogen, ranging from decrease to increase; a decrease of cardiac glycogen; and either no effect on muscle glycogen or an increase of the latter, which may involve both white and red muscle (cf. Epple 1969; Plisetskaya et al. 1976; Epple et al. 1980; Ablett et al. 1981a, b; Ottolenghi et al. 1982; Carneiro and Amaral 1983). The reasons for the varying responses of liver glycogen to insulin are thus far unexplained. Seasonal factors, temperature, and the nutritional state of the animal must be involved in some cases (L. Leibson and Plisetskaya 1969; Cowey et al. 1977a, b; Ablett et al. 1981a, b; Ottolenghi et al. 1982; Lewis and Epple 1984; Plisetskaya et al. 1984). In vitro, insulin-stimulated glucose incorporation into liver glycogen in *Opsanus tau* and *Notemigonus chrysoleucas* (Tashima and Cahill 1968; DeVlaming and Pardo 1975), but was ineffective in *Salmo gairdneri* (Morata et al. 1982a). On the other hand, there was both an increase in liver glycogen and glucose release in response to glucagon and epinephrine in the latter study, an effect than can only be explained by stimulation of gluconeogenesis. Hence, it appears that the effects of insulin on the carbohydrate metabolism of the teleost liver in vivo are strongly modified by the net simultaneous glycogenolytic-gluconeogenic impact of "catabolic" hormones (glucagon, epinephrine, cortisol), which may be sometimes additive. It remains to be seen if this hypothesis can indeed explain the varying effects of exogenous insulin on liver glycogen. In skeletal muscle, insulin stimulates glucose uptake, glycogen synthetase activity and the channeling of the metabolite into oxidative and/or lipogenic pathways (cf. L. Leibson et al. 1976b; Thorpe 1976; Ablett et al. 1981b).

The role of insulin in the lipid metabolism of teleosts has received less attention (for literature see Dave et al. 1979; Epple et al. 1980; Plisetskaya 1980; Ablett et al. 1981a, b; Murat et al. 1981). Isletectomy in *Myoxocephalus (Cottus) scorpius* was followed by increased deposition of liver lipid, which could be reversed by exogenous insulin (N. A. McCormick and Macleod 1925). In freshly captured eel, pancreatectomy caused an increase in liver FFA and a decrease in plasma FFA, but there was no effect on liver FFA in fasted animals. Serum cholesterol decreased temporarily in freshly captured eel but there were no changes in fasted ones; furthermore, there was no obvious impact on the adipose tissue of the eel in either study (Epple and Lewis 1977; Lewis and Epple 1984). However, without additional work, it is impossible to ascertain the extent to which absence of insulin in any of these studies was balanced by the simultaneous removal of the antagonistic actions of pancreatic glucagon and/or somatostatin. It is nevertheless obvious that, contrary to the situation in mammals, the overall impact of insulin on the adipose tissue of the eel must be very limited.

Exogenous insulin lowered the plasma FFA of teleosts in all pertinent studies, whereas effects on plasma cholesterol were less consistent (for literature see Dave et al. 1979; Plisetskaya 1980; Epple et al. 1980; Lewis and Epple 1984). The impact of insulin on hepatic lipids varies greatly; decrease, no effect, and a lipogenic action have been reported (for literature see Epple et al. 1980; Ablett et al. 1981a, b). In two studies, exposure of fishes to a high glucose-containing medium caused an increase in hepatic lipids, which could be reversed by insulin injections (Köhler 1963; Sterne et al. 1968), and the same effect of insulin was seen in a third study in which hepatic fat deposition had been caused by glucose injections (Yanni 1964). Insulin had either no effect on muscle lipids, or it was lipogenic (for literature see Epple et al. 1980; Ablett et al. 1981 a, b); and no lipogenic effect on heart muscle in vitro (Tashima and Cahill 1964, 1968). Thus, a lowering effect on plasma FFA seems to be the only clearcut insulin action of physiological significance so far identified in lipid metabolism in teleosts.

Pancreatectomy in freshly captured eel (*Anguilla rostrata*) caused a temporary increase in serum amino acid nitrogen (AAN) which was followed by a decrease 20 days after the surgery. On the other hand, there was no effect of the operation on this parameter in fasted eel (see Table 10.3). Serum urea was not affected in either experiment. These observations suggest that in fed eel insulin has an acute anabolic action on protein metabolism, which prevails over antagonistic (catabolic) mechanisms, a finding in good agreement with data on insulin responses to feeding and effects of exogenous insulin on teleost protein metabolism (see below). The net effects of exogenous insulin on the protein metabolism of teleosts are clearly anabolic: it lowers the circulating amino acids (citrulline being an exception, see Y. Inui et al. 1975), apparently by increasing their uptake into liver and muscle cells, and incorporation into proteins of both tissues; it stimulates hepatic secretion of newly formed proteins; furthermore it reduces hepatic gluconeogenesis (for literature see Ince and Thorpe 1978; Walton and Cowey 1979; Ablett et al. 1983; Y. Inui and Ishioka 1983 a, b; Plisetskaya et al. 1984). Circulating titers of teleost insulin have been determined by various investigators, now with increasing use of homologous radioimmunoassays (see Sect. 10.2). One study reports on daily variations of the insulin titer in a teleost. In sea bass (*Dicentrarchus labrax*) fed a "natural" diet (filleted fish) on a regular schedule, insulin showed a daily cycle with higher values during the dark period, and an inverse cycle of plasma glucose. This periodicity may have been entrained by the feeding pattern since the increase of insulin began after the normal feeding time, although the experiments were carried out 1 week after the last feeding. Interestingly, controls given a commercial diet with much lower protein and high carbohydrate content at the same feeding times did not display a daily cycle of plasma insulin (Gutierrez et al. 1984). Variations of plasma insulin during various stages of the life cycle have been demonstrated in a few species. In the hump-back salmon (*Oncorhynchus gorbuscha*), plasma levels of insulin become maximal before spawning but decrease precipitously thereafter. In the coho salmon (*Oncorhynchus kisutch*), plasma insulin levels were significantly lower in fasted than in fed animals (Plisetskaya et al. 1986 a). Very low plasma levels of insulin have also been found in postspawning coho salmons (see Sect. 10.2). In plaice (*Exophthalmus maeoticus*) and in scorpion fish (*Scorpaena porcus*) plasma insulin increases during the spawning period, a time of heavy food uptake in both species. In males of the plaice (*E. maeoticus*) a high insulinemia persists even after spawning (cf. Murat et al. 1981). In bream (*Abramis brama*), immature fish have a higher insulin titer than adults, which also seems to be related to a stronger food uptake (Murat et al. 1981). Thus, it appears that in teleosts protein-rich meals are associated with high insulin titers whereas starvation has the opposite effect (Murat et al. 1981); furthermore it is likely that amino acids are the most important insulin secretagogues (Ahmad and Matty 1975; Ince and Thorpe 1977; Ince 1979, 1980). On the other hand, Huth and Rapoport (1982) did not find an effect of leucine on insulin biosynthesis in carp islets in vitro. Glucose, which is a strong stimulator of insulin biosynthesis and release in mammals, seems to have a different role in teleosts. As in *Myxine glutinosa* (Emdin and Falkmer 1977), it has been thus far impossible to show a stimulatory effect of glucose on insulin biosynthesis (for literature see Huth and Rapoport 1982). However, in almost all in vivo and in vitro studies, glucose stimulated the release of insulin though the effect was weak when compared with certain amino acids (for literature see Huth and Rapoport 1982; Ince and So 1984). Fletcher et al. (1978) were unable to demonstrate a glucose effect on catfish islets in vitro but found that galactose acts as a stimulant. In the same species (*Ictalurus punctatus*), insulin release from the perfused Brockmann body was stimulated by a glucose concentration which was much higher than that required for somatostatin release (Ronner and Scarpa 1984). Stimulated insulin release by amino acids or glucose in teleosts (Ince and Thorpe 1977; Ince 1979, 1980; Ronner and Scarpa 1982) is similar to that in a lizard (Rhoten 1973a, b), chicken (King and Hazelwood 1976), and mammals (cf. Gerich et al. 1976), both in rate and biphasic pattern. However, as in other poikilotherms, the onset of insulin-induced glycemic and other changes in blood metabolites is much later than in homeotherms (for literature see Epple et al. 1980).

Any interpretation of the role of insulin in teleosts must consider the largely ignored fact that in vivo several control mechanisms modify the actions of islet hormones. Most notably, the impact of the very strong peptidergic innervation (see Chap. 7) is a virtually unknown factor. Second, the possible role of gastrointestinal secretagogue(s) remains to be identified although exogenous secretin has been shown to stimulate insulin release in vivo (Ince 1983). There is no proof of a physiological role of catecholamines in insulin release despite several experimental studies (see Chaps. 7 and 8). Another factor, ambient temperature, has a great impact on both biosynthesis (Huth and Rapoport 1982) and plasma clearance of insulin (Ince 1982). Furthermore, diet-induced changes in the concentrations of insulin receptors, at least in the liver, may play a role (Ablett et al. 1983). However, receptor binding of heterologous insulin may not be necessarily indicative of the biological effects of the hormone. Findings in mammalian tissues show that the latter seem to be dependent on a post-receptor phenomenon that discriminates among different types of insulin. Finally, it must be noted that fishes with low affinity insulins have been found to possess higher concentrations of insulin receptors and an increased sensitivity to insulin (cf. Muggeo et al. 1979; Leibush and Bondareva 1981).

Conclusions. In teleosts, insulin is an anabolic hormone. Its primary function seems to be in protein metabolism in which it fosters uptake of amino acids and their incorporation into tissue proteins. Stimulation of insulin release is particularly sensitive to circulating amino acids, whereas glucose has a decidedly weaker effect. Results of pancreatectomy suggest that the overall metabolic impact of endogenous insulin is much smaller than in mammals. However, injections of insulin antibodies cause hyperglycemia; and exogenous insulin lowers blood sugar, FFA, and amino acids, whereas its effects on liver and muscle glycogen are variable. Similarly, the data on its impact on tissue lipids do not allow generalizations. Many findings in teleosts indicate that the physiological state of the animals strongly affects the actions of insulin; it appears that the role of the hormone may be greater in states of increased metabolism and food uptake. As in other groups of poikilotherms, a better understanding of the physiology of insulin in teleosts requires the replacement of experiments with excessive doses of mammalian hormone by the measurement of the variations and responses of endogenous insulin. Fortunately, the number of such studies is now increasing.

10.4.5 Dipnoi

Considering the phylogenetic position of the lungfishes close to the base of the tetrapodes, and their ability to accumulate large quantities of urea in their plasma during estivation, it is surprising that their islet physiology has received so little attention. To the best of our knowledge there is only one single paper on insulin in lungfishes, and only one other study on glucagon (see Chap. 11.4.5). Using a bioassay method, El Hakeem and Babiker (1983) estimated the pancreatic insulin content of free-swimming and estivating African lungfish (*Protopterus annectens*), to be 1.27 and 0.82 U/mg dry weight, respectively.

10.4.6 Amphibia

There are no data on the islet physiology of the Gymnophiona and only a few studies on the Urodela. However, the Anura have been studied frequently since the early days of islet physiology (for literature see Penhos and Ramey 1973; Epple et al. 1980).

Urodela. Pancreatectomy has been performed only in one species, *Taricha torosa.* The operation was "radical", involving removal of part of the duodenum and interruption of the bile duct; blood sugar was measured up to 24h only. Hence, the etiology of the observed hyperglycemia is uncertain (cf. Wurster and Miller 1960; M. R. Miller 1961). In the same species, exogenous insulin lowered the blood sugar for several days, depending on the dose, and animals receiving high doses of insulin (50 U/kg) eventually died (Wurster and Miller 1960). In the pedogenic mudpuppy (*Necturus maculosus*) insulin (100 IU/kg) significantly affected the disappearance of a glucose load from blood, as early as 3 h after infusion. However, the glycemia continued to decrease, reaching very low values between 24 and 48 h. Since some animals had no measurable blood sugar levels at all, but did not show hypoglycemic symptoms, Copeland and DeRoos (1971) conclude that plasma glucose is not a required source of energy in this species. In liver organ cultures of *Amphiuma means* (D. Brown et al. 1975), insulin decreased GOT (glutamate-oxaloacetate transaminase) and GPT (glutamate-pyruvate transaminase). Pancreas organ cultures from the same species released insulin during culture periods up to 16 days; insulin secretion was stimulated, inter alia, by glucose and glucagon but reduced by epinephrine (Gater and Balls 1977). From their in vitro studies on the liver of the axolotl (*Ambystoma mexicanum*) Janssens and Maher (1986) concluded that insulin counteracts the glycogenolytic effect of glucagon by inhibiting the increase of cyclic AMP stimulated by glucagon.

Anura. Pancreatectomy in adult frogs and toads resulted in all cases in hyperglycemia, but the impact of removal of B-cells on this effect is difficult to assess. One of the major problems is the possible oversight of small pancreas remnants rich in islet tissue; also the regenerative ability of such tissues is substantial (cf. Epple and Lewis 1973; Penhos and Ramey 1973; Epple et al. 1980). Another problem may arise with the high, seasonally varying sensitivity of frogs to stress (C. L. Smith 1954) which may make it most difficult to separate specific and unspecific effects of pancreatectomy, especially when hyperglycemia is used as a metric. Furthermore, strong seasonal variations in fuel reserves, as well as in serum and pancreas concentrations of insulin and glucagon in serum pancreas (Schlaghecke and Blüm 1981) should inevitably affect the results of islet removal. In general, the survival period of pancreatectomized frogs and toads seems to be a week or less, while at the same time rapid and progressive hyperglycemia and ketonemia develop (cf. Penhos and Ramey 1973). Insulin injections cause a slow-starting and long-lasting hypoglycemia, and it appears that these animals can survive, at least under certain conditions, for days without measurable blood sugar levels (cf. M. R. Miller 1961; Hanke and Neumann 1972; Penhos and Ramey 1973; Hanke 1974a, b; Epple et al. 1980; DeRoos and Parker 1982). In *Xenopus laevis* the hypoglycemic insulin effect appears to be stronger in postmetamorphic than in larval animals (Hanke 1974c). Just as in trout, insulin passes the skin of the leopard frog (*Rana pipiens*) and lowers the plasma glucose (Greenwald and Munford 1976). Also as in teleosts (see Sect. 10.4.4), the reports on the effects of exogenous insulin on liver and muscle glycogen in vivo vary greatly; however, in vitro insulin stimulates the uptake of glucose into liver, and skeletal and heart muscle, and glycogen formation in skeletal muscle (for literature see Hanke 1974a, b; L. Leibson and Plisetskaya 1973; L. Leibson et al. 1976b; Epple et al. 1980). Gourley et al. (1969) found seasonal differences in the insulin effect on pyruvate uptake, oxidation, and synthesis into glycogen in skeletal muscle of a frog. Seasonal differences of exogenous insulin were also seen in studies on the lipid metabolism of frogs; these differences were compounded by variations in the temperature. Insulin lowered plasma FFA concentration of only in *Rana temporaria* that were acclimated to 5 °C and injected in September. No effect was seen in a parallel group acclimated to 25 °C, or in warm- and cold-acclimated frogs injected in winter. On the other hand, insulin caused an increase in plasma glycerol, but only in the cold-acclimated summer animals. Harri and Puuska (1973) suggest from these experiments that insulin lowers the titer of the plasma FFA by stimulation of re-esterification and/or oxidation. Gunesch (1974) observed an increased total lipid content of the body and a decreased lipid concentration in the extracellular fluid after injections of insulin in larval and juvenile *Xenopus laevis*. A lowering effect on FFA in the extracellular fluid was obvious, but not statistically significant, in animals from stage 60 on. In vitro, insulin stimulates incorporation of leucine into liver proteins of *Rana catesbeiana* (Penhos and Krahl 1962, 1963). In primary cultures of hepatocytes of the same species, insulin maintained high levels of synthesis and secretion of protein, the release of serum albumin being particularly sensitive to the presence of the hormone in the culture medium (Stanchfield and Yager 1979). It seems that the only data on endogenous insulin titers have been reported by Schlaghecke and Blüm (1981), who observed with a heterologous assay high levels of insulin in both blood and pancreas during the warm season, followed by a decline with a nadir in March and April.

In conclusion, the available information on the role of insulin in amphibians is insufficient to evaluate the spectrum of its biological activities. Like fishes, amphibians respond to insulin with hypoglycemia and may not depend on glucose for the fuel supply of their nervous tissues, at least under certain conditions. In general, insulin seems to have its usual anabolic net effects, but information on its impact on metabolism of lipid and protein is very scarce. Great seasonal variations of fuel stores and circulating islet hormones may explain in part the varying results on carbohydrate and lipid metabolism thus far reported. Application of species-specific insulin and measurement of endogenous insulin responses to various stimuli in intact animals are needed, but no such studies seem to exist.

10.4.7 Reptilia

Despite the key phylogenetic position of this group, physiology of reptilian islets continues to be neglected. Since the previous reviews of Penhos and Ramey (1973) and Epple et al. (1980), only a few aspects of the physiology of reptilian insulin seem to have been studied; apparently *Sphenodon punctatus* has never been used in work on islet physiology or biochemistry.

Chelonia. There are three studies on pancreatectomized turtles (Aldehoff 1891; Nishi 1910; Foglia et al. 1955), all of which applied radical surgery. Hence, the bulk of the recently discovered "open" insulin cells of the turtle intestine (Gapp et al. 1986) must also have been removed. The animals developed severe hyperglycemia and either were killed early, or showed a very poor survival rate. In the latter case, most animals died within 3 days following surgery, and only one specimen survived for 10 days (Foglia et al. 1955). From these findings, it is impossible to decide if the observed hyperglycemia was indeed a specific result of B-cell removal. Injections of mammalian insulin cause slowly developing and long-lasting hypoglycemia (cf. M. R. Miller 1961; Penhos and Ramey 1973; Marques et al. 1982). The results of Marques et al. (1982, 1985) suggest the presence of specific insulin binding sites in liver, adipose tissue, pituitary, adrenals, and thyroid of *Chrysemys dorbigni.*

Crocodilia. It seems that pancreatectomy was performed in only one species, *Alligator mississippiensis* (Penhos et al. 1967a). Here, the operation was followed by a progressive hyperglycemia and hyperketonemia. One animal survived for 15 weeks with a blood sugar level of 489 mg% (normal range: about 90 mg%) and a ketonemia of 134 mg% (normal range: 3.1–4.3 mg%). Since the ketonemia in these pancreatectomized alligators was the highest ever recorded for any vertebrate (Penhos and Ramey 1973), one wonders what the pH of their blood may have been. Crocodilia respond to insulin injections with long-lasting hypoglycemia; high doses of this hormone may cause a fatal shock unless glucose is given (cf. Penhos and Ramey 1973). In the caiman, insulin lowered plasma amino acids and fostered protein synthesis (Hernandez and Coulson 1968).

Squamata. In all lizards and snakes studied so far, with one exception, pancreatectomy caused an initial hypoglycemia which was followed by a hyperglycemia (cf. Penhos and Ramey 1973). The water snake (*Natrix piscator*) was the exception to this rule since there was a continuous increase of blood sugar without a hypoglycemic phase. In the same species, both hepatic and muscle glycogen were lowered by pancreatectomy, but there were considerable variations. Fifteen days after the operation the animals also showed hypercholesterolemia (Rangneker and Padgaonkar 1972). Insulin causes hypoglycemia in all species studied, but lizards appear particularly insulin-resistent when excessive doses are given (cf. M. R. Miller 1961; Penhos and Ramey 1973). On the other hand, *Natrix piscator* became aglycemic 24 h after injection of 40 IU/kg of mammalian insulin, and the blood sugar level had not completely returned to normal 7 days after the injection. Pancreatectomized specimens were less insulin-sensitive than controls, but hypophysectomized-pancreatectomized water snakes were aglycemic within 3 h following the injections (Padgaonkar and Rangneker 1975). In the lizard *Uromastix hardwickii*, insulin had no effect on blood cholesterol (Suryawanshi and Rangneker 1971). So far, there seems to be only one study on the variations of plasma insulin in intact reptilians. In their studies on the pancreas insulin content of the Sahelian lizard (*Varanus exanthematicus*), Dupé-Godet and Adjovi (1981 b) observed the highest values during the rainy season, i.e., the time of food uptake, and the lowest values during the dry season when the animals were fasting. The opposite was seen in a turtle, *Chrysemys dorbigni* (Marques and Kraemer 1968), and a snake, *Bothrops jararaca* (Prado and Prado 1946) whose pancreas contained the highest insulin titers during the winter season. Thanks to Rhoten's (1973 b; 1974 a, b; 1984) studies in the lizard, *Anolis carolinensis,* we have at least some insight into the mechanisms of insulin release. Using the perifused splenic pancreas lobe of this species, the author found, *inter alia*, that the B-cells respond to very high titers of glucose (500 mg/ml), or of arginine (10 mM) or leucine (10 mM) with a mammalian-like biphasic release of insulin. Stimulation of insulin release by glucose was not accompanied by an increase in proinsulin synthesis (Rhoten 1983).

Conclusions. The dearth of data makes it difficult to arrive at general conclusions on the role of insulin in reptiles; one hopes that the recent progress in isolation and characterization of reptilian hormones will be followed by in vivo studies with homologous insulin RIAs. While the islet physiology of the turtles remains largely unknown, it appears that the role of insulin in the majority of the Squamata resembles that of the birds (see Sect. 10.4.8), with whom they are only distantly related. On the other hand, the few data on crocodilians, the closest living relatives of the birds, suggest that the glycemia of these reptiles depends in a mammalian-like fashion on a well balanced insulin titer. These phylogenetic paradoxes also extend to the physiology of glucagon (see Chaps. 11.4.7 and 11.4.8). However, the enormous tolerance to hyperketonemia may set the crocodilians apart from all other vertebrates.

10.4.8 Aves

Since the early days following the discovery of pancreatectomy diabetes in mammals (Von Mehring and Minkowski 1889), there have been many attempts to obtain a similar result in birds. Unfortunately, it is likely that most investigators overlooked the splenic lobe with its large accumulation of A-cells (cf. Mialhe 1958; Cieslak and Hazelwood 1986a, b). Consequently, it remains an open question if the supposedly greater, mammalian-like dependence of carnivorous birds on insulin is a biological phenomenon or simply the result of incomplete surgery (for literature see Mialhe 1958; F. G. Young 1963; Sirek 1969). However, it appears from the results of total pancreatectomy in domestic ducks and geese (for literature see Sitbon et al. 1980) and from the removal of the A-cell-rich third and splenic lobes of the chicken pancreas (Mikami and Ono 1962) that at least in these three species the metabolic impact of glucagon prevails over that of insulin. This is also reflected in the higher number of A-cells (see Chap. 4.2) and the ratio of circulating insulin: glucagon (Sitbon et al. 1980; Hazelwood 1984). Nevertheless, a comparison of the responses of endogenous islet hormones to metabolites and of the effects of pancreatectomy between a truly granivorous species, such as the pigeon (cf. Mihail et al. 1963), and carnivorous/piscivorous birds (e.g., vultures and penguins) may be required before we can more safely speak of *the* role of avian insulin. A unique feature of the birds, which they seem to share only with the Squamata, is their very high blood sugar level (usually more than twice that of mammals). In duck, goose, and chicken, pancreatectomy results in a drastic decrease in blood sugar, again resembling the situation in Squamata, although birds seem to be unable to recover from the early hypoglycemia. An interesting problem arises with the persistence of measurable titers of plasma insulin immunoreactivity in both totally pancreatectomized chickens and geese (Colca and Hazelwood 1982; Sitbon and Mialhe 1978), which requires further investigation with respect to origin and significance. In both duck and chicken, infusion of anti-insulin antibodies causes hyperglycemia (Mirsky et al. 1964; Bondareva 1970), suggesting that in these species insulin has a role in the maintenance of glycemia despite the overpowering impact of glucagon. Exogenous insulin causes hypoglycemia, but birds seem to have a lizard-like (see Sect. 10.4.7) ability to resist excessive doses of this hormone (cf. Hazelwood 1986). The latter phenomenon may be linked to the absence of a hepatic effector system despite the presence of high-affinity receptors (Cramb et al. 1982b). If this interpretation is correct, then the weak glycemia response of birds to insulin would be due to an extrahepatic mechanism. The effect of exogenous insulin on the glycogen content of liver, skeletal muscle and heart varies with nutritional state and dose employed (cf. Hazelwood 1965, 1986). In vivo, the results of insulin injections are certain to be modified by the immediate, hypoglycemia-induced glucagon release and probably other mechanisms as well. However, in studies on isolated chicken hepatocytes, insulin failed to show any effect on glycogenolysis, glycogen synthesis, gluconeogenesis and lipogenesis, an observation that Cramb et al. (1982b) explained by the above-mentioned absence of a post-receptor mechanism. Similar to these – when compared with mammals – reduced effects of insulin in carbohydrate metabolism, the role of this hormone is also a minor one in the avian lipid metabolism. In pancreatectomized ducks there is a slow decrease of plasma FFA and triglycerides (Desbals et al. 1967) even when normoglycemia is maintained (cf. Mialhe 1976). Exogenous insulin stimulates, via hypoglycemia, the release of glucagon, the strong lipolytic action of which explains varying accounts of the effect of exogenous insulin on the avian lipid metabolism. Since insulin has apparently no direct antilipolytic effect on avian adipose tissue (Goodridge 1968; Grande 1969; Langslow and Hales 1969), the "paradoxical" increase of plasma FFA in adults of most avian species (only in the horned owl, there was no effect) after injections of this hormone may be mediated by glucagon. The latter suggestion has been confirmed by studies in pancreatectomized duck, in which the hyperlipemic effect of insulin is lacking (Desbals 1972). Nevertheless, findings in the duck suggest that insulin has a glucose-dependent hypolipoacidemic effect (Gross and Mialhe 1982). In a system of perifused chicken pancreas fragments, long-chain fatty acids, but not volatile fatty acids, stimulated insulin release (Colca and Hazelwood 1981). The biological significance of a small increase in hepatic FFA synthesis in neonatal chickens after insulin application in vitro remains to be established (Goodridge 1973). In contrast to its impact on carbohydrates and lipids, insulin seems to play its "traditional" anabolic role in avian protein metabolism. Subtotal and total pancreatectomy in the duck and goose, and 99% pancreatectomy in the chicken cause an increase in plasma AAN (Samsel 1973; Baker and Hazelwood 1977; cf. F. Laurent et al. 1985). Tolbutamide, which stimulates insulin release, caused a decrease of plasma AAN in the chicken (Baker and Hazelwood 1977), and so did insulin in the diabetic goose (cf. F. Laurent et al. 1985). Vice versa, oral amino acids caused an increase in the plasma insulin of the chicken (Langslow et al. 1970). Intravenous as well as oral administration of both alanine and arginine stimulate insulin release in the goose, but i.v. infusion of alanine is without effect in the duck (Khemiss and Sitbon 1982). These differences between closely related species are rather startling. Arginine infusion in the Gentoo penguin (*Pygocellis papua*) caused an immediate insulin release (Farina et al. 1975), while injection of glucose (1 g/kg) had no effect on plasma insulin (Chieri et al. 1972). In general, glucose is a poor stimulant for insulin secretion from the avian pancreas, and as in lizards, high glucose levels are required; furthermore, the second phase of insulin release, seen in the eel, lizard and mammals (see preceding sections), is either weak or absent in the chicken (for literature see Epple et al. 1980; Colca and Hazelwood 1981). Nevertheless, increased levels of glucose in the medium caused insulin release from pancreas fragments of older chickens (Colca and Hazelwood 1981) and from cultured splenic lobes of both embryonic and 2- or 9-day-old chickens (Foltzer et al. 1982).

From the preceding, it seems that insulin retains its anabolic role in avian protein metabolism, but "loses" much of its impact on lipid and carbohydrate metabolism to the overpowering actions of glucagon. The parallel to the situation in the Squamata is striking despite the phylogenetic distance between both groups. However, most of our information is based on data from two types of birds, i.e., Anatidae (duck, goose) and Phasianidae (chicken); hence, we know very little about the role of insulin in the birds as a group.

10.4.9 Conclusions

It appears that insulin and the IGFs evolved from a common ancestral molecule that was primarily a growth factor. While the IGFs largely retained the growth-promoting, slow-acting function, insulin increasingly assumed the complementary role of controller of metabolic processes. However, growth-promoting capabilities of insulin seem still to be required during prenatal development, even of mammals, and a related specific function of proinsulin at this stage appears also possible. In all adult vertebrates, insulin appears to be an anabolic hormone in protein metabolism. Its anabolic role in carbohydrate metabolism may also be ancient, but its homeostatic function in blood sugar regulation in nonhomeotherms requires further study. Our knowledge of the evolution of its impact on lipid metabolism is very poor. A lowering effect of exogenous insulin on plasma FFA is present in lampreys, teleosts, and tetrapods (except postnatal birds), but absent in hagfishes. The situation in the chondrichthys is unknown. One may suspect that a lipogenic effect of insulin appeared before its anticatabolic action, but there is still no proof that endogenous insulin directly affects the adipose tissue in fishes, amphibians, and reptiles. The development of an excessive ketonemia in the pancreatectomized alligator may be related to an anti-catabolic role of insulin in the lipid metabolism of this reptile. It remains to be seen if the dominating role of insulin in the control of mammalian adipose tissue is present in any other group of vertebrates. In general, the available data suggest that islet-produced insulin was originally a slow-acting hormone of "plenty", supporting the utilization and storage in nutrients during periods of feeding. Perhaps, its first post-hepatic target was skeletal muscle of the body wall, i.e., the major storage organ of many fishes, although the insulin-insensitivity of the lamprey muscle (in terms of metabolite uptake) is rather discordant with this view. However, insulin may play the ancient role of a growth factor in this organ, a possibility not yet tested. The ubiquitous sensitivity of insulin release to increases in plasma amino acids could have evolved as a signal for *anabolic* actions during times of feeding, and for *anti-catabolic* actions during times of fasting, preventing, in the latter case, inordinate tissue breakdown.

Chapter 11

Glucagon

Until three decades ago, glucagon received very limited attention. Perhaps the greatest obstacles in early glucagon research were conceptual, and based on mammalian bias. The main arguments against the existence of an insular hyperglycemic-glycogenolytic hormone were that (a) no related deficiency syndromes were noted after pancreatectomy, and (b) there appeared to be no need for a hormone duplicating the functions of epinephrine. However, it now appears that the overlapping (but not redundant!) actions of glucagon and epinephrine may provide a critical safety mechanism against the devastating effect of hypoglycemia in mammals (see e.g., P. E. Cryer et al. 1984; Sperling et al. 1984), and that some vertebrate groups may depend more on glucagon than on insulin in their regulations of metabolism of carbohydrate (birds and Squamata) and lipid (birds) (see below). Nevertheless, even in mammals some basic questions as to the role of glucagon in glycemic regulations of mammals are still disputed (cf. Tan et al. 1985). There seems to have been no account of the comparative physiology of glucagon since the reviews by Plisetskaya (1975) and Epple et al. (1980). On the other hand, many important contributions on the basic and mammalian physiology of glucagon are contained in two recent volumes edited by Lefèbvre (1983a, b).

11.1 Molecular Structure and Related Compounds

Like insulin, glucagon is synthesized as a part of a larger precursor molecule (cf. Bell et al. 1983; Lopez et al. 1983; Tager 1984). The amino acid sequence of pancreatic preproglucagon from the anglerfish (*Lophius americanus*) and four mammals (hamster, rat, ox, human) has been deduced from DNA sequencing of the glucagon gene or cDNA (for literature see Schmidt et al. 1985; Plisetskaya et al. 1986b). In the mammals so far studied, a single glucagon gene codes for a preproglucagon which contains three glucagon-related peptides: glucagon, glucagon-like peptide-1 (GLP-1), glucagon-like peptide-2 (GLP-2). The latter two peptides are arranged in tandem in close proximity to the C-terminal of the glucagon precursor (Fig. 11.1). The two mammalian GLPs consist of 37 and 35 amino acid residues, respectively, and they show different degrees of homology to pancreas glucagon (GLP-1: 48% identical residues; GLP-2: 38%). It seems that the mammalian A-cell releases both GLPs in a single peptide (Patzelt and Schiltz 1984). In the anglerfish and the catfish (*Ictalurus punctatus*), two different mRNAs give rise to two different preproglucagons which contain the sequence of

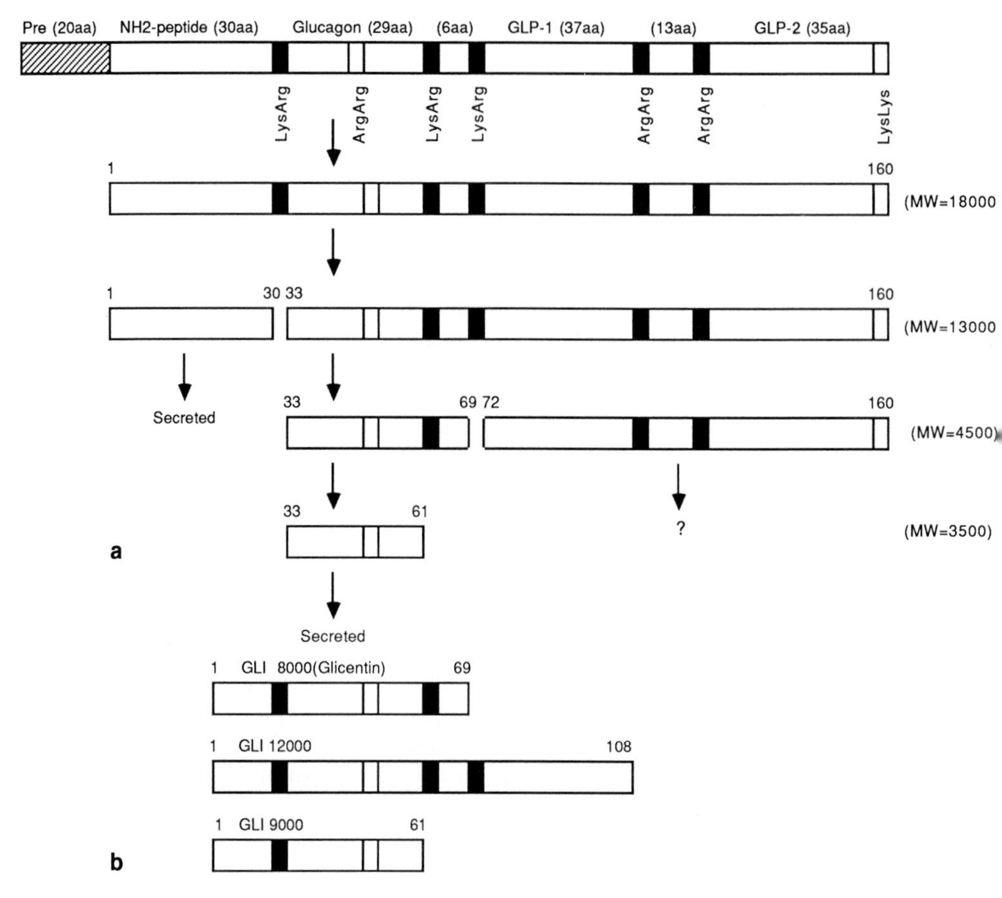

Fig. 11.1. Proposed ways of processing of mammalian preproglucagon (*a*), and the structure of nonpancreatic glucagon-containing peptides (*b*). In a, basic dipeptides are indicated; those that are potential sites for cleavage are indicated by *dark boxes*. *Numbers in parentheses* at the end of each line are the sizes of the glucagon-containing intermediates. *Numbers above the lines* are the amino acids at the ends of the polypeptide in relation to the sequence of preproglucagon; (*b*) structure of nonpancreatic glucagon-containing polypeptides. GLI 8000 and GLI 12000 are major polypeptides in the intestine; GLI 9000 accumulates in the serum of animals with renal failure (Bell et al. 1983)

glucagon and, in contrast to the mammalian precursor, a single GLP (Lund et al. 1983; Andrews and Ronner 1985; see Fig. 11.2). The GLPs from catfish and coho salmon (*Oncorhynchus kisutch*) islets have been isolated and sequenced (Andrews and Ronner 1985; Plisetskaya et al. 1986b). Catfish GLP contains 34 amino acid residues as opposed to 31 in the salmon and anglerfish. When compared with salmon glucagon, 11 out of 29 positions are matches, and 14 are single base exchanges. Table 11.1 (see pp. 126 and 127) compares the amino acid sequences of the glucagons and GLPs of three teleosts with those of the human. The structural similarities between the GLPs and glucagon suggest that the GLPs may have physiological functions, a notion supported by a potent stimulation of adenylate

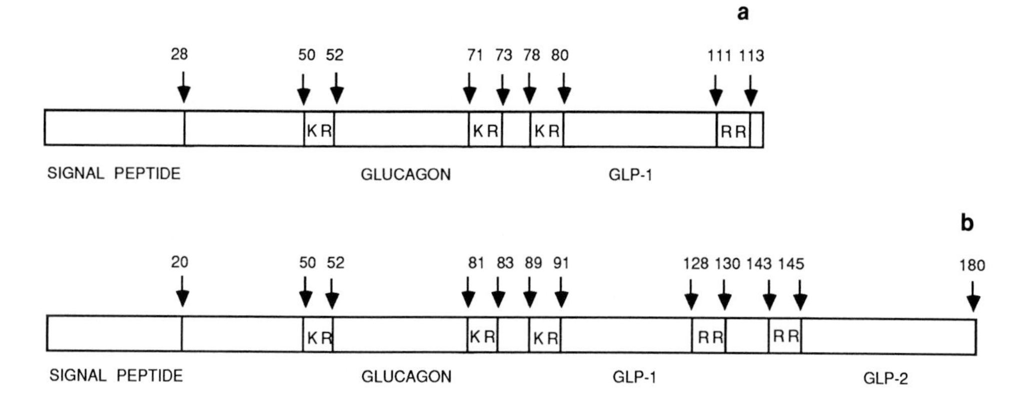

Fig. 11.2. Comparison of (*a*) anglerfish and (*b*) human preproglucagon. The putative proteolytic processing sites (two adjacent basic amino acid residues) are indicated in *black*. Note that in contrast to Fig. 11.1, the numbering includes the amino acids of the pre-(signal) peptides. (After Andrews and Ronner 1985)

cyclase in rat hypothalamus and pituitary (Hoosein and Gurd 1984), a weak stimulation of exocrine pancreas secretion (Uttenthal et al. 1984) and a glucose-dependent insulinotropic effect on isolated rat islets (Schmidt et al. 1985). Since the GLPs do not inhibit glucagon binding to brain and liver membranes (Hoosein and Gurd 1984), their possible physiological role must differ from that of glucagon. Thus, it appears that the pancreatic A-cell contains two types of "standard peptides", glucagon and GLP, in its secretory cocktail. A variety of larger forms with glucagon-like immunoreactivity have been isolated from a variety of tissues, and from different species (cf. Lopez et al. 1983). Among these forms is glicentin, a putative proglucagon fragment which consists of 69 amino acid residues. Of these, residues 33–61 contain in mammals the amino acid sequence of glucagon (Thim and Moody 1981; see Fig. 11.1).

Most mammalian glucagons are 29 amino acid residues long with a molecular weight of 3485 daltons. The porcine, bovine, rat, hamster, and human molecules are sequentially identical (cf. Huang et al. 1986), and while the primary structure of rabbit (Sundby and Markussen 1971 a, b), camel (Sundby et al. 1974) and sheep (Jackson 1981, cited by Bromer 1983) glucagons have not been determined, these peptides are compositionally identical to the human hormone (Table 11.2, see p. 128). Guinea pig glucagon, like insulin, differs considerably from that of other mammals (Conlon et al. 1985 c; Huang et al. 1986). Whether other hystricomorph glucagons are similar to that of the guinea pig is unknown; also there seems to be no information on the South American primates whose insulin must be very different from the typical mammalian insulin (see Sect. 10.2). Avian glucagons also have 29 residues and differ only slightly in primary structure from the mammalian hormone. Turkey glucagon has one substitution at position 28 (Markussen et al. 1972) while duck glucagon differs in positions 16 and 28 (Sundby et al. 1972). Chicken and turkey glucagons are probably identical (Pollock and Kimmel 1975); alligator glucagon is probably identical with the duck molecule (Lance et al. 1984). The gastrointestinal tract of the axolotl contains large quantities of peptides that seem to be strongly homologous with mammalian pancreatic glucagon (Conlon et al. 1985b).

The four known teleost glucagons (Table 11.1) have 13 identical amino acid residues at the N-terminal. Salmon glucagon differs at positions 14, 16, and 18 from catfish and the two anglerfish glucagons, the substitution of the glutamine for a leucine residue in position 14 being unique for all glucagon-like peptides so far sequenced. In addition, salmon glucagon differs in position 27 from anglerfish II glucagon and in positions 21, 24, and 29 from anglerfish I glucagon. Together with the findings on GLPs, this suggests that the salmon and catfish glucagon genes share a closer ancestral relationship with the anglerfish glucagon gene II than gene I (Plisetskaya et al. 1986b). As can be seen from Tables 11.1 and 11.2, the differences between the teleost glucagons and the known glucagons of the other vertebrates are remarkable. On the other hand, selachian glucagon(s) are compositionally closer to those of mammals. This observation, first reported for the shark (*Squalus acanthias*) by Sundby (1976), was fully confirmed for a ray (*Torpedo marmorata*). Conlon and Thim (1985) found that pancreas glucagon from the latter differs in only three positions from the known mammalian glucagons. Thus, it appears that with the exception of the guinea pig (or hystricomorphs?) and the teleosts (or actinopterygians?), the primary structure of islet glucagon was strongly conserved. In mammals, the (29-residue) islet type of

Table 11.1. Comparison of the known teleost glucagons (A) and GLP's (B) with human GLP's, and comparison between salmon glucagon and GLP. Note that position 1 in the human GLP-I shown here corresponds to number 7 according to Bell et al. (1983). (After Plisetskaya et al. 1986b)

	1				5					10				
Glucagon (A)														
Coho Salmon	His	Ser	Glu	Gly	Thr	Phe	Ser	Asn	Asp	Tyr	Ser	Lys	Tyr	Gln
Catfish	–	–	–	–	–	–	–	–	–	–	–	–	–	Leu
Anglerfish – II	–	–	–	–	–	–	–	–	–	–	–	–	–	Leu
Anglerfish – I	–	–	–	–	–	–	–	–	–	–	–	–	–	Leu
Human	–	–	Gln	–	–	–	Thr	Ser	–	–	–	–	–	Leu
GLP (B)														
Coho Salmon	His	Ala	Asp	Gly	Thr	Tyr	Thr	Ser	Asn	Val	Ser	Thr	Tyr	Leu
Catfish	–	–	–	–	–	–	–	–	Asp	–	–	Ser	–	–
Anglerfish – II	–	–	–	–	–	–	–	–	Asp	–	–	Ser	–	–
Anglerfish – I	–	–	–	–	–	Phe	–	–	Asp	–	–	Ser	–	–
Human – I	–	–	Glu	–	–	Phe	–	–	Asp	–	–	Ser	–	–
Human – II	–	–	–	–	Ser	Phe	Ser	Asp	Glu	Met	Asn	–	Ile	–
Salmon glucagon and GLP (C)														
Glucagon	His	Ser	Glu	Gly	Thr	Phe	Ser	Asn	Asp	Tyr	Ser	Lys	Tyr	Gln
GLP	–	Ala	Asp	–	–	Tyr	Thr	Ser	Asn	Val	–	Thr	–	Leu
Base changes		1	1			1	1	1	1	2		2		1

glucagon occurs in the pancreas and gastric mucosa (cf. Hatton et al. 1985), whereas two larger forms with NH_2- and COOH-terminal extensions are found in the intestinal mucosa and hypothalamus (Tager 1984; Conlon et al. 1985c; see also Chap. 2.2.2).

Glucagon is one member of a family of nearly a dozen peptides that includes secretin (Mutt et al. 1970), VIP (Mutt and Said 1974), GIP (Moody et al. 1984), the growth hormone-releasing factor (Guillemin et al. 1984), peptide PHI-27 (Tager 1984), and it shares some structural homology with prealbumin (Jörnvall et al. 1981).

11.2 Antigenicity, Assays, and Endogenous Titers

The development of a glucagon radioimmunoassay (Unger et al. 1959) closely followed that of insulin (Berson et al. 1956). Since glucagon is weakly antigenic and attempts to raise antisera against the molecule have often resulted in nonspecific or low-affinity antibodies, specificity is attained through carboxyterminally directed antisera, that is the 22–29 amino acid sequence (Heding 1983). Glicentin has enough carboxy-terminal amino acids beyond the glucagon sequence to mask the glucagon binding sites of such antisera. Antisera directed

15					20					25				29		
Glu	Glu	Arg	Met	Ala	Gln	Asp	Phe	Val	Gln	Trp	Leu	Met	Asn	Ser		
–	Thr	–	Arg	–	–	–	–	–	–	–	–	–	(–	–)		
–	Thr	Arg	–	–	–	–	–	–	–	–	–	Lys	–	–		
–	Asp	–	Lys	–	–	Glu	–	–	Arg	–	–	–	–	Asn		
Asp	Ser	–	Arg	–	–	–	–	–	–	–	–	–	–	Thr		
Gln	Asp	Gln	Ala	Ala	Lys	Asp	Phe	Val	Ser	Trp	Leu	Lys	Ser	Gly	Arg	Fla
–	–	–	–	–	–	–	–	Ile	Thr	–	–	–	–	–	Gln	Pro
–	–	–	–	–	–	–	–	–	–	–	–	–	Ala	–	–	Gly
Lys	–	–	–	Ile	–	–	–	–	Asp	Arg	–	–	Ala	–	Gln	Val
Glu	Gly	–	–	–	–	Glu	–	Ile	Ala	–	–	Val	Lys	–	–	Gly
Asp	Asn	Leu	–	–	Arg	–	–	Ile	Asn	–	–	Ile	Gln	Thr	Lys	Ile
Glu	Glu	Arg	Met	Ala	Gln	Asp	Phe	Val	Gln	Trp	Leu	Met	Asn	Ser		
Gln	Asp	Gln	Ala	–	Lys	–	–	–	Ser	–	–	Lys	Ser	Gly	Arg	Ala
1	1	1	2		1				2			1	1	1		

Table 11.2. Amino acid sequences in islet glucagons in vertebrates other than teleosts. (For literature and data, see Pollock and Kimmel 1975; Lance et al. 1984; Conlon and Thim 1985; Conlon et al. 1985c; Huang et al. 1986)

	1				5					10					15					20					25				29
Human and other mammals studied so far; except the guinea pig	His	Ser	Gln	Gly	Thr	Phe	Thr	Ser	Asp	Tyr	Ser	Lys	Tyr	Leu	Asp	Ser	Arg	Arg	Ala	Gln	Asp	Phe	Val	Gln	Trp	Leu	Met	Asn	Thr
Guinea pig	–	–	–	–	–	–	–	–	–	–	–	–	–	–	–	–	–	–	–		Gln	–	Leu	Lys	–	–	Leu	–	Val
Turkey, chicken?	–	–	–	–	–	–	–	–	–	–	–	–	–	–	–	–	–	–	–	–	–	–	–	–	–	–	–	Ser	–
Duck, alligator?	–	–	–	–	–	–	–	–	–	–	–	–	–	–	–	Thr	–	–	–	–	–	–	–	–	–	–	–	Ser	–
Torpedo (Selachian)	–	–	Glu	–	–	–	–	–	–	–	–	–	–	–	–	Asn	–	–	–	Lys	–	–	–	–	–	–	–	–	–

toward the amino-terminal end of the molecule cross-react nonspecifically with gut glucagon-like immunoreactivity (GLI) or glicentin. According to Heding (1983), the best antisera are obtained from rabbits injected subcutaneously with glucagon conjugated with albumin, although a few workers have successfully obtained good antisera from guinea pigs. In rat and guinea pig liver membrane assays, guinea pig glucagon is tenfold less potent than porcine glucagon, which suggests that the glucagon receptor is well conserved in mammals. Taken together with the high glucagon concentration in the guinea pig pancreas, this suggests furthermore that circulating glucagon (just like insulin) is elevated in the guinea pig (Huang et al. 1986). It seems that all glucagon radioimmunoassays reported to date in nonmammalian vertebrates have employed antisera to mammalian glucagon, some of which were nonspecific (amino-terminally directed) and others specific for pancreatic glucagon. Quantifying plasma baseline values of pancreatic glucagon is difficult for two reasons: (1) antiserum specificity and subsequent cross reactivity with various GLIs may give false-positive results; and (2) pancreatic glucagon may emanate, in addition to the stomach, from several nonpancreatic sources (see Chap. 2.2.2). Holst et al. (1983) have found all molecular forms of glucagon circulating in pancreatectomized humans, which corroborates earlier observations of glucagonemia in pancreatectomized dogs (Matsuyama and Foà 1974; Vranic et al. 1974) and birds (Colca and Hazelwood 1982). Therefore, depending on the species, any in vivo measurements of circulating "pancreatic" glucagon, no matter how specific the antiserum, may be affected by extrapancreatic glucagon. The degree to which these extrapancreatic sources contribute to a glucagon assay, and the factors that control their release are largely unknown. In fasted birds, variations of plasma glucagon are mainly due to pancreatic glucagon (cf. Strosser et al. 1984), a situation helpful in the design of experiments. However, there are differences in the response to i. v. amino acids even between closely related birds. Arginine causes release of both gut and pancreas glucagon in the duck, but is without effect on gut glucagon in the goose (Khemiss and Sitbon 1982).

Plasma glucagon values are sketchy among sub-mammals. Hazelwood (1984; Cieslak and Hazelwood 1986a) states that avian plasma contains four to ten times more glucagon than mammals, and Rhoten (1976) described the plasma values of the lizard *Anolis carolinensis* to be similar to the avian situation – high plasma glucagon in the presence of high plasma glucose. In its annual cycle the frog *Rana esculenta* displays its highest plasma glucagon values late in hibernation (Schlaghecke and Blüm 1981), very similar to the mammalian hedgehog (*Erinaceus europaeus*), whose plasma glucagon titers peak upon arousal from cold-induced lethargy (Hoo-Paris et al. 1982). Simple cold exposure of sheep (0° for 4–19 days) had no effect on plasma glucagon (Sasaki et al. 1982), so hibernation may involve metabolic processes that demand glucagon, and indeed the A-cells of hibernating animals may have special adaptations for hibernation (Assan et al. 1983). Interestingly, Umminger and Bair (1973) observed a relationship between cold-induced hyperglycemia and degranulation of A-cells in the killifish *Fundulus heteroclitus*. Catfish (*Ictalurus nebulosus*) glucagon exhibits a high degree of similarity with porcine glucagon by RIA (Andrews and Ronner 1985), whereas salmon (*Oncorhynchus kisutch*) glucagon cross-reacted very poorly in this system (Plisetskaya et al. 1986b). Recently, Gutierrez et al. (1986) estimated the plasma levels of 11 teleosts and the dogfish (*Scyliorhinus canicula*) with an adaptation of a technique described for mammals. In all teleosts with the exception of *Sparus aurata*, the average value was below 1 ng/ml. The lowest value (0.14 ± 0.04 ng/ml) was found in the dogfish. Glucagon immunoreactivity has been found in the plasma of the hagfish *E. stouti* (Stewart et al. 1978) and in islet extracts of the lamprey *Lampetra fluviatilis* (Zelnik et al. 1977); since there is no immunohistochemical support for an islet glucagon in these animals (see Chap. 4.1) it may be that the finding of Zelnik et al. (1977) has been due to contamination with mucosa material.

In summary, much work remains to be done before meaningful baseline glucagon levels are available for comparison among the poikilotherms. Because the interfering assay factors listed above probably apply to all vertebrates, this will be a very difficult task. Furthermore, with liver as the primary target organ (see below) only portal vein levels may be of any significance in many species. Yet, contamination by various gut glucagons and GLI's will be greatest in the portal vein.

11.3 Glucagon Before Birth

Almost nothing seems to be known about the physiological role before birth in mammals of glucagon, which appears early in islet development, perhaps even simultaneously with insulin and somatostatin (see Chap. 3). Stefan et al. (1982b) have reported that glicentin precedes glucagon in the human A-cell, but the functional significance of this phenomenon is unclear. Girard and Sperling (1983) reviewed the quantitative aspects of fetal glucagon, concluding that all species studied (rat, rabbit, sheep, and human) display a mid-gestational zenith in pancreatic glucagon content. Glucagon is also present in the fetal circulation, but glucagon receptor number or affinity is always lower in fetal than in adult liver, a situation which contrasts with that of the insulin receptors. At least in sheep and humans, virtually all plasma glucose is derived from the mother via transplacental transfer; only in chronic hypoglycemia may fetal glucose production be activated. Together with the normal absence of glycogenolysis and gluconeogenesis in utero (see also McCormick et al. 1985) these findings suggest that glucagon has a very limited, if any, role in the metabolism of fetal mammals. On the other hand, its immediate postnatal significance is shown by an increase of glucagon receptors, and a three- to fivefold increase of its plasma titer at birth (for literature see Sperling et al. 1984). Studies on the splenic lobe of the embryonic chicken pancreas in vitro (Foltzer et al. 1984) revealed that glucagon release occurs on day 15 of incubation, and that it is three times higher in explants from day 19. Further, very strong increases in glucagon release were seen in explants taken 2 and 9 days after hatching. In the 15-day-old embryo, but not in the 18-day-old embryo and the 2-day-old chicken, 30 mM of glucose enhanced glucagon release. A similar, stimulatory glucose effect was seen in the 9-day-old chicken material. These paradoxical findings require further study.

11.4 The Physiology of Islet Glucagon in Postnatal Vertebrates

When compared with insulin, the overall spectrum of the actions of islet glucagon appears rather limited. In all gnathostomes so far studied, with the possible exception of the chondrichthyes, the liver seems to be the primary target of glucagon, and at least one function, glycogenolysis, may be ubiquitous. It is possible that also hepatic gluconeogenesis is generally controlled by glucagon, but the information is incomplete. Hepatic ketogenesis is also under glucagon control in some taxa, but evidence from two groups with particularly interesting patterns of lipid metabolism is either nonexistent (Chondrichthyes) or uncertain (birds). Besides the liver, the only unequivocally proven extra-insular target organ of islet glucagon in vivo seems to be the avian adipose tissue. However, chances are good that glucagon will also prove to be a hypoosmotic hormone, at least in teleosts (see Chap. 14). In the homeotherms, glucagon is essential for the maintenance of blood-sugar homeostasis, a task that it achieves in close antagonistic cooperation with insulin (see e.g., Unger and Orci 1981a, b; Unger 1983; Hazelwood 1984). As in the case of insulin, it may be useful to look at the mammalian situation before attempting to trace the evolution of islet glucagon, beginning with the cyclostomes.

11.4.1 Mammalia

In this group, or better perhaps, in nonhibernating Placentalia, the maintenance of an appropriate glycemia per se is one of the most important regulatory functions, and in this the ratio of circulating insulin: glucagon plays a pivotal role; the higher the percentage of glucagon, the greater the release of hepatic fuel. Consequently, there is always a relatively or absolutely increased glucagon level when, in postprandial states, increased mobilization of metabolites occurs. Under *normal resting conditions*, the human liver must provide approximately half of the extracellular glucose pool. This basal hepatic glucose production is about 75 percent glucagon-mediated, which suggests that glucagon may be essential for life. However, there is so far only one published case in which it seems likely that specific glucagon-deficiency led to fatal hypoglycemia (cf. Unger and Orci 1981a, b; Unger 1983). During *acute "emergencies"* such as "fight-or-flight" situations, various forms of stress or heavy exercise, the glucose utilization may increase more than tenfold. Under these conditions there is a sharp increase in plasma glucagon which, in concert with decreased insulin and increased catecholamine levels, causes increased hepatic glucose output. This mechanism helps to provide sufficient fuel for the appropriate organs, especially brain and skeletal muscle. During *extended starvation*, survival depends on the economic utilization of additional substrates. Again, the titer of glucagon is increased. However, since the hepatic glycogen reserves are largely consumed after 24 h, other sources of fuel must be utilized. This is mediated via a simultaneous decrease in plasma insulin (see Chap. 10.4.1), which enhances mobilization of amino acids from protein, mainly from muscle. This substrate is now used for hepatic gluconeogenesis, a process fostered by the change of the insulin: glucagon ratio in favor of the latter.

However, since this type of fuel production would be incompatible with a reasonable chance of eventual survival, another source of energy must be utilized. Therefore, after a few days of starvation, gluconeogenesis is largely replaced by ketogenesis. In this case, again a lower level of insulin provides mobilization of metabolites, i.e., release of free fatty acids from adipocytes, while glucagon stimulates the hepatic oxidation of the fatty acids to ketones. The latter now substitute for glucose as a fuel in the brain, and utilization of both fatty acids and ketones by peripheral tissues further reduces the need for gluconeogenesis. On the other hand, increased blood sugar level after a *meal rich in carbohydrates* coincides with an increased insulin: glucagon ratio, since under this condition plasma glucose, somatostatin (?) and insulin, and some other factors, together suppress release of pancreatic glucagon. However, this "herbivore" (or "fructivore") constellation differs from the situation after a *meal rich in proteins* and poor in carbohydrates. Now, the suppression of glucagon release by increased titers of insulin (and somatostatin?) could lead to a dangerous hypoglycemia, if the A-cell were not to receive an overriding signal. This signal arrives in form of amino acids that create a postprandial "carnivore" constellation. The latter consists of a balanced insulin: glucagon ratio that assures an appropriate glycemia. Finally, the state of *insulin deficiency* in type-I diabetes may be considered a cruel, albeit most instructive experiment on the role of islet glucagon. Apparently due to a lack of A-cell control by insulin, a marked hyperglucagonemia causes an excessive hepatic production of both glucose and ketones. However, if glucagon is also absent, the insulin deficiency by itself does not cause the hepatic overproduction of metabolites (for literature see Itoh et al. 1981; Unger and Orci 1981a, b; Weir 1981; Lefèbvre 1983a, b; Unger 1983). Unger (1985) suggests that the major effect of insulin at the hepatocyte level is to oppose the effect of glucagon by reducing the cyclic AMP-dependent protein kinase activity. Admittedly, this picture is largely based on the human condition, and it is simplified because it ignores the impact of several other "metabolic" hormones (see Chap. 8), variations in the state of the substrate (which is supposed to be readily available), and the impact of nervous factors. Nevertheless, it contains the basic facets for the following consideration of the role of glucagon in nonmammalian species.

It is clear that in mammals pancreatic glucagon is in several aspects a direct antagonist of insulin. However, in addition to the intrainsular interactions with insulin and somatostatin (see Chap. 9.3), and the role of ions (Leclercq-Meyer and Malaisse 1983), the knowledge on the direct impact of circulating metabolites on glucagon release is not satisfactory, either. While there is general agreement that the secretory activity of the mammalian A-cell is stimulated by amino acids, and inhibited by glucose (Weir 1981; Itoh et al. 1981; Gerich 1983b; Assan et al. 1983; Luyckx and Lefèbvre 1983; Asplin et al. 1984; Unger 1985), the precise role of leucine in vivo is open to question (cf. Leclercq-Meyer et al. 1985a); and so is the physiological significance of the inhibitory effect of fatty acids (Steffens and Strubbe 1983). The roles of glucagon in mammalian mineral metabolism (Kolanowski 1983) as well as in the control of renal (Lefèbvre and Luyckx 1983) and cardiovascular functions (Farah 1983; Tanaka et al. 1983) are also far from clear. Finally, it must be recalled that nothing seems to be known about the functions of glucagon in marsupials and monotremes.

11.4.2 Cyclostomata

Barring surprises with the unidentified hormone of the argyrophil cells in the islets of adult lampreys (see Chap. 4.1), it appears that cyclostome islets do not produce glucagon. Indirect evidence for the absence of glucagon in cyclostome islets comes from experiments that demonstrated that exogenous glucagon does not affect lamprey hepatic glycogen, and that this hormone even causes a decrease in blood sugar, an effect that may be due to stimulation of insulin release (cf. Hardisty and Baker 1982). Vice versa, anti-glucagon serum seems to lower plasma insulin (Murat et al. 1981). This raises the question if, perhaps, glucagon from the "open" gut cells of cyclostomes acts as a secretagogue for insulin release. Furthermore, the reasonably well-contained blood sugar level of the lampreys (see Chap. 10.5.2) demonstrates that, at the agnathian level, mechanisms other than effects of glucagon, and also of catecholamines (Dashow and Epple 1983; Plisetskaya et al. 1984b) control the production of glucose.

11.4.3 Chondrichthyes

In selachians, exogenous glucagon is hyperglycemic (Grant et al. 1969; Patent 1970) but it is not clear whether this effect reflects a physiological role of this hormone; no accompanying changes of liver glycogen thereafter were apparent in *Squalus acanthias* (Patent 1970). In the holocephalian *Hydrolagus colliei*, exogenous glucagon was totally ineffective, although epinephrine caused hyperglycemia (Patent 1970). It is possible that the low glycogen content of the chondrichthyan liver limited the impact, or at least the *acute* impact, of exogenous glucagon in the above-cited studies though, of course, other factors such as species specificity of the hormone cannot be discounted (Patent 1973). In this connection, it is noteworthy that the holocephalian islets produce, apparently in addition to glucagon, a glucagon-related peptide of unknown function (see Chap. 4.2). The virtual lack of information on the glucagon physiology in the chondrichthyes is especially regrettable because of the peculiarities of their lipid metabolism (Zammit and Newsholme 1979).

11.4.4 Actinopterygii

It appears that among this group only teleosts have been studied. In species of this group, injections of apparently large quantities of mammalian glucagon are followed by a hyperglycemia of varying degree whereas the effects on liver and muscle glycogen vary greatly (for literature see Epple et al. 1980; Murat et al. 1981; Carneiro and Amaral 1983). Recently, Plisetskaya (pers. commun.) found that in salmon *physiological* doses of salmon glucagon rarely caused hyperglycemia; frequently, however, the hormone caused a slight hypoglycemia that was not mediated by insulin. In a study on *Pimelodus maculatus* hepatic glycogen actually increased (Carneiro and Amaral 1983). There is no question that glucagon-stimulated hyperglycemia in several teleosts is due to gluconeogenesis, but it seems that the glycemic contributions of glycogenolysis and gluconeogenesis vary greatly with the physiological state and probably also with the species (for literature see Walton and Cowey 1979; Renaud and Moon 1980; Murat et al. 1981; Morata et al. 1982b; Inui and Ishioka 1983b). In the teleost liver, gluconeogenesis from amino acids, pyruvate, and lactate has been demonstrated (cf. Phillips and Hird 1977a; Walton and Cowey 1979; Renaud and Moon 1980; Morata et al. 1982b; cf. Mommsen 1986); gluconeogenesis from ketone bodies appears a distinct possibility (cf. DeRoos et al. 1985). The general picture that emerges is that interactions among insulin, glucagon and epinephrine control the utilization of glucogenic substrates in the liver (for literature see Epple et al. 1980; Murat et al. 1981; see also Chap. 10.4.4), and thus also, indirectly, glycogen deposition in skeletal muscle. However, many details of the hepatic interactions of these hormones (and probably others as well) remain to be clarified. The role of glucagon in the lipid metabolism of teleosts is unclear. Exogenous glucagon was found to exert no impact on plasma FFA in the European eel (*Anguilla anguilla*) (A. L. Larsson and Lewander 1972), but increased plasma FFA in the pike (*Esox lucius*); it had no effect on plasma cholesterol (Thorpe and Ince 1974; Ince and Thorpe 1975). Farkas (1969) found no effect on the release of FFA or glycerol from adipose tissue of *Lucioperca lucioperca*. The few data on the role of glucagon in protein metabolism of teleosts agree, in general, with its gluconeogenic role in the liver, although single injections in the pike (Thorpe and Ince 1974) and Japanese eel, *Anguilla japonica* (Y. Inui and Yokote 1977) did not change the blood level of AAN. However, repeated injections lowered the AAN level, an effect that was abolished by hepatectomy (Y. Inui and Yokote 1977). In vitro, glucagon stimulated uptake of amino acid by liver, but not opercular muscle of the Japanese eel (Y. Inui and Ishioka 1983a, b). Very little is known about the impact of metabolites on glucagon release in teleosts. Using a heterologous assay system and a perfused pancreas preparation, Ince and So (1984) found that acute increases of glucose content resulted in a monophasic, dose-dependent decline in glucagon secretion over 30 min. When compared with the simultaneously measured somatostatin (S-14) release, the response of glucagon occurred at a lower glucose titer (8.3 mM vs. 16.7 mM, respectively). Perfusion of Brockmann bodies of catfish (*Ictalurus punctatus*) with 10 mM glucose had no effect on glucagon release, but 10 mM arginine caused a significant elevation (Ronner and Scarpa 1982). Considering that the presence of glucagon in the Brockmann bodies of *Scorpaena* species was demonstrated more than 30 years ago (Audy and Kerly 1952), the progress on its physiology in the teleosts is less than impressive.

11.4.5 Dipnoi

The only physiological investigation of glucagon in lungfishes seems to be a recent study on the Australian lungfish (*Neoceratodus forsteri*), which showed that glucagon (10^{-6} M) stimulates glycogen breakdown and glucose release from hepatic tissue in vitro (Hanke and Janssens 1983).

11.4.6 Amphibia

Nothing appears to be known about the physiology of glucagon in the Gymnophiona; furthermore, information on the Urodela is almost nonexistent. In *Taricha torosa*, intraperitoneal injections of amorphous glucagon (1000 µg/kg!) had no glycemic effect, whereas intravenous injections of 100 µg/kg of crystalline glucagon caused a threefold rise in blood sugar (cf. M. R. Miller 1961). In liver organ cultures of *Amphiuma means*, glucagon stimulated gluconeogenesis from pyruvate and alanine (Brown et al. 1975). In the same species, insulin release was increased by glucagon (Gater and Balls 1977). In vitro, glucagon increases hepatic glycogenolysis in the axolotl, *Ambystoma mexicanum* (Janssens and Maher 1986).

Glucagon elevated glycemia in all studies on Anura; it has a hepatic glycogenolytic action (for literature see Farrar and Frye 1979; Epple et al. 1980). Since the hepatic effect is not necessarily associated with decrease in the glycogen content (Hanke 1974b; Farrar and Frye 1979), it is likely that gluconeogenesis is stimulated by glucagon. However, there is no direct proof for this suggestion. As expected, in contrast to epinephrine, glucagon did not cause an increase in blood lactate in *Rana pipiens* (Farrar and Frye 1979). In two northern species, *Rana temporaria* and *Rana pipiens*, the glycemic effect of glucagon varies with the season; in the latter twice as much glucagon was needed in winter frogs to duplicate the results with summer animals (Hanke 1974b; Farrar and Frye 1979). In *Rana temporaria*, glucagon injections had no effect on the titer of plasma FFA (Harri and Puuska 1973). In a liver preparation of a toad (*Bufo arenarum*), glucagon decreased the levels of the total lipids and cholesterol in the perfusate, but greatly enhanced its ketone content (Penhos et al. 1967b). In late tadpoles of *Xenopus laevis* (Hanke and Neumann 1972) and *Rana pipiens* (Farrar and Frye 1979), exogenous glucagon is also hyperglycemic, apparently due to hepatic glucose release. The only study on endogenous glucagon in Amphibia seems to be a report by Schlaghecke and Blüm (1981), who found no clear seasonal variations in pancreatic glucagon in *Rana esculenta* despite the annual cycle in the plasma titer (see Chap. 11.2). Finally, it may be noted that in frogs anti-glucagon serum seems to decrease serum insulin (Murat et al. 1981). As in the lamprey (see Chap. 11.4.2), this raises the question of a stimulatory effect of glucagon, possibly of gut origin, on the release of insulin. From the preceding, the dearth of data on the physiology of glucagon in the Amphibia is obvious, so that general conclusions appear unwarranted.

11.4.7 Reptilia

The hyperglycemia following pancreatectomy in turtles and crocodilians, and the temporary hypoglycemia after this operation in several species of Squamata, suggest considerable variations in the role of glucagon between these groups. The "paradox" of similarity between the islet physiology of Squamata and Aves (see below), in obvious contrast to their phylogenetic relationship, invites speculation as to the diphyletic, independent origin of the high glucagon and blood sugar titers in both groups (see Chap. 10.4.7). Glucagon is hyperglycemic in all reptiles studied so far (for literature see Penhos and Ramey 1973; Epple et al. 1980). In a turtle species, chronic glucagon infusion decreased hepatic glycogen, but not muscle glycogen (Marques 1967). In pancreatectomized, but not in intact alligators, exogenous glucagon was ketogenic (Penhos et al. 1967a); in an intact lizard (*Uromastix hardiwickii*) glucagon caused hypocholesterolemia (Suryawanshi and Rangneker 1971). The only data on endogenous glucagon responses in reptiles are those of Rhoten (1976, 1978). Like birds, the lizard *Anolis carolinensis* has a physiological "hyperglycemia" and "hyperglucagonemia". When the splenic lobe of the anolian pancreas was exposed to a basal glucose concentration of 225 mg% in vitro, a subsequent 56% decline in glucose concentration had no significant effect on glucagon release. Only a further decrease to 50 mg% stimulated glucagon secretion. On the other hand, an increase of glucose to 600 mg% failed to decrease glucagon release significantly, as also did arginine (10 mM), a potent glucagon secretagogue in mammals. During the annual cycle of a Sahelian lizard (*Varanus exanthematicus*), Dupé-Godet and Adjovi (1981a) found the highest pancreatic concentrations of glucagon in the fasting season, and the lowest ones during times of food intake. Apart from their interesting phylogenetic position in general, reptiles invite, in particular, studies on the glucagon physiology of the Crocodilia and Squamata. The discrepancies in the phylogenetic relationship and islet physiology between the latter two groups and birds have been pointed out previously; one wonders how these differences are reflected in the lipid metabolism. For example, is the ketonemia in the pancreatectomized alligator due to pancreatic-type glucagon from the gut, which is now unbalanced by islet insulin? Is adipose tissue of Squamata as sensitive to glucagon as that of the birds?

11.4.8 Aves

Various studies on pancreatectomized ducks and geese, and A-cell-deficient chickens, have shown that the metabolism of these species depends strongly on pancreatic glucagon; however, as pointed out in Chap. 10.4.8, it may be premature to draw conclusions on birds in general, since these three species represent only two families in a single order of a nutrition-wise and ecologically most heterogeneous class of vertebrates. In these three species, glucagon is essential to life by preventing a fatal hypoglycemia, and in close coordination with insulin, in control of blood sugar level. Pancreatectomized ducks and geese can be maintained alive by application of a proper mixture of insulin and glucagon (cf. Sitbon and Mialhe 1978). Glucagon has been found to be hyperglycemic in all species studied so far (for literature see Langslow and Hales 1971; Epple et al. 1980; Sitbon et al. 1980; Hazelwood 1984), including carnivorous species and the emperor penguin (cf. Grande 1970; Groscolas and Bezard 1977). In vivo and in vitro this hormone strongly stimulates hepatic glycogenolysis, and fosters gluconeogenesis (cf. Fister et al. 1983). The presence of the latter process is also suggested by observations on fasting chickens, in which hepatic glycogen stores are quickly depleted but partially replenished after a few days (Hazelwood and Lorenz 1959; Langslow and Hales 1971). In a comprehensive study on isolated chicken hepatocytes, Dickson and Langslow (1978) found that glucagon stimulates gluconeogenesis from all substrates used; alanine, glycerol, pyruvate, dihydroxyacetone, fructose, and lactate. In the duck, alanine seems to be the major substrate in glucagon-stimulated gluconeogenesis (Laurent et al. 1985). Strong hyperglycemia suppresses the concentration of circulating glucagon in the goose (Sitbon and Mialhe 1979); suppression of glucagon release from chicken pancreas fragments was observed when the glucose levels of the perifusate were elevated (Colca and Hazelwood 1981). The effect of circulating glucose on the A-cell seems to be mediated via insulin-dependent intracellular glucose metabolism (Laurent et al. 1981). In all, data from various laboratories suggest that glucagon exerts its essential glucoregulatory function in the birds studied so far by a glucose-mediated feedback mechanism. However, it remains to be seen if the essential part of this feedback mechanism, i.e., a high glucose sensitivity of the A-cell, also exists in insectivorous, carnivorous, and piscivorous birds.

Triglycerides, which are of utmost importance for the metabolism of birds in general, can be deposited in huge amounts in the adipose organs of migratory birds (cf. J. R. King 1972; Dolnik 1975). Like many nonmammalian vertebrates, birds also store conspicuous amounts of lipids in liver. The liver is also the major site of lipogenesis from whence triglycerides are transported to the adipose tissue as low-density or beta-lipoproteins (O'Hea and Leveille 1969). The adipose tissue itself seems to play only a minor role in fatty acid synthesis. However, it supplies FFA that seem to be the main energy source of birds. In contrast to the insignificant role of insulin, avian lipid metabolism is largely controlled by glucagon. This hormone seems to suppress the release of triglycerides from the liver (DeOya et al. 1971); it also induces a strong lipolysis in adipose tissue, which, at variance with that of other vertebrates so far studied, is highly sensitive to physiological levels of this hormone. The impact of glucagon on the adipose tissue is antagonized by a direct inhibitory action of somatostatin, a situation also so far known for birds only (see Chap. 13.3). The glucagon effect causes a rapid increase in plasma FFA. In excessive doses, glucagon produces fatty livers, probably by providing substrate (FFA) from the adipose tissue that is used for triglyceride production, in combination with inhibition of hepatic triglyceride release (cf. Langslow and Hales 1971). The importance of glucagon for the avian lipid metabolism is demonstrated by the decrease in plasma FFA in pancreatectomized ducks, and the suppression of the glucagon titer by oleic acid infusions. Thus, at least in the duck, plasma FFA level seems to be controlled overwhelmingly by a glucagon-FFA feedback mechanism (Sitbon et al. 1980), which, however, is mediated by somatostatin (Gross and Mialhe 1986). Results obtained with fragments of chicken pancreas (Colca and Hazelwood 1981) are compatible with this interpretation.

Unlike in mammals, studies in the duck revealed no clear correlation between the plasma titers of beta-hydroxybutyrate (virtually the only circulating ketone body in this species) and FFA. Glucagon infusions, which increase the titer of the FFA, have no impact on that of ketone bodies; and when the FFA titer declines after total pancreatectomy, plasma level of beta-hydroxybutyrate does not change. On the other hand, the plasma level of the ketone body increases strongly in the fasted duck, and also temporarily after subtotal pancreatectomy (cf. Sitbon et al.

1980). Thus, while the basic role of glucagon in the triglyceride-FFA metabolism of birds, at least the species thus far studied, appears rather clear, its impact on the ketone metabolism is uncertain. Information on effects of glucagon on avian protein metabolism is limited. In addition to a gluconeogenic action in the liver (see above), a feedback mechanism with circulating amino acids appears possible. In the duck, infusion of arginine provokes an immediate release of glucagon (cf. Sitbon et al. 1980). In the goose, intravenously or orally administered alanine and arginine were equally effective in provoking secretion of pancreatic glucagon (Khemiss and Sitbon 1982). Finally, it may be noted that both stress and exercise stimulate glucagon secretion in birds (cf. Harvey et al. 1982), whereas a possible role of the hormone in thermoregulation needs further studies (Palokangas et al. 1973).

The data presented in the preceding section indicate that glucagon is the dominant hormone of avian carbohydrate and lipid metabolism, whereas its role in protein metabolism is probably less prominent. In the commonly studied domestic birds the maintenance of glycemia involves interaction with insulin (insulin: glucagon molar ratio), whereas interactions of both hormones in the control of lipemia remain uncertain. It seems that the antilipolytic effect of insulin in mammals is "replaced" by a direct inhibitory effect of somatostatin (see Chap. 13.3) in adipose tissue. Glucagon prevents fatal hypoglycemia by hepatic glycogenolytic and/or gluconeogenic functions; it provides FFA, the major substrate for muscle metabolism, via lipolysis in the adipocytes. Both hepatic glycemic and FFA-related functions apparently involve feedback mechanisms. On the other hand, no relationship between plasma ketones and glucagon could be identified. However, it must be emphasized that most of the above generalizations require confirmation from work with carnivorous and piscivorous species.

11.4.9 Conclusions

The presence of A-cells in all vertebrates above the cyclostomes suggests that pancreas-released glucagon functions in the metabolism of all gnathostomes. Perhaps, it is *the* hyperglycemic hormone for the nonstress situations, whereas epinephrine plays an overlapping role mainly in fight-or-flight responses. In all gnathostomes thus far studied, with the exception of the chondrichthyes, a hepatic glycogenolytic and/or gluconeogenic action has been demonstrated. A confirmation of the biological significance of several of these observations by use of species-specific glucagon in physiological doses, and in vivo, is desirable. Furthermore, the phylogenetic occurrence of the ketogenic action of this hormone is largely unknown. In birds, glucagon is the dominating hormone in both carbohydrate and lipid metabolism, and it plays a similarly important role in the carbohydrate metabolism of the Squamata. In both groups of homeotherms (birds and mammals), glucagon is directly involved in the control of blood sugar homeostasis via a feedback mechanism. The contribution of pancreatic-type glucagon from the gastrointestinal tract must be considered in all studies on the titers of the endogenous hormone in vivo.

Chapter 12

Pancreatic Polypeptide

12.1 Molecular Structure and Related Compounds

Pancreatic polypeptide, a 36-amino acid peptide with a molecular weight of 4250, was first isolated from chicken pancreas (Kimmel et al. 1971) and named avian pancreatic polypeptide (aPP). Subsequent preparations of the homologous hormone from bovine and human pancreas were termed bPP and hPP, respectively. With the exception of two teleosts, *Dicentrarchus labrax* (Thorpe and Duve 1985), and *Oncorhynchus kisutch* (Kimmel et al., manuscript), PP has not been detected by RIA in islets below the amphibians (Greeley et al. 1984). Immunohistological findings suggest that it occurs in the pancreas of all vertebrates above the cyclostomes (see Chap. 4.2). In their report on the primary structure of rat PP, Kimmel et al. (1984) compiled the sequences of nine PPs, from mammals, birds, and reptiles. Of the 36 amino acids, 10 are invariant, the lowest percentage of invariant residues among the four major islet hormones (Table 12.1). Additional sequence changes might then be expected in amphibian and fish PPs. Indeed, in a recent study Kimmel and coworkers (manuscript) report that the salmon pancreas contains a 36-residue peptide that is more homologous with porcine NPY (83%) and PYY (75%) than any of the previously characterized pancreatic peptides; and they mention a similar peptide also for *Lepisosteus spatula* (see below). Another homolog of PP, peptide YG, was recently identified in the PP cells of the islets of the anglerfish by Noe et al. (1986). Compared with hPP, the sequence variations of the different PPs range from two amino acids in the dog and pig to 23 in the goose. As might be predicted, alligator PP is more akin to aPP than to its mammalian homologs (Lance et al. 1984). Thus, there is extensive variation in the primary structure of PP, but with sufficient invariant residues to maintain proper conformation. Schwartz et al. (1980) identified a putative pancreatic polypeptide precursor in dog islets, the peptide having a molecular weight of 9000 and being cleaved into PP and a smaller component during synthesis; Schwartz and Tager (1981) further isolated a 20-amino acid peptide from the canine pancreas which probably represents a COOH-terminal fragment of the PP precursor.

Several investigators have reported PP-like immunoreactivity in extra-pancreatic locations, particularly the gut and nervous system. Recently, however, those reactivities (Table 12.1) have been attributed to peptide YY (PYY) and neuropeptide Y (NPY), respectively, both so named because of the single-letter designation of Y for tyrosine residues found at the amino- and carboxy-terminal ends (see review by Emson and de Quidt 1984).

Table 12.1. Comparison of the amino acid sequence of 11 peptides of the PP family. (After Kimmel et al. 1984)

					5					10					15		
Human	Ala	Pro	Leu	Glu	Pro	Val	Tyr	Pro	Gly	Asp	Asn	Ala	Thr	Pro	Glu	Gln	Met
Bovine	–	–	–	–	–	Glu	–	–	–	–	Asn	–	–	–	–	–	–
Porcine	–	–	–	–	–	–	–	–	–	–	Asp	–	–	–	–	–	–
Canine	–	–	–	–	–	–	–	–	–	–	Asp	–	–	–	–	–	–
Ovine	–	Ser	–	–	–	Glu	–	–	–	–	Asn	–	–	–	–	–	–
Rat	–	–	–	–	–	Met	–	–	–	–	Tyr	–	–	His	–	–	Arg
Turkey	Gly	–	Ser	Gln	–	Thr	–	–	–	–	Asp	–	Pro	Val	–	Asp	Leu
Goose	Gly	–	Ser	Gln	–	Thr	–	–	–	Asn	Asp	–	Pro	Val	–	Asp	Leu
Alligator	Thr	–	–	Gln	–	Lys	–	–	–	–	Gly	–	Pro	Val	–	Asp	Leu
Invariant	–	–	–	–	Pro	–	Tyr	Pro	Gly	–	–	Ala	–	–	Glu	–	–
PYY	Tyr	Pro	Ala	Lys	Pro	Glu	Ala	Pro	Gly	Glu	Asp	Ala	Ser	Pro	Glu	Glu	Leu
NPY	Tyr	Pro	Ser	Lys	Pro	Asp	Asn	Pro	Gly	Glu	Asp	Ala	Pro	Ala	Glu	Asp	Leu

12.2 PP Antigenicity, Assays, and Endogenous Titers

PP antigenicity has been very problematic in investigations using both ICC and RIA. While there is sufficient cross-reactivity between the commonly used PP antisera and most PP molecules to allow the immunocytochemical identification of PP and related compounds, the cross-reactivity is too nonspecific to allow accurate RIA determinations in heterologous systems. For example, there is less than 14 percent cross-reactivity between avian PP antisera and mammalian PP (Hazelwood 1981). The problem is further depicted by the early failure to detect PP in amphibian islets by RIA (Langslow et al. 1973), which has only recently been resolved (Greeley et al. 1984). PP could not be assayed in four teleost species despite the use of carboxy-hexapeptide-specific antiserum (Greeley et al. 1984), suggesting that amino acid differences might occur in regions of the fish molecule that are invariant in "higher" forms. Furthermore, Kimmel et al. (1984) observed that while rat PP differs from hPP and bPP by only eight amino acids there is very little reactivity of the rat hormone with antisera to the other two. They conclude that PP antigenic sites must be restricted to very short residue sequences, perhaps related to slight conformational differences. Clearly, until PP, NPY and PYY can be specifically distinguished in RIA systems, little can be said about circulating PP in most vertebrates.

Floyd and Vinik (1981) noted that the mean circulating level of PP in healthy human subjects with an average age of 25 years was 54 pg/ml. As age increased, so did PP titers, exceeding 200 pg/ml in the seventh decade. This correlates with the age-related increases in F-cell volume observed by Stefan et al. (1982a). In fetal lambs plasma PP levels were 20 pM at 105 days gestation (145 day gestation period), increasing to adult levels of >200 pM one week post partum (Shulkes and Hardy 1982). Fasting levels of chicken PP are 2–3 ng/ml, or roughly 40 times values in mammals (Kimmel et al. 1978). In contrast to insulin and glucagon, PP is eliminated from the circulation after pancreatectomy in chickens (Colca and

		20					25					30					35	
Ala	Gln	Tyr	Ala	Ala	Asp	Leu	Arg	Arg	Tyr	Ile	Asn	Met	Leu	Thr	Arg	Pro	Arg	Tyr
–	–	–	–	–	Glu	–	–	–	–	–	–	–	–	–	–	–	–	–
–	–	–	–	–	Glu	–	–	–	–	–	–	–	–	–	–	–	–	–
–	–	–	–	–	Glu	–	–	–	–	–	–	–	–	–	–	–	–	–
–	–	–	–	–	Glu	–	–	–	–	–	–	–	–	–	–	–	–	–
–	–	–	Glu	Thr	Gln	–	–	–	–	–	–	Thr	–	–	–	–	–	–
Ile	Arg	Phe	Tyr	Asn	–	–	Gln	Gln	–	Leu	–	Val	Val	–	–	His	–	–
?	Arg	Phe	Tyr	Asp	Asn	–	Gln	Gln	–	Arg	Leu	Val	Val	Phe	–	His	–	–
Ile	–	Phe	Tyr	Asn	–	–	Gln	Gln	–	Leu	–	Val	Val	–	–	–	–	Phe
–	–	–	–	–	–	Leu	–	–	Tyr	–	–	–	–	–	Arg	–	Arg	–
Ser	Arg	Tyr	Tyr	Ala	Ser	Leu	Arg	His	Tyr	Leu	Asn	Leu	Val	Thr	Arg	Gln	Arg	Tyr
Ala	Arg	Tyr	Tyr	Ser	Ala	Leu	Arg	His	Tyr	Ile	Asn	Leu	Ile	Thr	Arg	Gln	Arg	Tyr

Hazelwood 1982). In mammals and birds, the rapid plasma clearance rate of PP is within the range of many other peptide hormones (Pollock and Kimmel 1981); it appears that the kidneys play a major role in this process (Adamo et al. 1983). No data on circulating PP are yet available for other vertebrate classes.

12.3 The Physiology of Islet PP

Despite considerable efforts during the last decade, the precise functions of PP have not been elucidated. The major difficulties appear to be due to the following: (1) PP release in mammals is probably regulated by a combination of neural, hormonal and metabolic signals with critical, highly complex interactions; (2) interspecific differences are considerable; (3) data obtained in vivo and in vitro are often conflicting; (4) bolus injections and infusions of the same secretagogue give different results; (5) oral and intravascular applications of secretagogues also give different results. Most of the pertinent data have been reviewed or summarized by Floyd et al. (1977), Chance et al. (1979), Hazelwood (1981, 1984), Floyd and Vinik (1981). In man, perhaps the best-studied species, PP is released in response to a meal in a biphasic manner. The first, "cephalic" phase, which seems to be almost or completely under neural (vagal, cholinergic) control, occurs within minutes. A second phase occurs later and may last for hours. It seems to be both neurally and hormonally controlled. The first phase can be elicited by gastric distention and sham feeding, while the total biphasic response occurs after meals rich in proteins or fats. Carbohydrate meals seem to be less effective, although oral glucose is a strong stimulus for PP release, and this response is exaggerated in diabetics. On the other hand, both bolus injections and i. v. infusions of glucose cause a significant decrease of plasma PP. The picture that seems to emerge is that under physiological conditions PP release is strongly affected by nutrients within the gastrointestinal tract, which here act on both afferent neurons and endocrine cells. In contrast to this situation in man, the intact dog responds to a meal with a slow progressive increment of PP release with a maximum after 60–90 min (Tasaka et al. 1984). However, intravenous application of amino acids or glucose causes a decline of circulating PP in the intact dog, while preceding vagotomy reverses this effect (A. Inui et al. 1986); and to add to the confusion, the perfused, isolated canine pancreas shows a biphasic PP response to arginine (Hermansen and Schwartz 1979; Weir et al. 1979). In the chicken, feeding elicits a biphasic PP response similar to that in the human (Hazelwood 1980, 1984). This response occurs after feeding of protein, fat, or carbohydrate diet (Kimmel et al. 1978), while the most effective nutrient stimulus in vitro is a mixture of amino acids in relative proportion to that of commercial chicken feed (Hazelwood 1980). Leucine, isoleucine, phenylalanine and valine, which stimulate PP release from perifused fragments of chicken pancreas, are ineffective in vivo (Colca and Hazelwood 1982). In the same system, volatile fatty acids did not affect PP release while change from a 1-mM linoleate medium to control medium caused an increased PP release as an off-response (Colca and Hazelwood 1981). At this point, it appears that the stimulation of proventricular secretions is a major function of PP in the chicken. However, receptor sites in gut and pancreas suggest additional gastrointestinal functions (Kimmel and Pollock 1981), while specific binding of aPP to membrane preparations of the avian cerebellum (Adamo et al. 1983) could point to a further target tissue (for discussion, see Adamo and Hazelwood 1984).

The biological effects of plasma PP in man appear paradoxical, since they may involve inhibition of both contractions of gall bladder and secretions by exocrine pancreas. Obviously, this is difficult to reconcile with the stimulatory effects on the gastric secretions, which could be physiological, since a pure PP tumor was found to be associated with peptic ulcer (Bordi et al. 1978). Another possible role of PP is indicated by its stimulation of net fluid uptake from rat small intestine (Mitchenere et al. 1981). In addition to the stimulation of gastrointestinal functions, avian PP seems also to have metabolic functions. In young chickens, it has been found to be glycogenolytic (without being hyperglycemic) and glycerolemic. In nonfasted chickens, it stimulates synthesis of liver lipids with subsequent transfer of these lipids to depot sites, resulting in transient hypertriglyceridemia. These findings are in discord with reports on the absence of PP receptors on chicken hepatocytes (cf. Cramb et al. 1982a, b). However, they are compatible with the report of Adamo et al. (1983) on very specific, but low affinity binding of PP to chicken hepatocyte membranes. Considering the high "portal" concentrations of PP arriving at the liver, low affinity may be no obstacle to a physiological PP action on the avian liver. On the other hand, in fasted chickens PP injections cause a reduction of plasma levels of free fatty acids and glycerol, an observation in agreement with an antilipolytic effect at the isolated adipocyte (cf. Hazelwood 1980). No information on the role of PP in nonhomeotherms seems to exist with the exception of one in vitro study on the islets of a teleost (*Dicentrarchus labrax*). As in mammals, glucose was without effect while arginine stimulated PP release (Thorpe and Duve 1985).

12.4 Conclusions

So far, the study of the physiology of PP has been a most frustrating task. The known or suspected reasons for this are manifold: great interspecific variations of the molecule, together with a high "unspecific" antigenicity (see also Chap. 1); highly complex control of PP release in the few species so far studied; great structural and physiological variations of its presumed major target system, the digestive tract, even among mammals (e.g., human, dog, rat, ruminant); absolute ignorance as to what might constitute a reasonably basic model for pertinent studies. Perhaps, a breakthrough in the understanding of the functions of islet PP may require a comparative approach, using "lower" vertebrates, or its study in a mammal without islet innervation, such as *Acomys cahirinus* (see Chap. 7). At this point, it appears that islet PP is primarily involved in the control of digestive functions, particularly in the release of gastric secretions.

Chapter 13

Somatostatin

Somatostatin, which occurs in the islet organ of all vertebrates thus far studied (for literature see Dupé-Godet and Adjovi 1983; Conlon et al. 1985a; Plisetskaya et al. 1986c) with the probable exception of larval lampreys, was the last of the major islet hormones to be discovered. However, there is already an enormous volume of pertinent data, most of which are compiled in recent reviews by Arimura (1981), Gerich (1981, 1983a), Bethge et al. (1982) and Reichlin (1983a, b). A striking feature of both insular and extra-insular somatostatin is its almost uniformly inhibitory effect, which contrasts with its wide distribution as well as with the multitude of pathways by which it achieves this effect. The latter include: (probable) autocrine inhibition; neurotransmission and neuromodulation; cytotransmission; paracrine secretion; functions as portal hormone, telehormone, and CSF hormone, as well as "lumone" (exocrine hormone) activity. One wonders if a few reports on stimulation by somatostatin can also ultimately be explained by inhibitory mechanisms. According to Bethge et al. (1982), somatostatin causes histamine release from mast cells in vitro, and Szabo (1983) suggests that it activates von Kupffer cells and possibly other macrophages as well. Thus, it remains to be seen if the term "panhibin", as proposed by McCann et al. (1980) will be more appropriate than the restrictive names "somatostatin" or "SRIF" (somatotropin release inhibiting factor). On the other hand, the molecular heterogeneity of somatostatin may play an important role in the modifications of its effects since the receptors for the different forms of somatostatin seem to vary (see Chap. 9.1.1 and below). Furthermore, an apparently rapid disappearance rate (1–2 min in mammals: Arimura 1981) may help to restrict the effects of somatostatin, when released in smaller quantities, to nearby targets.

13.1 Molecular Structure and Related Compounds

The first form of somatostatin to be identified was a tetradeca-peptide (S-14) with a molecular weight of 1639 from the ovine hypothalamus (Brazeau et al. 1973). Meanwhile, additional molecular forms (cf. Plisetskaya et al. 1986c; Varndell et al. 1986) have been detected (Table 13.1), and the sequences of the somatostatin precursors (Fig. 13.1) have been deduced from the nucleotide sequence of the cloned somatostatin cDNAs for anglerfish (*Lophius americanus*), catfish (*Ictalurus punctatus*), rat, and human (cf. Shen and Rutter 1984). Disagreement exists as to the precise synthetic pathways of the two forms so far clearly established for mammals (S-14 and S-28). Both are probably products of a 10,348-dalton pro-

```
                                                   -102     -100
                                                   met  leu  ser
UAGAGUUUGACCAGCCACUCUCCAGCUCGGCUUUCGCGGCGCCGAG     AUG  CUG  UCC

                                   -90
cys  arg  leu  gln  cys  ala  leu  ala  ala  leu  ser  ile  val  leu  ala
UGC  CGC  CUC  CAG  UGC  GCG  CUG  GCU  GCG  CUG  UCC  AUC  GUC  CUG  GCC

               -80                                                 -70
leu  gly  cys  val  thr  gly  ala  pro  ser  asp  pro  arg  leu  arg  gln
CUG  GGC  UGU  GUC  ACC  GGC  GCU  CCC  UCG  GAC  CCC  AGA  CUC  CGU  CAG

                                             -60
phe  leu  gln  lys  ser  leu  ala  ala  ala  ala  gly  lys  gln  glu  leu
UUU  CUG  CAG  AAG  UCC  CUG  GCU  GCU  GCC  GCG  GGG  AAG  CAG  GAA  CUG

               -50                                                 -40
ala  lys  tyr  phe  leu  ala  glu  leu  leu  ser  glu  pro  asn  gln  thr
GCC  AAG  UAC  UUC  UUG  GCA  GAG  CUG  CUG  UCU  GAA  CCC  AAC  CAG  ACG

                                             -30
glu  asn  asp  ala  leu  glu  pro  glu  asp  leu  ser  gln  ala  ala  glu
GAG  AAU  GAU  GCC  CUG  GAA  CCU  GAA  GAU  CUG  UCC  CAG  GCU  GCU  GAG

               -20                                                 -10
gln  asp  glu  met  arg  leu  glu  leu  gln  arg  ser  ala  asn  ser  asn
CAG  GAU  GAA  AUG  AGG  CUU  GAG  CUG  CAG  AGA  UCU  GCU  AAC  UCA  AAC

                                              1
pro  ala  met  ala  pro  arg  glu  arg  lys  ala  gly  cys  lys  asn  phe
CCG  GCU  AUG  GCA  CCC  CGA  GAA  CGC  AAA  GCU  GGC  UGC  AAG  AAU  UUC

               10                  14
phe  trp  lys  thr  phe  thr  ser  cys  AM
UUC  UGG  AAG  ACU  UUC  ACA  UCC  UGU  UAG  CUUUCUUAACUAGUAUUGUCCAUA

UCAGACCUCUGAUCCCUCGCCCCCACACCCCAUCUCUCUUCCCUAAUCCUCCAAGUCUUC

AGCGAGACCCUUGCAUUAGAAACUGAAAACUGUAAAUACAAAAUAAAAUUAUGGUGAAAU

UAU(A)n
```

Fig. 13.1. The sequence of human preprosomatostatin mRNA deduced from that of the composite sequences of the cDNA clones. The predicted amino acid sequence is indicated and numbered by designating the first residue of somatostatin-14 as 1. The amino acids toward the NH_2 terminus and COOH terminus of the peptide are numbered negatively and positively, respectively. A possible signal peptide extends from Met at position −102 to Gly at −79; the propeptide extends from Ala at −78 to Lys at −1; and somatostatin-28 extends from Ser at −14 to Cys at 14 (Shen et al. 1982)

peptide (Shen et al. 1982); furthermore it appears that at least in some cases, S-28 is the precursor of S-14 (Zingg and Patel 1983; Gluschankof et al. 1984). However, there is also evidence that the two forms arise independently (Patzelt et al. 1980; Wu et al. 1983); consistent therewith, the presence of three cDNA sequences coding for mRNA and somatostatin-like peptides in anglerfish islets (Goodman et al. 1980a, b; Hobart et al. 1980; Shields 1980) suggests that S-14 and S-28 could be produced from two different prohormones. The islets and intestine of the catfish (*Ictalurus punctatus*) contain S-14, but, predominantly, a related 22-residue peptide (S-22), which may be restricted to teleosts (Fletcher et al. 1983; Andrews et al. 1984). In rat assay systems, S-22 has the same activity spectrum as S-14, but reduced potency (Oyama et al. 1981). The islets of another teleost, the coho salmon (*Oncorhynchus kisutch*) contain, besides "typical" S-14 (S-14-I) a second type of S-14 (S-14-II) and large quantities of a S-25; S-14-II has an amino acid composition identical to the 14 amino acids of the C-terminal of S-25 (Plisetskaya

Table 13.1. The primary structures of the currently known islet somatostatins. (Compiled from various sources; data on salmon somatostatins from Plisetskaya et al. 1986c)

	"Pro-chain"														S-14-I													
	−14		−12		−10		−8		−6		−4		−2			2		4		6		8		10		12		14
S-14-I (mammals; pigeon; salmon, catfish, anglerfish; *Torpedo*)															Ala	Gly	Cys	Lys	Asn	Phe	Phe	Trp	Lys	Thr	Phe	Thr	Ser	Cys
S-22 (catfish)							Asp	Asn	Thr	Val	–	Ser	Lys	Pro	Leu	Ala	–	Met	–	Tyr	–	–	–	Ser	Ser	–	Ala	Cys
S-25 (salmon)					Ser	Val	Asp	Asn	Asp	–	Pro	Gln	Glu	–	–	–	–	–	–	–	Tyr	–	–	Gly	–	–	–	–
S-28-I (anglerfish)	Ala	–	Ser	Gly	Gly	–	Leu	Leu	–	–	–	–	–	–	–	–	–	–	–	–	–	–	–	–	–	–	–	–
S-28-II (anglerfish)	–	Val	Asp	–	Thr	Asn	Asn	Leu	Pro	–	–	–	–	–	–	–	–	–	–	–	Tyr	–	–	Gly	–	–	–	–
S-28 (rat, pig, human)	Ser	Ala	Asn	Ser	Asn	Pro	Ala	Met	Ala	Pro	Arg	Glu	Arg	Lys	–	–	–	–	–	–	–	–	–	–	–	–	–	–

et al. 1986c). In contrast to the variability of somatostatin in the teleosts is the report of Conlon et al. (1985a) on an elasmobranch (*Torpedo marmorata*). In the pancreas, various brain and gastrointestinal regions of this species, the authors found only typical S-14. Considering the unusual cellular heterogeneity of the elasmobranch islet organ (see Chap. 4.2) this observation is surprising. A probable somatostatin homolog, the dodecapeptide urotensin II has been found in several molecular forms in the urophysis of teleosts and elasmobranchs. Here, it is released as a caudal neurosecretion; cardiovascular, osmoregulatory, and reproductive functions have been attributed to it (cf. Ichikawa et al. 1984; Bern et al. 1985; Owada et al. 1985).

The histologic distribution of the different forms of somatostatin varies. In the digestive tract of cat and man (Chayvialle et al. 1980; Penman et al. 1980), S-14 predominates in the antrum, duodenum, pancreas, and external intestinal muscle layer (myenteric neurons?), while S-28 prevails in the ileal and colonic mucosa. In the rat, there are three forms of somatostatin immunoreactivity in the portal vein blood, and in tissues. The two smaller forms of these are probably identical with S-14 and S-28. Pancreas and stomach of the rat contain mainly presumptive S-14, while the larger forms prevail in the intestine (Patel et al. 1981). Baskin and Ensinck (1984) suggest that the intestine of this species is a major source of circulating S-28. In guinea pig tissues Conlon (1984b) found only one form of somatostatin (S-14). In the duck, S-14 predominates in the pancreas, while S-28 predominates in the gut (DiScala-Guenot et al. 1984). The intrasplenic islet of a lizard (*Varanus exanthematicus*) seems to produce only a large somatostatin, possibly identical with S-28, while blood from the pancreas proper (other islets?) and from the abdominal organs contains both S-28- and S-14-like moieties (Dupé-Godet 1984). In the islet and intestine of catfish (*Ictalurus punctatus*), S-22 prevails over S-14, while in the brain the ratio is reversed; in the islet of the anglerfish (*Lophius americanus*) almost all somatostatin is S-14, with a very small admixture of S-22 (Fletcher et al. 1983). Immunostaining of salmonid islets with homologous antisera showed that S-25 prevails over S-14-I (M. Nozaki, A. Gorbman and E. Plisetskaya, pers. commun.). Considering the presence of more than one circulating somatostatin moiety (Baldissera et al. 1983; cf. Reichlin 1983b; Gerich 1983a; Dupé-Godet 1984), it is interesting that tissue binding may involve receptors specific for different types of somatostatin (cf. Tran et al. 1985). Taken together with the differing effects of S-14 and S-28 on target organs, these observations make it clear that the identification of both origin and action spectra of the individual somatostatins is of great urgency.

Both mammalian S-14 and S-28, as well as a number of homologs, have been synthesized (cf. Gerich 1983a; Marco et al. 1983; Taparel et al. 1985). Unfortunately, the naturally occurring forms of somatostatin appear to be of limited use as therapeutic tools because of their short half-lives (see above). In particular, their application in the treatment of diabetes mellitus (Type I) appears to hold little promise (Long 1983). However, in the treatment of acute gastrointestinal disorders, encouraging results have been reported. Thus, large quantities of exogenous somatostatin curtail hemorrhage by gastric ulcers in the majority of patients, probably by lowering of blood flow to the stomach and suppression of secretion of acid and gastrin. Furthermore, somatostatin-induced suppression of exocrine pancreas secretion seems to improve considerably the survival of patients with pancreatic surgery (cf. Reichlin 1983b). On the other hand, use of long-acting somatostatin analogs with specific inhibitory effects may soon lead to spectacular successes in the treatment of hyperfunctional or tumoral tissues (cf. Long 1983). Recently, Redding and Schally (1984) found that [L-5-Br-Trp^8] somatostatin decreased weight and volume of two different types of pancreatic carcinomas in rodents. It is hoped that the development of long-acting somatostatin analogs with preferential effects on individual types of endocrine cells (e.g., insular A- or B-cells) will provide clinicians and reseachers with highly specific tools for "chemo-ectomies" in the near future (cf. Gerich 1983a).

13.2 Antigenicity, Assays, and Endogenous Titers

Because immunological measurement of somatostatin is fraught with great difficulties and pitfalls, estimates of plasma levels of somatostatin vary among laboratories. The difficulties are partly due to heterogeneity of circulating somatostatins, the binding of most of circulating S-14 to plasma proteins, rapidly acting, degrading plasma enzymes, and incomplete recovery during extraction procedures (for literature and discussion, see Baldissera et al. 1983; Gerich 1983a; Reichlin 1983b; Dupé-Godet 1984). However, it is clear that there is a strong transhepatic gradient of plasma somatostatin, at least in mammals and a lizard, *Varanus exanthematicus* (Dupé-Godet 1984). In mammals, the highest titers seem to occur in hypophysial portal blood where, contrary to the situation in peripheral blood, they are affected by anesthetics (Chihara et al. 1979). On the other hand, no significant gradient among titers in jugular, hepatic, and abdominal veins of *Varanus exanthematicus* was found (Dupé-Godet 1984). The somatostatin titers in the peripheral plasma of normal mammals (man, dog, rat) are usually well below 500 pg/ml, but in patients with somatostatinoma, they have been reported to reach 3000–25000 ng/ml (cf. Reichlin 1983b). In the chicken, titers are 200–400% higher than in normal mammals (Hazelwood 1984; Cieslak and Hazelwood 1986a, b); in *V. exanthematicus*, they are approximately 1100 pg/ml. Very high plasma titers (3–4 ng/ml) have been estimated for the channel catfish, *Ictalurus punctatus* (Ronner, pers. commun.). Probably they are due to S-14 and/or S-28, since S-22 does not cross-react with antibodies to S-14 (Ronner and Scarpa 1984). In a homologous RIA system for catfish S-22, the sensitivity for this hormone was at least 1000-fold greater than to S-14-I (Oyama et al. 1982). Similarly, salmon S-25 seems to cross-react very poorly with homologous S-14-I (Plisetskaya et al. 1986c). From the above compilation, it seems that the number of insular D-cells (see Chap. 4) may be reflected in the titers of circulating somatostatin immunoreactivity. However, this possibility, even if confirmed by hard, comparable data, remains of limited significance until we understand the role of the various forms of plasma somatostatins.

13.3 The Physiology of Islet Somatostatin

The absence of somatostatin cells in the islets of larval lampreys (see Chap. 4.1) indicates that they do not need *islet* somatostatin at all, so that one might speculate that this is related to the relatively constant food supply of these filter feeders. In adult lampreys, the intrainsular distribution of B- and somatostatin cells in largely different follicles (see Chap. 4.1) indicates that paracrine interactions between somatostatin and insulin may not exist. Considering the very small number of somatostatin cells in the islet organ of the *Myxine glutinosa*, one wonders if the islet fraction of this hormone has any function at all in hagfishes; on the other hand, the biological significance of an inhibitory effect of very high doses of somatostatin on insulin release from islets of *Eptatretus* species in vitro remains to be proven (Stewart et al. 1978). An increase of FFA, after somatostatin injections into adult lampreys (Plisetskaya 1980), appears compatible with a suppression of insulin release. However, because this effect was only seen in January, and was absent in November, it may well have been due to an insulin-independent mechanism. It is easy to imagine that the irregular, but at times heavy food intake of the parasitic lampreys and scavenging hagfishes requires a more sophisticated control system of digestive and metabolic actions than that present in the ammocoete. This, perhaps, called for an islet hormone that counteracts excessive postprandial increases of *circulating* metabolites, a role for which the inhibitor somatostatin (at the intestinal level?) would be perfectly suited (see below). Almost no experimental data on somatostatin in the Chondrichthyes seem to exist. In the perfused rectal salt gland of *Squalus acanthias*, S-14 inhibited VIP-stimulated secretion (Silva et al. 1985). Among the Osteichthyes, the functions of somatostatin have been studied both with respect to pituitary control and "peripheral" actions in a few teleosts. In vitro, mammalian S-14 and S-28 inhibit growth hormone release from the goldfish pituitary (Marchant and Peter 1985), while catfish S-22 was ineffective. When compared with the impact of S-14-I on rat tissues, catfish S-22 had a reduced inhibitory action on release of growth hormone, insulin, and glucagon (Oyama et al. 1981). Paradoxically, despite substitutions in positions 21 and 24 (sites important for growth hormone inhibition: Morel et al. 1984) anglerfish S-28-II reduced strongly growth hormone release from rat pituitary in vitro (Noe et al. 1979). In juvenile coho salmon, low doses of homologous S-25 (4 μg/kg body weight) caused a transitory (2-h) decrease in plasma insulin, but increased plasma FFA, liver lipase activity and hepatic glycogen (Plisetskaya et al. 1986c). It is noteworthy that the effect on liver lipid metabolism also occurs in vitro (Sheridan et al. 1985), a reminder of the never-settled question of an islet hormone with a specific lipolytic effect on the liver (cf. Epple 1965, 1968b; see also Chap. 14). When given to unanesthetized cardiac-cannulated American eel, somatostatin (S-14) caused hyperglycemia. However, the effect varied with the dose. Low doses (0.1 μg/kg) were followed by a modest hyperglycemia after 160 min; higher doses up to 200 μg/kg were without effect, but an excessive dose (1 mg/kg) caused a transitory hyperglycemia after 20 min. Furthermore, antisomatostatin had essentially the same hyperglycemic effect as the low dose of the hormone (Epple et al. 1983). The interpretation of these data is difficult. The effect of the low dose is probably physiological and, perhaps, mediated via suppression of insulin release. The lack of glycemic response to higher doses could be due to a simultaneous inhibition of release of both insulin and glucagon. The rather rapidly developing hyperglycemia following excessive dose of somatostatin possibly reflects a direct interference with neuronal pathways. Hyperglycemia after antisomatostatin injections could be the net result of interferences at several (insular and extrainsular) antagonistic systems affecting glycemia; alternatively, it could be an unspecific effect of the antibody used (Plisetskaya, pers. commun.). In the coho salmon (*Oncorhynchus kisutch*), anti-S-25 caused an elevation of plasma insulin and an increase in hepatic glycogen (Plisetskaya et al. 1986c). In another teleost (*Heterochromis bimaculatus*) astronomical doses of S-14 (up to 12 mg/kg body weight!) evoked *hypoglycemia* and behavioral changes (Christ 1985). These studies demonstrate vividly the difficulties in interpretation of exogenous manipulations of plasma somatostatin.

In the perfused pancreas of the European silver eel (*Anguilla anguilla*) somatostatin inhibited amino acid-stimulated insulin secretion (Ince 1980). In the same system, perfusion with a high physiological dose of glucose (8.3 mM) was without an effect on somatostatin secretion, whereas higher doses (16.7 and 33.3 mM) caused a dose-dependent biphasic somatostatin release, with a rapid return to control levels after an acute change to baseline concentrations of glucose (Ince and So 1984). If our experience with American eels (*A. rostrata*) can be applied to the European species, blood glucose values above 13.5 mM are only encountered under extreme experimental conditions, such as prevail immediately after pancreatectomy (Lewis and Epple 1972; Epple and Lewis 1977) and excessive doses of epinephrine (Epple and Nibbio 1985). Hence, the physiological significance of the glucose-induced somatostatin release from the perfused eel pancreas appears uncertain. Perhaps the threshold of the D-cell for glucose in the unanesthetized, intact eel is lower than in the pancreas preparation. In a perfused Brockmann body preparation of the catfish (*Ictalurus punctatus*) Ronner and Scarpa (1982) observed a clear biphasic somatostatin secretion in response to 10 mM glucose, whereas arginine caused only a small release. In a further study, these authors found that half-maximal somatostatin release occurred at a concentration of 5 mM glucose, whereas higher concentrations (9 mM) were needed to stimulate equivalent insulin secretion (Ronner and Scarpa 1984). Based on data of Ronner and Scarpa (1982), as well as of their own, Ince and So (1984) suggest that rapid release of somatostatin in response to glucose in teleosts may be "...the main determining factor in the magnitude of insulin secretion, and possibly of glucagon secretion, and hence their net physiological effect in peripheral tissues." However, they also recognize that this postulated effect of islet somatostatin is certain to be modified in vivo by neural and gastrointestinal factors. At first glance, the rapid responses (within a few minutes) in the perfusion studies of Ince (1980), Ronner and Scarpa (1982) and Ince and So (1984) appear to be discordant with the late metabolic responses after somatostatin injections in intact fishes (Plisetskaya 1980; Epple et al. 1983). Since such a late effect is also present after insulin and glucagon injections (see Chaps. 10.4 and 11.4), this discrepancy may be due to intrinsic differences in the response time between the islet cells, and the target tissues of islet hormones of poikilotherms. If so, this would parallel the situation with the catecholaminergic systems of poikilotherms (at least of fishes), in which epinephrine stimulates immediate release of other catecholamines as well as respiration, whereas its metabolic effects are expressed much more slowly (Dashow and Epple 1983; Macey et al. 1984; Epple and Nibbio 1985). No islet-related data on the effects of circulating somatostatin in amphibians appear available. However, the anuran pancreas contains very high quantities of somatostatin-like immunoreactivity (Vale et al. 1976; Falkmer et al. 1978; Tomita and Pollock 1981). In reptiles, the studies of Dupé-Godet (1984) on *Varanus exanthematicus* have laid promising foundations for further work. The presence of high titers of at least two circulating somatostatins of pancreatic origin, differing sites of their production (intrasplenic islet vs. pancreas proper; see Sects. 13.1 and 13.2), plus sufficient size of *V. exanthematicus* for repeated blood sampling, suggest that this species may provide an useful reptilian model for islet research in general. The high somatostatin content in the pancreas of *Alligator mississippiensis* (Lance et al. 1984) may indicate high plasma levels similar to *V. exanthematicus* and birds.

The peculiar islet structure of birds should make possible independent local interactions between somatostatin and glucagon, and somatostatin and insulin, respectively (see Chap. 14). Therefore, one wonders how the very high pancreatic and circulating concentrations of this peptide (Vale et al. 1976; Weir et al. 1976; Hazelwood 1984) fit into this picture. Strosser et al. (1983a, b) have shown that S-14 and S-28 are equipotent antilipolytic hormones in the duck, and that they act directly on the adipocytes. In the intact duck, the infusion of S-14 stimulates glucagon release indirectly via interference in the glucagon-FFA feedback system, while S-28 may inhibit glucagon release directly (Strosser et al. 1984). The increase in glucagon is insufficient to inhibit the decrease in FFA, but it prevents the hypoglycemia described for pancreatectomized ducks. S-28, however, causes a strong decrease in plasma glucose, possibly by suppression of glucagon release (Strosser et al. 1984). On the other hand, both exogenous S-14 and S-28 inhibit insulin secretion in this species (Sitbon et al. 1980; Strosser et al. 1984). In the isolated perfused chicken pancreas, somatostatin continuously inhibits basal rates of secretion of glucagon and insulin; moreover, it appears that the suppression of glucagon secretion by high glucose concentrations (42 mM) is at least partially dependent on local somatostatin release (Honey et al. 1980, 1981). In studies on 99% pancreatectomized chickens, Cieslak and Hazelwood (1986b) found no evidence of stimulation of glucagon release by somatostatin. Using organ culture Foltzer et al. (1982, 1983) confirmed an inhibitory effect of somatostatin also for the glucose-stimulated insulin release by the embryonic and post-hatching chicken pancreas, but found no effect of somatostatin on glucagon release in the embryonic pancreas. Thus, while the results of the investigations agree on a direct inhibitory effect of somatostatin on avian insulin secretion, the discrepancy between the reported somatostatin effects on glucagon release in duck and chicken remains to be explained. Nevertheless, it appears that pancreatic somatostatin could play an essential role in the regulation of the appropriate insulin/glucagon ratio in blood (Hazelwood 1984).

In mammals there is extensive information in possible actions of islet somatostatin. Unfortunately, most of these are based on injections of very high, probably unphysiological doses. Pertinent data on release of endogenous somatostatin are limited; a clear separation between insular and gastrointestinal fractions has been achieved only in a few studies. Pancreatic somatostatin is released following gastrointestinal loading with protein, fat, or glucose, and after infusions of glucose and amino acids. However, the gastrointestinal D-cells respond only to luminal nutrients, as might be expected from "open" cells (for details and discussion, see Schusdziarra 1980). Intra-insular function is suggested by the presence of somatostatin receptors on A-, B- and D-cells (Patel et al. 1982), so that it now seems that very small increases in somatostatin entering the islet arteries (10 or 20% of the basal concentrations leaving the islets) effectively inhibit secretion of both insulin and glucagon in the dog (Kawai et al. 1982b; Unger 1983). The implication is that telehormonal rather than portal-type or paracrine somatostatin controls canine islet activities. However, the possible existence of different types of somatostatin receptors on A- and B-cells (cf. Gerich 1983a) may modulate these effects; at least in the rat, somatostatin seems to inhibit preferentially the release of insulin (Tannenbaum et al. 1982). The liver may be the first target of the high somatostatin concentrations leaving the islets (Gerich 1981). Evidence for several possible somatostatin actions on this organ is equivocal (see e.g., Bone et al. 1984), although it may have a direct impact on hepatic lipid metabolism (Catalán et al. 1984). Since somatostatin can alter the hepatic extraction rates of insulin and glucagon (Ishida et al. 1980), the liver could be the site where somatostatin(s) profoundly affect(s) intermediary metabolism by changes in the ratio of circulating insulin: glucagon. Furthermore, in concert with gastrointestinal fractions, islet somatostatin may control the hepatic function by decreasing the blood flow through the hepatic portal vein (Jaspan et al. 1979). The possible list of posthepatic (telehormonal) functions that could involve islet somatostatin is considerable (Arimura 1981; Gerich 1981, 1983a; Bethge et al. 1982; Reichlin 1983a, b). In concert with a varying portion of gastrointestinal fractions, islet somatostatin may have two main targets in mammals: (1) the islet organ, on which it acts via negative feedback, and (2) other tissues of the digestive tract. In the latter, six major functional realms seem to be affected by inhibitory actions of somatostatin (Gerich 1981, 1983; Reichlin 1983b): gut hormone secretion; exocrine secretions of stomach, intestine, liver, and pancreas; motor activity of the digestive tube and gall bladder; absorption; blood flow; cell proliferation. Thus, as a part of a dual hormonal control system of gastrointestinal and metabolic functions (Epple et al. 1980), and in concert with gastrointestinal somatostatins, islet somatostatin may indeed be a major regulator of nutrient homeostasis (Schusdziarra 1980; Schusdziarra et al. 1985). This suggestion is supported by the release of islet somatostatin in response to increases in plasma glucose, amino acids, and fats (cf. Schusdziarra 1980).

13.4 Conclusions

The islet organ of all vertebrates, with the exception of the larval lamprey, seems to produce somatostatin. It is likely that the molecular heterogeneity of circulating somatostatin already known extends to most taxa, and that it reflects to some extent varying functional spectra. However, one wonders why the structure of S-14-I has been retained so stringently in all taxa so far studied, including the guinea pig, a species with unusual insulin, glucagon (see Chaps. 10 and 11), and other gastrointestinal hormones (cf. Huang et al. 1986) and proteins as well (Wriston 1984). In general, the plasma titers and pancreatic content of somatostatin seem to be higher in "lower" gnathostomes than in mammals, but the significance of this phenomenon is obscure. It is clear that the impact of islet somatostatin, at least on extra-insular targets, can only be interpreted in conjunction with the functions of gastrointestinal somatostatins. There is good evidence to believe that insular and gastrointestinal somatostatins jointly form a functional system, which participates in the control of the nutrient homeostasis via inhibitory actions. By an integrated response (release) to both gastrointestinal (luminal) and blood-borne stimuli, the circulating somatostatins seem to constitute the antagonists to stimulators of both digestion and increases in circulating metabolites. Perhaps PP and gastrin will prove to be the major antagonistic principles. No reasonably complete picture can emerge until the significance and functional spectra of circulating somatostatins outside the digestive system (e.g., effects on adipose tissue) are understood. It is not known if there are interactions among hypophysio-portal and systemic somatostatins in any taxon. While the contribution of hypophysio-portal somatostatin to the total pool of circulating somatostatins may be too small to have any impact beyond the anterior pituitary gland, the high local, portal titers (controlling the pituitary gland) may safeguard against interference by somatostatins from extrahypothalamic sources.

Chapter 14

Summary and Perspectives

Considered for decades an unique gland with an unique secretion (insulin), the islet organ has lost, in recent years, its special status. In the gnathostomes it is now recognized as a complex endocrine gland that produces four or more "major" hormones, and a varying number of co-secretions. None of these hormones and co-secretions appear to be specific for the islet organ; rather, the major islet secretions (or closely related messenger substances) existed long before the islet organ evolved from "open" mucosa cells. When, apparently in parallel with the complex organization of the vertebrates, the "need" for new endocrine glands developed, the originally "open" insulin cells shifted from the gut to a site ideally suited for the monitoring of the systemic blood. Here, with an arterial blood supply and a location *downstream* from the protopancreas, the insulin cells (now "closed" B-cells) are exposed to the metabolites in the same concentrations as they arrive at their tissue sites of uptake (Fig. 2.6). Teleologically, this arrangement is rational because it allows a reinforcement and coordination of messages from open mucosa cells, released by luminal stimuli, and the message from the B-cells, released by blood-borne stimuli. It is likely that before the ammocoete stage, there was a situation in which B-cells acted in concert with "open" insulin cells of the gut. Since gut cells producing glucagon, PP, somatostatin (or closely related peptides) still reside within the gastrointestinal mucosa of "higher" vertebrates (for literature see Chap. 2.2), we may assume that the corresponding islet cells left the gut for the same reason as the B-cells: reinforcement of, and/or coordination with, gut messages. Therefore, we proposed to interpret the islets and open gut endocrines as members of a dual control system of digestive and metabolic functions (Epple et al. 1980), consisting of agonists and antagonists (see Fig. 14.1).

Considering the enormous variations in digestive tract and nutrition, sometimes even among closely related vertebrate species, one would certainly expect differences in islet structure, composition, and innervation. However, the degree of the structural and functional plasticity of the islet organ is truly amazing, although great phylogenetic variability seems to be a feature of the APUD-type endocrines in general (Epple and Brinn 1980). Many examples of this plasticity have been pointed out in previous chapters. Perhaps, some of the most notable are the X-cells of holocephalians, the flame-shaped cells of *Lepisosteus osseus*, the innervation of the teleost islets, and the lack of islet innervation in the spiny mouse. Why do sluggish predators of the deep, as the holocephalians are basically, produce an unusual hormone in 50 percent of their islet cells? What is the nature and role of another probably unique hormone of the common flame-shaped cells of *L. osseus*, a pike-like predator close to the teleosts? Considering that scores of chondrichthyans and thousands of actinopterygians with life styles

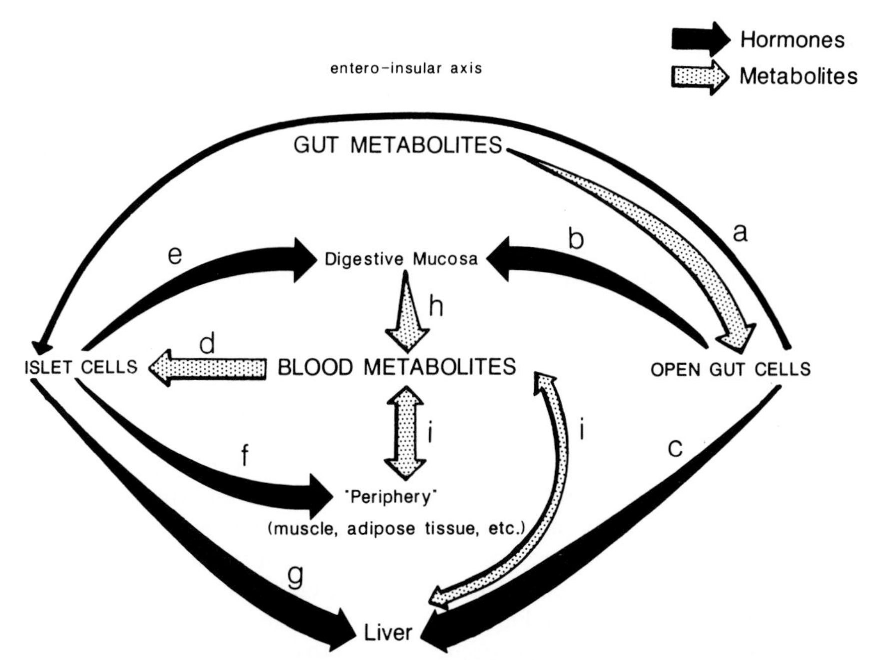

Fig. 14.1. Proposed dual endocrine control system of digestive and metabolic functions. Gut metabolites stimulate "open" gut cells, whose hormones convey messages to: (*a*) the islet cells (enteroinsular axis); (*b*) digestive mucosa; (*c*) liver. In turn, blood metabolites originating from these three targets; (*d*) stimulate the islet cells whose hormones affect (*e*) digestive mucosa; (*f*) peripheral tissues, and (*g*) liver. Integration of signals conveyed by gut and islet hormones results in the coordination of metabolite uptake from the gut (*h*) and metabolite fluxes (*i*) to and from other organs; thus excessive blood levels of metabolites are avoided. For simplicity, nervous controls and details of hormonal interactions are omitted

similar to the holocephalians and *L. osseus*, respectively, seem to manage perfectly without these special islet secretions, one cannot help but wonder. With respect to the heavy islet innervation of the teleosts, we have speculated that it may serve as an overriding control system in the integration of islet responses to metabolic and osmoregulatory stimuli (Epple and Brinn 1976). However, in the case of *Acomys cahirinus*, neither food nor life style seem to set this species so far apart from other rodents that they could be related to the, certainly secondary, lack of islet innervation. The possibility that this lack is connected with a (Type 2) diabetic tendency (cf. Dulin et al. 1983; Nesher et al. 1985) is fascinating from a clinical point of view, but it makes the biologist wonder why, in evolutionary terms, such a safeguard against the detrimental effects of overeating should have been lost so easily. Another question arises with the discovery of "open" insulin cells in a selachian (*Squalus acanthias*) (El-Salhy 1984) and a turtle, *Pseudemys scripta* (Gapp et al. 1986). Is this a primitive feature that has been abandoned by the extant cyclostomes and many gnathostomes (see above), or is it simply another demonstration of the plasticity of the ability of the APUD system to create components when biological advantages call for it? Is, in the Chondrichthyes, the presence of these cells somehow related to the great number of islet cells? Two other questions are almost inevitable: is the "insulin" of these gut cells identical with islet insulin? And how do the intestinal and pancreatic insulin cells cooperate in the digestive and metabolic regulations?

The structural variability of the islet organ and its possible functional implications (or lack thereof) have been discussed in Chap. 9.3. Suffice it to re-emphasize here that, on a phylogenetic scale, it is difficult to identify a morphological pattern consistent with a distinct downstream arrangement of islet cells along capillaries. The revised model of Unger (1983) is certainly rational for the rat, but it remains to be seen if nature applied it as frequently as it deserves from a logical point of view. For example, we have noted that neither the horse nor birds comply with the model, and that interactions between islet cells via gap junctions, co-secretions, and receptor modulations could be invoked to propose an equally efficient alternative. Perhaps the mutual regulation of islet secretions sometimes involves a combination of both models.

In Chap. 6, we have joined once more the disussion of insulo-acinar interactions. The available evidence suggests that the islet organ evolved as a structure *sui generis*, functionally independent from the equivalents of the exocrine pancreas (see Fig. 2.6); however, both organs share with the liver an origin from a restricted area of the gut endoderm. This close embryonic relationship explains, e. g., the peculiar intermingling of the three organs in the Polypteridae and *Lepisosteus osseus*, the possible occurrence of mixed endocrine-exocrine pancreas cells, and the proliferation of islet cells from the hepatic duct system of the cyclostomes. On the other hand, we have been unable to detect an evolutionary trend, or principle, that explains insulo-acinar systems as raison d'être for the intrapancreatic islets. We are not disputing the possible, though unproven, functional importance of such systems in some species; yet we reiterate that there is no reason to consider the impact of islet secretions on the surrounding acinar tissue – certainly in most cases – as anything but a byproduct of the topographic association. Perhaps mammals are the only group of vertebrates in which small islets have to some extent an even intrapancreatic distribution; but there are advantages other than insulo-acinar interactions to explain this phenomenon (see Chap. 6). On an overall phylogenetic scale, there is a strong trend to concentrate the A-, B- and D-cells in a limited number of large aggregations, a phenomenon easily explained by ontogenetic mechanisms (see Chap. 3).

While the close portal-type interaction between islet organ and liver may exist in all gnathostomes, although there are open questions especially with the Chondrichthyes, the situation in the cyclostomes is rather confusing. The liver of the extant forms is very small (see Chap. 10.4.2); furthermore, a specific impact of insulin and somatostatin on this organ has not yet been shown unequivocally. Perhaps the islet-hepatic axis of hagfishes and adult lampreys has degenerated, so that in search for a basic situation, we should direct our attention to the ammocoetes. Unfortunately, however, the latter is often considered a rather specialized form (cf. Hardisty 1982; Malatt 1985). However, if one were unaware of its presence and were to attempt to construct a missing link between *Branchiostoma lanceolatum* and adult lampreys, one would doubtless propose a model very similar to the ammocoetes of the Petromyzontidae. The same exercise, applied to the digestive system of *B. lanceolatum* and the gnathostomes, would yield a pancreas not too different from that of the ammocoete of *Mordacia mordax*. Thus, at least for pancreas research, the ammocoetes might be more useful than prevailing thinking suggests. At any rate, we are left with very few hard facts on the physiology of the islet-liver system at the cyclostome level, i.e., at the phylogenetically most critical stage. The pertinent information on the Chondrichthyes is also very meager, although their unusually fatty liver, lack of adipose tissue and heavy utilization of ketone bodies (cf. DeRoos et al. 1985) should make them most tempting research objects. Inevitably, the group for which insulo-hepatic axis is best known is the placental mammals in which insulin, glucagon, and somatostatin have strong mutual interactions with the liver (see Chaps. 10.4.1, 11.4.1 and 13.3). The hormones affect metabolism of the hepatocytes, and hepatocytes determine the quantities of islet hormones released into the systemic circulation. Nevertheless, many uncertainties, particularly with respect to the differential actions of different somatostatins, and with the cooperation between gut and islet secretions at the hepatocyte, level remain to be clarified. Finally, a basic difference between the mammals (Rothwell and Stock 1985) including man (Merklin 1974) and most of the other vertebrates raises questions: Is there a relationship between the absence of hepatic lipid storage and the presence of brown adipose tissue in mammals? Is it possible that the low-melting-point lipids of the brown adipose tissue represent a rapidly mobilizable "substitution" for the lack of intracellularly stored hepatic lipids during nonshivering thermogenesis? What are the precise interactions between insulin, to which brown adipose tissue seems to be most sensitive (Nedergaard and Lindberg 1982), and other hormones and neurons in this tissue?

The remarkably strong conservation of the molecular structure of insulin makes one wonder if, or how, this phenomenon is related to the function of the hormone. Unfortunately, a clear answer still escapes us since the information on the "lower" vertebrates is very limited. However, one could imagine that it may prevent confusion with messages from structurally similar growth factors (see Chap. 10.1.2). The occurrence of insulin-like material in the digestive system of invertebrates points to a possible ancient role of this hormone in the regulation of enteric functions (see Chap. 2.1). If so, this role must have diminished in adult vertebrates in favor of a metabolic function, a change perhaps signaled by the appearance of extramucosal insulin cells. However, insulin receptors in the gastrointestinal epithelium of the fetal rat (Sodoyez-Goffaux et al. 1985) suggest that insulin controls the prenatal gut even in mammals. Virtually nothing is known on insulin actions in enteric regulation in poikilotherms; perhaps the "open" insulin cells of the selachians indicate that at this evolutionary level insulin still has important gastrointestinal functions, but this is merely a guess. Of course, the presence of such open insulin cells in the gut of a turtle does not encourage attempts to construct phylogenetic connections between these observations.

The physiological stimuli causing insulin release in the ammocoete are unknown, but it is likely that the diet of these filter feeders (cf. Hardisty 1979) contains a higher proportion of carbohydrates than that of the adult lampreys and hagfishes. Hence, one wonders if (a) glucose is a physiological stimulus for insulin release at the ammocoete stage, and (b) persistence of this mechanism explains the glucose sensitivity of the B-cells in many carnivorous species, from the adult lamprey to the dog. For the biologist, it is puzzling that glucose should be such a widespread stimulator of insulin release although (1) the majority of lower vertebrates (e.g., all adult lampreys, hagfishes, Chondrichthyes, Polypteridae, sturgeons, *Lepisosteus*, *Amia*, lungfishes) are carnivorous or scavengers, and (2) there is no closely regulated blood sugar level in most of these species. On the other hand, amino acids seem to be a strong signal for insulin release in all vertebrates with the possible exception of the Myxinidae (Emdin 1982a), including species in which B-cells do not respond to changes in plasma glucose such as the penguin, *Pygocellis papua* (Chieri et al. 1972). This phenomenon, plus the anabolic action of insulin in the body wall of almost all species studied, make it likely that islet insulin became involved early in the regulation of protein metabolism in the following way: During times of food uptake, gut-derived amino acids stimulate secretion of insulin, which then fosters their uptake by various tissues, in particular, skeletal muscle; during times of fasting, muscle-derived amino acids again stimulate insulin release, resulting in an anti-catabolic effect of the hormone which prevents inordinate tissue breakdown. This hypothesis is consistent with the variations of endogenous insulin titers thus far demonstrated in poikilotherms, the long-lasting effect of insulin in these species, the suppression of plasma AAN by exogenous insulin, and the increase of plasma AAN found after islet- or pancreatectomy (see Chap. 10.5). Because the reported insulin insensitivity of the body wall of adult lampreys (Chap. 10.5.2) is at variance with this picture, one wonders if the hormone acts in this case as a growth factor. On the other hand, it remains to be seen if the phenomenon can be explained by the use of mammalian insulin, or the peculiar physiological state of the prespawning lampreys.

The available data on cyclostomes and, in particular, the contradictory findings in hagfishes make it impossible to obtain a reasonably clear picture of the role of endogenous insulin in the carbohydrate metabolism of early vertebrates. The issue is even more clouded in the Chondrichthyes (at least the elasmobranchs), which apparently depend to only a very limited degree on circulating glucose and instead use extensively ketone bodies (cf. DeRoos et al. 1985). Thus, the Polypteridae (cf. Bjerring 1985) and lungfishes should be among the most suited extant forms to trace the evolution of the insulin-carbohydrate interactions. Unfortunately, no pertinent information seems to exist on either group, while a deluge of data from the teleosts makes it most frustrating to separate "the grains from the chaff" in this evolutionary "side-branch". One is tempted, in paraphrase of Shakespeare, to offer a whole kingdom of teleost data for a single thorough study on *Polypterus*. There is no proof of a mammalian-like, overpowering role of insulin in regulation of blood sugar in any species "below" the homeotherms; nevertheless, it remains to be seen if this statement must be modified for some forms once we know more about the impact of higher temperatures and states of activity. In the homeotherms, survival requires a critical minimum level of blood sugar. However, the contribution of insulin in the maintenance of glycemia differs considerably between mammals and birds, insulin dominating over glucagon in the former, and the opposite situation characterizing the latter.

Our knowledge on the evolution of the role of insulin in the lipid metabolism is at best sketchy. Exogenous insulin lowers plasma FFA in lampreys, teleosts and tetrapodes (except postnatal birds), but not in hagfishes. The situation in the Chondrichthyes is unknown, and there is still no proof that endogenous insulin affects the adipose tissue in fishes, amphibians and reptiles. The presence of insulin-binding sites in turtle adipose tissue (Marques et al. 1982), and the development of an excessive ketonemia in the pancreatectomized alligator suggests an anti-catabolic role of insulin in the lipid metabolism of these reptiles (see below). However, it remains to be seen if the dominant role of insulin in the control of mammalian adipose tissue is paralleled in any other group of vertebrates.

Considering the large number of studies on the comparative physiology of insulin, the overall result is meager. Some reasons for this deplorable state of affairs are obvious: (1) The frequent use of high, unphysiological doses of mammalian insulin in studies on poikilotherms; such experiments must be a sledgehammer approach to a delicate system that normally acts over extended periods of time. (2) Mammalian bias, which caused many investigators to restrict their scope to "easy" mammalian-type parameters in species with very different organizations. (3) Dearth of information on the ammocoete and the parasitic stage of the lamprey. Almost all data on adult lampreys are from prespawners, which a cynic may describe as "fasting sex maniacs in various stages of their demise". (4) Uncertainty as to the biological status of the extant hagfishes. How degenerate is their islet organ? (5) Dearth or absence of experimental studies on virtually all gnathostomes close to phylogenetic "key positions" (selachians, Polypteridae, lungfishes, turtles, crocodiles, monotremes, marsupials). (6) Uncertainty as to the biological significance of data obtained with heterologous insulins in lower forms.

It is widely believed that somatostatin was the second gut hormone to become an islet secretion. However, the appearance of argyrophil cells simultaneously with the somatostatin cells in the islets of metamorphosing lampreys, plus additional islet cells in prespawning *Petromyzon marinus* indicate that the issue is not yet settled (Chap. 4.1). Furthermore, the tinctorial differences between the somatostatin cells of the cyclostomes and the D-cells are so marked that the identity of both cell types is by no means assured. Considering the plasticity of the APUD system, there is no reason to believe that a transformation from intestinal to insular somatostatin cells could not have occurred more than one time. Of course, we have no information yet on the nature of the somatostatin-like immunoreactivity in the cyclostome islets. The molecular heterogeneity of somatostatin, the immunological differences among the different forms, the already known differential effects of S-14, S-22, and S-28, and the presence of different somatostatin receptor types all leave no doubt that any generalization concerning occurrence and function of this substance must be made with great caution (Chap. 13). One can only hope that the availability of synthetic somatostatin (S-14) does not lead to widespread and indiscriminate use before its presence and plasma levels in the species under study have been ascertained. Otherwise, we could be burdened with another pile of dead animals, killed in the pursuit of wrong parameters with a foreign drug. Nevertheless, the heterogeneity of circulating somatostatin makes this substance one of the most fascinating, but also difficult, hormones with which to work. So far, we have only sketchy information on its functions in "lower" vertebrates. In lampreys, the general, though not complete, restriction of

B- and somatostatin cells to follicles of their own makes a paracrine action of somatostatin on the B-cells unlikely, as does the largely extra-insular occurrence of D-cells in the turtle pancreas. The very small number of somatostatin cells in *Myxine glutinosa* may indicate a reduced or lack of physiological impact in these dwellers of the deep, whereas the cytological criteria in prespawning lampreys point to an important role of somatostatin even after months of fasting (Hilliard et al. 1985). Perhaps the findings on somatostatin release in teleost preparations (Chap. 13.3) indicate a homeotherm-like role in postprandial regulations. However, the recently reported effects of S-25 on the salmon liver (Chap. 13.3) raise two old, largely ignored issues (cf. Epple 1965, 1968b): (1) Is there a specific islet hormone that is involved in the catabolic regulation of liver lipids? (2) How do islet hormones affect the hepatic lipid stores of nonmammalian vertebrates which, contrary to the fatty liver of mammals, are a physiological phenomenon? While it is generally accepted that islet-related hepatic fat deposition of mammals is due to insulin deficiency (which causes breakdown of the adipose tissue stores: see Chap. 10.4.1), it is difficult, actually impossible, to explain the regulation of the greatly fluctuating liver lipids in nonmammalian species by the same mechanism. For instance, chondrichthyes have an extremely fatty liver (cf. Patent 1970, 1973) and no adipose tissue; and in birds, the hepatic lipids must be affected strongly by an interplay between glucagon and somatostatin (see Chaps. 11.4.8 and 13.3, and below). Obviously, mammalian bias would not be helpful in the approach to these questions.

In birds, D-cells are associated with both types of islets. In the A-islets, the D-cells are intermingled with A-cells, with which they form strands along capillaries. This arrangement is compatible with a communication via gap junctions as well as paracrine or portal mechanisms. However, it would obviate any telehormonal effect of somatostatin, provided it is of the same type as produced in the A-islets. In the avian B-islets, D-cells often form a tangentially oriented outer layer on B-cells which are concentrically arranged around a capillary. Interspecifically, this D-cell rim varies greatly, being particularly strong in the pigeon. By this arrangement, at least communication via gap junctions and paracrine secretions appears possible, but the capillary connections within the B-islets have not yet been clarified. The significance of the different location of the D-cells in the two types of avian islets is unknown but, interestingly, Van Campenhout and Cornelis (1954) obtained in some species a silver reaction in the D-cells of the B-islets only. Since it was later shown that argyrophil D-cells occur in the A-islets of species in which the Belgian authors obtained negative results, the findings of Van Campenhout and Cornelis were thought to be merely due to technical problems (Epple 1965). With new information on different action spectra of the somatostatins in the duck (Strosser et al. 1984), it may be worthwhile to reconsider the case. Are the cytological differences between the D-cells related to the production of different somatostatins? Differences in the tissue contents of S-14 and S-28 may point in this direction. Another important and unexpected finding on avian somatostatin is the direct antilipolytic effect of both S-14 and S-28 on the adipocyte of the duck (Strosser et al. 1983b). Is this a general mechanism in birds? How widespread is it in other taxa? Does the "paradox" similarity of the islet physiology between birds and Squamata include this phenomenon? Is lack of islet somatostatin involved in the excessive ketonemia of the pancreatectomized alligator? And last, but not least: is it possible to trace evolutionary trends or interactions related to the antilipolytic actions of somatostatin and insulin? In mammals, islet somatostatin is an important factor in the control of nutrient homeostasis. Together with gut somatostatin, it seems to provide the classical example of the dual control system of digestive and metabolic functions (see above), insular and gut somatostatin jointly moderating the uptake of nutrients at the gut level (cf. Schusdziarra 1980), in direct opposition to gastrin from "open" gut cells, and to PP from "closed" islet cells. Obviously, this basic concept, which must be subject to species-related variations in detail, ignores for simplicity the multitude of possible or known interactions with hormonal and nervous factors. In particular, modulation of releases of pancreatic insulin and glucagon via telehormonal somatostatin and the modulation of the systemic insulin: glucagon ratio at hepatic level via the combined portal effects of gut and islet somatostatin must be investigated in additional species. Perhaps the basic function of plasma somatostatin in all chordates is the coordination of nutrient entry from the gut, whereas a role in regulation of blood metabolites followed later with the evolving islet organ. On the other hand, a direct somatostatin action on the adipocyte may be a phylogenetically late event, perhaps restricted to birds. Nevertheless, the structural variability of the somatostatin molecule and receptors (Chap. 13) may foreshadow a wider functional spectrum than anticipated on the basis of the available data.

Unless a *Latimeria*-like miracle (cf. Forey 1980; Locket 1980) presents us with an extant ostracoderm, we may never know if the absence of glucagon in the cyclostome islets is a truly primitive condition. However, one wonders if there is a relationship between the absence of islet glucagon and the metabolically rather insignificant role of the cyclostome liver (see Chap. 10.4.2). For the time being, it appears that the Chondrichthyes are the phylogenetically oldest of the extant vertebrates with islet A-cells, yet we know virtually nothing about the role of glucagon at this level since the weak hyperglycemic effect after large doses of mammalian glucagon does not allow many conclusions (see Chap. 11.4.3). As previously pointed out, the high lipid content of the chondrichthyan liver, and the likelihood that ketone bodies are a major fuel, at least in selachians, should make these animals fascinating models for the study of the role of hormones in hepatic ketogenesis. At the latest, from the actinopterygian level on, glucagon seems to affect three functions of the liver: ketogenesis, gluconeogenesis and glycogenolysis. This speculation is based on sketchy data from teleosts, and several tetrapods (Chap. 11). However, major, completely unstudied groups include the Polypteridae, Gymnophiona, Monotremata, and Marsupialia. Among the reptiles, the strong ketogenesis in the pancreatectomized alligator makes one wonder if gut glucagon, unbalanced by pancreas insulin and/or somatostatin, is involved (see Chap. 11.4.7). Similar to the case of somatostatin, further questions arise with the glucagon sensitivity of the avian adipose tissue: Does the similarity in islet physiology between saurians and birds also extend to the role of glucagon in the lipid metabolism? Or is glucagon sensitivity of the adipose tissue a general phenomenon among reptiles, the line leading to the mammals being secondarily deprived of this mechanism? Finally, it must be noted that our lack of information on islet physiology in the early mammals leaves us without an answer to the question of a role of glucagon in the homeostatic control of glycemia in Monotremata and/or Marsupialia.

Pancreatic polypeptide remains the most enigmatic of the four major islet secretions. The available evidence from mammals and birds suggests a function in the control of the digestive system, in which its stimulatory effect on the stomach indicates a gastrin-like action. The spectrum of its actions in homeotherms requires further investigation, whereas its role in poikilotherms is completely unknown. At least in teleost islets, PP-like immunoreactivity may be due to a related peptide, aPY (Noe et al. 1986).

There can be no question that the last two decades have been a period of good progress in comparative physiology of the islets. However, especially in the case of insulin, the need for careful experimental design has also become most obvious since data are of little, if any, value if they are not biologically meaningful. One must, therefore, hope that in future islet research more emphasis will be placed on the use of physiological doses and species-specific hormones (and anti-hormones). Furthermore, in vivo studies should be carried out, whenever possible, in animals with a sufficient body size for cardiovascular cannulation and repeated blood sampling. It would be worthwhile to identify "model" species, and to concentrate on these in cooperative research programs, beginning with the isolation and preparation of species- or taxon-specific hormones and hormone assays. Considering the now known structural and immunological variations of the islet hormones, this type of research may be more profitable than a shotgun approach, using species just because they are locally available. For some taxa, "model" species may be difficult to identify, since, in addition to the above-mentioned features, the animals should be hardy in surgery, reasonably stress-resistent, and easy to keep in captivity. Furthermore, they should have anatomical features favorable for endocrinectomies. From personal experience, we would recommend the early anadromous stage of Southern Hemisphere lamprey (*Geotria australis*), and the eels of the genus *Anguilla*. In both cases, all islet tissue can be removed surgically, the body size is sufficient for repeated blood collection via cardiovascular cannulas, and the animals prefer tubes as hiding places, which reduces stress of captivity. In addition, both forms are very hardy in surgery and do not require feeding (for literature and details, see Epple and Nibbio 1985; Epple et al. 1983; Macey et al. 1984). Similarly useful species should be found among the bottom-dwelling selachians. Although their islet organs are surgically inaccessible in vivo (Chap. 4.2), both the larger species of *Polypterus* and all three forms of extant lungfishes seem to have otherwise favorable features. Among the amphibians, the large *Bufo marinus*; and among the reptiles, especially *Varanus exanthematicus* (cf. Dupé-Godet 1984) come to mind.

We have referred to mammalian bias as a factor in the limited progress made in some aspects of comparative islet research. One example is the neglect of the study of noncarbohydrate parameters, although there is no question that most poikilotherms are basically carnivorous species. In particular, the importance of gluconeogenesis and ketogenesis in these species (see e.g., Phillips and Hird 1977a, b; Leech et al. 1979; Zamnit and Newsholme 1979; Savina and Derkachev 1983) should have directed more attention to parameters of lipid and protein metabolism. Another example is the osmoregulatory role of plasma urea in the Chondrichthyes, lungfishes, and even a frog, *Rana cancrivora* (cf. G. W. Brown and Brown 1985), whose possible control by islet secretions apparently has never been studied. Similarly, the role of free amino acids in the intracellular osmoregulations of pancreatectomized aquatic animals has been addressed in only one study in the American eel (Epple and Kocsis 1980). Eels (*A. rostrata*) deprived of their islet secretions, survive well in freshwater, but very poorly when transferred to seawater (cf. Epple and Lewis 1975; Epple and Miller 1981). The death of these animals in seawater seemed to be connected with an inability to correct the normally transitory serum hyperosmolality and muscle dehydration. Recent studies on other teleost species (*Gillichthys mirabilis*, *Sarotherodon mossambicus*, *Platichthys flesus*) indicate that glucagon may be the islet hormone in question, and that this substance modulates the branchial ion excretion (Foskett et al. 1982, 1983; Davis and Shuttleworth 1985). Obviously, these findings are well compatible with our suggestion that the endocrine pancreas was originally an organ dually involved in osmoregulation and intermediary metabolism, whose function shifted in the tetrapods to predominantly metabolic regulations (Epple and Brinn 1976). Unfortunately, it seems that a large number of in vitro studies on the effect of insulin on the ion transport across anuran epithelia has never been extended to work with intact animals (for literature see Crabbé 1981; Moore 1983); and it remains to be seen if a few scattered data on osmoregulatory effects of somatostatin can be related to the islet fraction of this hormone (cf. Loretz et al. 1983; Silva et al. 1985). Thus, the evolution of the osmoregulatory effects of islet hormones, and their integration with processes of intermediary metabolism are one more promising area for further studies. Clearly, this example emphasizes that comparative islet physiology is still in a pioneer stage, and that there should be no lack of challenging questions for many years to come.

References

Abad ME, Agulleiro B, Rombout JHWM (1986) An immunocytochemical and ultrastructural study of the endocrine pancreas of *Sparus auratus* L. (Teleostei). Gen Comp Endocrinol 64:1–12

Abel JJ (1926) Crystalline insulin. Proc Nat Acad Sci USA 12:132

Ablett RF, Sinnhuber RO, Holmes RM, Selivonchick DP (1981 a) The effect of prolonged administration of bovine insulin in rainbow trout (*Salmo gairdneri*). Gen Comp Endocrinol 43:211–217

Ablett RF, Sinnhuber RO, Selivonchick DP (1981 b) The effect of bovine insulin on [^{14}C] glucose and [^{3}H] leucine incorporation in fed and fasted rainbow trout (*Salmo gairdneri*). Gen Comp Endocrinol 44:418–427

Ablett RF, Taylor MJ, Selivonchick DP (1983) The effect of high-protein and high carbohydrate diets on [^{124}I] iodoinsulin binding in skeletal muscle, plasma membranes and isolated hepatocytes of rainbow trout (*Salmo gairdneri*). Br J Nutr 50:129–139

Adamo ML, Hazelwood RL (1984) Cerebellar binding of avian pancreatic polypeptide. Endocrinology 114:794–800

Adamo ML, Dyckes DF, Hazelwood RL (1983) In vitro binding and degradation of avian pancreatic polypeptide by chicken and rat tissues. Endocrinology 11:508–516

Adashi EY, Resnick CE, D'Ercole AJ et al. (1985) Insulin-like growth factors as intraovarian regulators of granulosa cell growth and function. Endocrine Rev 6:400–420

Adrian TE, Bloom SR, Edwards AV (1983) Neuroendocrine responses to stimulation of the vagus nerves in bursts in conscious calves. J Physiol (Lond) 344:25–35

Agulleiro B, Ayala AG, Abad ME (1985) An immunocytochemical and ultrastructural study of the endocrine pancreas of *Pseudemys scripta elegans* (Chelonia). Gen Comp Endocrinol 60:95–103

Ahmad MM, Matty AJ (1975) Insulin secretion in response to high and low protein ingestion in rainbow trout (*Salmo gairdneri*). Pakistan J Zool 7:1–6

Ahrén B, Lundquist I (1982) Interaction of vasoactive intestinal peptide (VIP) with cholinergic stimulation of glucagon secretion. Experientia 38:405–406

Ahrén B, Lundquist I, Järhult J (1984) Effects of α_1-, α_2-, and β-adrenoreceptor blockers on insulin secretion in the rat. Acta Endocrinol 105:78–82

Albert SG (1982) Insulin biosynthesis in isolated channel catfish (*Ictalurus punctatus*) islets: the effect of glucose. Gen Comp Endocrinol 48:167–173

Aldehoff G (1891) Tritt auch bei Kaltblütern nach Pankreasexstirpation Diabetes mellitus auf? Z Biol 28:293

Ali-Rachedi A, Varndell IM, Adrian TE et al. (1984) Peptide YY (PYY) immunoreactivity is co-stored with glucagon-related immunoreactants in endocrine cells of the gut and pancreas. Histochemistry 80:487–491

Alumets J, Håkanson R, Sundler F (1978 a) Distribution and ultrastructure of pancreatic polypeptide (PP) cells in the pancreas and gut of the chicken. Cell Tissue Res 194:377–386

Alumets J, Håkanson R, O'Dorisio T et al. (1978 b) Is GIP a glucagon cell constituent? Histochemistry 58:253–257

Andrew A (1977) Pancreatic D cells in very young chick embryos. Gen Comp Endocrinol 31:463–465

Andrew A (1982) The APUD concept: where has it led us? Br Med Bull 38:221–225

Andrew A (1984) The development of the gastro-entero-pancreatic neuroendocrine system in birds. In: Falkmer S, Håkanson R, Sundler F (eds) Evolution and tumour pathology of the neuroendocrine system, Chap 6. Elsevier, Amsterdam, New York pp 91–109

Andrew A, Rawdon BB (1980) The ultrastructural identity of pancreatic polypeptide cells in chicks. Histochemistry 65:261–267

Andrews PC, Pubols VN, Hermodson MA et al. (1984) Structure of the 22-residue somatostatin from catfish. An O-glycosylated peptide having multiple forms. J Biol Chem 259:13267–13272

Andrews PC, Ronner P (1985) Isolation and structure of glucagon and glucagon-like peptide from catfish pancreas. J Biol Chem 260:3910–3914
Andries JC, Tramu G (1985) Ultrastructural and immunohistochemical study of endocrine cells in the midgut of the cockroach *Blaberus craniifer* (Insecta, Dictyoptera). Cell Tissue Res 240:323–332
Arimura A (1981) Recent progress in somatostatin research. Biomed Res 2:233–257
Arquilla ER, Stenger DP (1985) Immunology of islet cells. In: Volk BW, Arquilla ER (eds) The diabetic pancreas, second edition. Plenum Press, New York, pp 493–512
Arquilla ER, Miles PV, Morris JW (1972) Immunochemistry of insulin. In: Freinkel N, Steiner DF (eds) Endocrine pancreas. Handbook of physiology, sect 7, vol 1. Williams and Wilkins, Baltimore, pp 159–174
Ashcroft SJH (1981) Metabolic controls of insulin secretion. In: Cooperstein SJ, Watkins D (eds) The islets of Langerhans. Biochemistry, physiology and pathology, Chap 6. Academic Press, New York, pp 117–148
Asplin CM, Hollander PM, Palmer JP (1984) How does glucose regulate the human A cell *in vivo*? Diabetologia 26:203–207
Assan R, Marre M, Gormley M (1983) The amino acid-induced secretion of glucagon. In: Lefébvre PJ (ed) Glucagon II, Chap 24. Springer, Berlin New York Tokyo, pp 19–41
Atz JW, Epple A, Pang PKT (1980) Comparative physiology, systematics and the history of life. In: Pang PKT, Epple A (eds) Evolution of vertebrate endocrine systems. Texas Tech Press, Lubbock, Texas, pp 7–15
Audy G, Kerly M (1952) The content of glycogenolytic factor in pancreas from different species. Biochem J 52:77–78
Ayer-LeLievre C, Fontaine-Perus J (1982) The neural crest: its relation with APUD and paraneuron concepts. Arch Hist Jap 45:409–427
Baetens D, Malaisse-Lagae F, Perrelet A, Orci L (1979) Endocrine pancreas: three-dimensional reconstruction shows two types of islets of Langerhans. Science 206:1323–1325
Bahnsen M, Burrin JM, Johnston DG et al. (1984) Mechanisms of catecholamine effects on ketogenesis. Am J Physiol 247:E173–E180
Bajaj M, Blundell TL, Pitts JE et al. (1983) Dogfish insulin. Primary structure, conformation and biological properties of an elasmobranchial insulin. Eur J Biochem 135:535–542
Baker LA, Hazelwood RL (1977) Chicken cerebrospinal fluid as affected by fasting, sulfonylurea administration and pancreatectomy. Comp Biochem Physiol 56A:385–393
Baldissera FGA, Holst JG (1984) Glucagon-related peptides in the human gastrointestinal mucosa. Diabetologia 26:223–228
Baldissera FGA, Munoz-Perez MA, Holst JG (1983) Somatostatin 1-28 circulates in human plasma. Regul Pept 6:63–139
Banks I, Sloan JM, Buchanan KD (1980) Glucagon-like (GLI) and secretin-like immunoreactivity (SLI) within the digestive system of the bivalve mollusc *Cerastoderma edule*. Regul Pept 1 (Suppl): 7
Bannister JV, Parker MW (1985) The presence of a copper/zinc superoxide dismutase in the bacterium *Photobacterium leiognathi*: a likely case of gene transfer from eukaryotes to prokaryotes. Proc Nat Acad Sci USA 82:149–152
Banting FG, Best CH (1922) The internal secretion of the pancreas. J Lab Clin Med 7:251–256
Bargmann W (1939) Die Langerhansschen Inseln des Pankreas. In: Von Möllendorf W (ed) Handbuch der mikroskopischen Anatomie des Menschen. Springer, Berlin, Bd VI/2, S 197–288
Baron H (1934) Insel- und Zymogengewebe in ihren gegenseitigen Beziehungen bei *Gasterosteus aculeatus* und einigen anderen Teleostiern. Z Wiss Zool 146:1–40
Barrington EJW (1942) Blood sugar and the follicles of Langerhans in the ammocoete larva. J Exp Biol 19:45–55
Barrington EJW (1972) The pancreas and intestine. In: Hardisty MW, Potter IC (eds) The biology of lampreys. Academic Press, London, pp 135–169
Baskin DG, Ensinck JW (1984) Somatostatin in epithelian cells of intestinal mucosa is present primarily as somatostatin 28. Peptides (NY) 5:615–621
Baskin DG, Gorray KC, Fujimoto WY (1984) Immunocytochemical identification of cells containing insulin, glucagon, somatostatin, and pancreatic polypeptide in the islets of Langerhans of the guinea pig with light and electron microscopy. Anat Rec 208:567–578

Bassas L, de Pablo F, Lesniak MA, Roth J (1985) Ontogeny of receptors for insulin-like peptides in chick embryo tissues. Early dominance of insulin-like growth factor over insulin receptors in brain. Endocrinology 117:2321–2329

Bassett JM (1972) Plasma glucagon concentrations in sheep: their regulation and relation to concentrations of insulin and growth hormone. Aust J Biol Sci 25:1277–1287

Bautz A, Schilt J (1986) Somatostatin-like peptide and regeneration in planarians. Gen Comp Endocrinol 64:267–272

Baxter-Grillo D, Blázques E, Grillo TAI et al. (1981) Functional development of the pancreatic islets. In: Cooperstein SJ, Watkins D (eds) The islets of Langerhans. Biochemistry, physiology and pathology, Chap 2. Academic Press, London New York, pp 35–49

Beischer W, Schmid M, Kerner W, Keller L, Pfeiffer EF (1978) Does insulin play a role in the regulation of its own secretion? Horm Metab Res 10:168–169

Bell GI, Santerre RF, Mullenbach GT (1983) Hamster preproglucagon contains the sequence of glucagon and two related peptides. Nature (London) 302:716–719

Bencosme SA (1955) The histogenesis and cytology of the pancreatic islets in the rabbit. Am J Anat 96:103–152

Bendayan M (1982) Contacts between endocrine and exocrine cells in the pancreas. Cell Tissue Res 222:227–230

Bennett-Clarke CA, Joseph SA (1986) Immunocytochemical localization of somatostatin in human brain. Peptides 7:877–884

Benzo CA, Green TD (1974) Functional differentiation of the chick endocrine pancreas: insulin storage and secretion. Anat Rec 180:491–496

Berelowitz M, Frohman L (1983) Immunoreactive neurotensin in the pancreas of genetically obese and diabetic mice. Diabetes 32:51–54

Berelowitz M, LeRoith D, von Schenk H et al. (1982) Somatostatin-like immunoactivity and biological activity is present in *Tetrahymena pyriformis*, a ciliated protozoan. Endocrinology 110:1939–1944

Bern HA, Pearson D, Larson BA, Nishioka RS (1985) Neurohormones from fish tails – the caudal neurosecretory system. I. "Urophysiology" and the caudal neurosecretory system of fishes. Rec Prog Horm Res 41:533–552

Bernstein HA, Dorn A, Hahn H-J, Kostmann G, Ziegler M (1980) Cellular localization of insulin-like immunoreactivity in the central nervous system of spiny mice, C57Bl/6J and C57Bl/KsJ mice. Acta Histochem Cytochem 13:623–626

Berson SA, Yalow RS, Baumann A et al. (1956) Insulin ^{131}I metabolism in human subjects: demonstration of insulin-binding globulin in the circulation of insulin treated subjects. J Clin Invest 35:170–190

Berthoud HR (1984) The relative contribution of the nervous system, hormones, and metabolites to the total insulin response during a meal in the rat. Metabolism 33:18–25

Bethge N, Diel F, Usadel KH (1982) Somatostatin – a regulatory peptide of clinical importance. J Clin Chem Clin Biochem 20:1–11

Bevis PJR, Thorndyke MC (1979) A cytochemical and immunofluorescence study of endocrine cells in the gut of the ascidian *Styla clava*. Cell Tissue Res 199:139–144

Bevis PJR, Thorndyke MC (1981) Stimulation of gastric enzyme secretion by porcine cholecystokinin in the ascidian *Styla clava*. Gen Comp Endocrinol 45:458–464

Bhaumick B, Bala RM (1985) Ontogeny and characterization of basic somatomedin receptors in rat placenta. Endocrinology 116:492–498

Birch NP, Christie DL, Renwick GC (1984) Immunoreactive insulin from mouse brain cells in culture and whole rat brain. Biochem J 218:19–27

Bishop AE, Polak JM, Green IC et al. (1980) The location of VIP in the pancreas of man and rat. Diabetologia 18:73–78

Bjerring HC (1985) Facts and thoughts on piscine phylogeny. In: Foreman RE, Gorbman A, Dodd JM, Olsson R (eds) Evolutionary biology of primitive fishes. Plenum Press, New York London, pp 31–57

Bjorenson JE (1985) Effect of initial developmental stage on future morphology of transplanted chick pancreas. Cell Tissue Res 240:367–373

Bliss M (1982) The discovery of insulin. The University of Chicago Press, Chicago

Bloom SR, Polak JM (1981) Introduction. In: Bloom SR, Polak JM (eds) Gut hormones, second edition. Churchill Livingston, Edinburgh, London, pp 3–9

Bloom SR, Adrian TE, Barnes AJ, Polak JM (1979) Autoimmunity in diabetics induced by hormonal contaminants of insulin. Lancet i:14–16
Bloom SR, Edwards AV, Ghatei MA (1983) Endocrine responses to exogenous bombesin and gastrin releasing peptide in conscious calves. J Physiol 344:37–48
Blum F (1927) Zur Geschichte der hyperglykämisierenden Substanz der Pancreasdrüse. Z inn Med 28:923
Blundell T, Humbel RE (1980) Hormone families: pancreatic hormones and homologous growth factors. Nature (London) 287:781–787
Blundell T, Dodson G, Hodgkin D, Mercola D (1972) Insulin: The structure in the crystal and its reflection in chemistry and biology. Adv Prot Chem 26:279–402
Blundell T, Wood S (1982) The configuration, flexibility, and dynamics of polypeptide hormones. Ann Rev Biochem 51:123–154
Boarder MR, McArdle W (1984) Characterization of metenkephalen (arg6, phe7) immunoreactivity in human adrenal and human pheochromocytoma. Regul Pept 9:187–197
Bobbioni E, Marre M, Helman A, Assan R (1983) The nervous control of rat glucagon secretion. Horm Metab Res 15:133–138
Bolaffi JL, Reichlin S, Goodman DBP, Forest JN Jr (1980) Somatostatin: occurrence in urinary bladder epithelium and renal tubules of the toad, *Bufo marinus*. Science 210:644–646
Bone AJ, Younan SIM, Conlon JM (1984) Effects of somatostatin, glucagon, and insulin on glycogen synthesis in isolated rat hepatocytes. Horm Metab Res 15:513–515
Bondareva VM (1970) Effect of insulin deficiency upon the carbohydrate metabolism in chick embryos. Ontogenes 1:282–285
Bondareva VM, Soltitskaya LP, Nusacov YI (1980) Immunobiological pecularities of salmon, *Onchyrhynchus gorbuscha*, insulin and species specific radio-immune system for its assay. Zh Evol Biochim Physiol 16:518–521
Bonner JT (1971) Aggregation and differentiation in the cellular slime molds. Ann Rev Microbiol 25:75–92
Bonner-Weir S, Orci L (1982) New perspectives on the microvasculature of the islets of Langerhans in the rat. Diabetes 31:883–889
Bonner-Weir S, Weir G (1979) The organization of the endocrine pancreas: A hypothetical unifying view of the phylogenetic differences. Gen Comp Endocrinol 38:28–37
Bordi C, Togni R, Baetens D, Ravazzola M, Malaisse-Lagae F, Orci L (1978) Human islet cell tumor storing pancreatic polypeptide: A light and electron microscopic study. J Clin Endocrinol 46:215–219
Bosman FT, Blankenstein M, Daxenbichler G (1985) What‘s new in endocrine factors of tumor growth? Path Res Pract 180:81–92
Bratusch-Marrain PR (1983) Insulin-counteracting hormones: their impact on glucose metabolism. Diabetologia 24:74–79
Brazeau P, Vale W, Burgus R et al. (1973) Hypothalamic polypeptide inhibits the secretion of immunoreactive pituitary growth hormone. Science 179:77–79
Brinn JE (1973) The pancreatic islets of bony fishes. Am Zool 13:653–665
Brinn JE (1975) Pancreatic islet cytology of Ictaluridae (Teleostei). Cell Tissue Res 162:357–365
Brinn JE, Epple A (1976) New types of islet cells in a cyclostome, *Petromyzon marinus* L. Cell Tissue Res 171:317–329
Brinn JE, Burden HW, Schweisthal MR (1977) Innervation of the cultured fetal rat pancreas. Cell Tissue Res 182:133–138
Brockman RP, Laarveld B (1985) Effects of insulin on net hepatic metabolism of acetate and β-hydroxybutyrate in sheep (*Ovis aries*). Comp Biochem Physiol 81A:255–257
Bromer WW (1983) Chemical characteristics of glucagon. In: Lefébvre PJ (ed) Glucagon I, Chap 1. Springer, Berlin Heidelberg New York, Tokyo pp 1–22
Bromer WW, Sinn LG, Behrens OK (1957) The amino acid sequence of glucagon. V. location of amide groups, acid degradation studies and summary of sequential evidence. J Am Chem Soc 79: 2807–2810
Brown GW, Brown SG (1985) On urea formation in primitive fishes. In: Foreman RE, Gorbman A, Dodd JM, Olsson R (eds) Evolutionary biology of primitive fishes. Plenum Press, New York London, pp 321–337

Brown D, Fleming N, Balls M (1975) Hormonal control of glucose production by *Amphiuma means* liver in organ culture. Gen Comp Endocrinol 27:380–388
Bruni JF, Watkins WB, Yen SSC (1979) β-endorphin in the human pancreas. J Clin Endocrinol Metab 49:649–651
Buchan AMJ (1984) An immunocytochemical study of endocrine pancreas of snakes. Cell Tissue Res 235:657–661
Buchan AMJ, Lance V, Polak JM (1982) The endocrine pancreas of *Alligator mississippiensis*. Cell Tissue Res 224:117–128
Budd GC, Pansky B, Cordell B (1983) Insulin or insulin-like peptides in the pituitary gland. J Cell Biol 97:404A
Buffa R, Solcia E, Fiocca R, Crivelli O, Pera A (1979) Complement-mediated binding of immunoglobulins to some endocrine cells of the pancreas and gut. J Histochem Cytochem 27:1279–1280
Bünzli HF, Glatthaar B, Kunz P et al. (1972) Amino acid sequence of the two insulins from mouse (*Mus musculus*). Hoppe-Seyler's Z Physiol Chem 353:451–458
Burnstock G (1972) Purinergic nerves. Pharmacological Reviews 24:509–581
Bussolati G, Capella C, Vassallo G, Solcia E (1971) Histochemical and ultrastructural studies on pancreatic A cells. Evidence for glucagon and non-glucagon components of the a granule. Diabetologia 7:181–188
Campfield LA, Smith FJ (1983) Neural control of insulin secretion: interaction of norepinephrine and acetylcholine. Am J Physiol 244:R629–R634
Cantin M, Genest J (1986) The heart as an endocrine gland. Sci Amer 254:76–81
Cantor P, Rehfeld JF (1984) The molecular nature of cholecystokinin in the feline pancreas and related nervous structures. Regul Pept 8:199–208
Carillo M, Zanuy S, Duve H, Thorpe A (1986) Identification of hormone-producing cells of the endocrine pancreas of the sea bass, *Dicentrarchus labrax*, by ultrastructural immunocytochemistry. Gen Comp Endocrinol 61:28–301
Carneiro NM, Amaral AD (1983) Effects of insulin and glucagon on plasma glucose levels and glycogen content in organs of the freshwater teleost *Pimclodes maculatus*. Gen Comp Endocrinol 49:115–121
Catalàn RE, Martinez AM, Aragonès MD (1984) Evidence for a role of somatostatin in lipid metabolism of liver and adipose tissue. Regul Pept 8:147–159
Cegrell L (1968) The occurrence of biogenic monoamines in the mammalian endocrine pancreas. Acta Physiol Scand Suppl 314:1–60
Cerasi E (1975) Mechanism of glucose stimulated insulin release in health and in diabetes: some reevaluations and proposals. Diabetologia 11:1–13
Chan VG, Fontaine A (1971) Is there a B-cell homolog in starfish? Gen Comp Endocrinol 16:183–191
Chan DKO, Woo NYS (1978) Effect of cortisol on the metabolism of the eel, *Anguilla japonica*. Gen Comp Endocrinol 35:205–215
Chance RE, Johnson MG, Hoffmann JA, Lin TM (1979) Pancreatic polypeptide: a newly recognized hormone. In: Baba S, Kancho T, Yanaihara N (eds) Proinsulin, insulin, c-peptide. Excerpta Medica, Amsterdam, Oxford, pp 419–425
Chang AY, Diani AR (1985) Chemically and hormonally induced diabetes mellitus. In: Volk BW, Arquilla ER (eds) The diabetic pancreas, second edition. Plenum Press, New York, pp 415–438
Chan-Palay V, Nilaver G, Palay SL et al. (1981) Chemical heterogeneity in cerebellar Purkinje cells: existence and coexistence of glutamic acid decarboxylase-like and motilin-like immunoreactivities. Proc Nat Acad Sci USA 78:7787–7791
Chan-Palay V, Engel AG, Wu J-Y, Palay SL (1982) Coexistence in human and primate neuromuscular junctions of enzymes synthesizing acetylcholine, catecholamine, taurine, and γ-aminobutyric acid. Proc Nat Acad Sci USA 79:7027–7703
Charles MA (1981) Pathophysiology of insulin secretion in normal and diabetic humans. In: Brownlee M (ed) Diabetes mellitus, vol I. Garland STBP Press, New York, London, pp 123–206
Chayvialle JA, Paulin C, Dubois PM, Descos F, Dubois MP (1980) Ontogeny of somatostatin in the human gastro-intestinal tract, endocrine pancreas and hypothalamus. Acta Endocrinol 94:1–10
Cheng H, Leblond CP (1974) Origin differentiation and renewal of the four main epithelial cell types in the mouse small intestine V. Unitarian theory of origin of the four epithelial cell types. Am J Anat 141:537–562

Cherksey B, Altszuler N, Zadunaisky J (1981) Preponderance of β-adrenergic binding sites in pancreatic islet cells of the rat. Diabetes 30:172–174
Cherrington AD, Fuchs H, Stevenbon RW et al. (1984) Effect of epinephrine on glycogenolysis and gluconeogenesis in conscious overnight-fasted dogs. Am J Physiol 247:E137–E144
Chieri RA, Basabe JC, Farina JMS, Foglia VG (1972) Studies on carbohydrate metabolism in penguins (*Pygocellis papua*). Gen Comp Endocrinol 18:1–4
Chidambaran S, Meyer RK, Hasler AD (1973) Effect of hypophysectomy, isletectomy and ACTH on glycemia and hematocrit in the bullhead, *Ictalurus melas*. J Exp Zool 184:75–80
Chihara K, Arimura A, Schally AV (1979) Immunoreactive somatostatin in rat hypophysial portal blood: effects of anesthetics. Endocrinology 104:1434–1441
Ch'ng JLC, Polak JM, Bloom SR (1984) Clinical and pathological manifestations of pancreatic endocrine tumours. Front Horm Res 12:136–147
Christ H (1985) Effect of met-enk, substance P and SRIF on the behaviour of *Hemichromis binaculatus*. Peptides 6:139–148
Cieslak SR, Hazelwood RL (1986a) The role of the splenic pancreatic lobe in regulating metabolic normally following 99% pancreatectomy in chickens. Gen Comp Endocrinol 61:476–489
Cieslak SR, Hazelwood RL (1986b) Effectiveness of somatostatin in reuglating pancreatic splenic lobe hormone secretion following 99% pancreatectomy in adult chickens. Gen Comp Endocrinol 63:284–294
Clark A, Grant AM (1983) Quantitative morphology of endocrine cells in human fetal pancreas. Diabetologia 25:31–35
Clark SA, Stumpf WA, Sar M et al. (1980) Target cells for 1,25-dihydroxyvitamin D_3 in the pancreas. Cell Tissue Res 209:515–520
Clutter WE, Bier DM, Shah SD, Ayer PE (1980) Epinephrine plasma metabolic clearance rates and physiologic thresholds for metabolic and hemodynamic actions in man. J Clin Invest 66:94–101
Cohn DV, Langerle R, Fischer-Colbrie R et al. (1982) Similarity of secretory protein-I from parathyroid gland to chromogranin A from adrenal medulla. Proc Nat Acad Sci USA 79:6056–6059
Cohn DV, Elting JJ, Frich M, Elde R (1984) Selective localization of the parathyroid secretory protein-I/adrenal medulla chromogranin A protein family in a wide variety of endocrine cells of the rat. Endocrinology 114:1963–1974
Colca JR, Hazelwood RL (1976) Pancreatectomy in the chicken: does an extra-pancreatic source of insulin exist? Gen Comp Endocrinol 28:151–162
Colca JR, Hazelwood RL (1981) Insulin, pancreatic polypeptide, and glucagon release from the chicken pancreas *in vitro*: responses to changes in medium glucose and free fatty acid content. Gen Comp Endocrinol 45:482–490
Colca JR, Hazelwood RL (1982) Persistence of immunoreactive insulin, glucagon and pancreatic polypeptide in the plasma of depancreatomized chickens. J Endocrinol 92:317–326
Collip JB (1923) The original method as used for the isolation of insulin in semipure form for the treatment of the first clinical case. J Biol Chem 55:xl–xli
Conlon JM (1983) Biosynthesis and molecular forms of somatostatin. In: Mngola EN (ed) Diabetes 1982. Excerpta Medica, Amsterdam, pp 243–249
Conlon JM (1984b) Isolation and structure of guinea pig gastric and pancreatic somatostatin. Life Sci 35:213–220
Conlon JM, Thim L (1985) Primary structure of glucagon from an elasmobranchian fish, *Torpedo marmorata*. Gen Comp Endocrinol 60:398–405
Conlon JM, Thim L (1986) Primary structure of insulin and a truncated C-peptide from an elasmobranchian fish, *Torpedo marmorata*. Gen Comp Endocrinol 64:199–205
Conlon JM, Agoston DV, Thim L (1985a) An elasmobranchian somatostatin: primary structure and tissue distribution in *Torpedo marmorata*. Gen Comp Endocrinol 60:406–413
Conlon JM, Ballmann M, Lamberts R (1985b) Regulatory peptides (glucagon, somatostatin, substance P, and VIP) in the brain and gastrointestinal tract of *Ambystoma mexicanum*. Gen Comp Endocrinol 58:150–158
Conlon JM, Hansen HF, Schwartz TS (1985c) Primary structure of glucagon and a partial sequence of oxyntomodulin (glucagon-37) from the guinea pig. Regul Pept 11:309–320
Conlon JM, Dobbs RE, Orci L, Unger RH (1979) Glucagon-like polypeptides in canine brain. Diabetes 28:700–702

Cooke PS, Nicoll CS (1984) Role of insulin in the growth of fetal rat tissues. Endocrinology 114:638–643

Cooperstein SJ, Watkins D (1981) Action of toxic drugs on islet cells. In: Cooperstein SJ, Watkins D (eds) The islets of Langerhans. Biochemistry, physiology and pathology, Chap 15. Academic Press, London, New York, pp 387–425

Copeland DL, deRoos R (1971) Effect of mammalian insulin on plasma glucose in the mudpuppy (*Necturus maculosus*). J Exp Zool 178:35–44

Coupland RE (1979) Catecholamines. In: Barrington EJW (ed) Hormones and evolution, vol 1. Academic Press, New York, London, San Francisco, pp 309–340

Coutinho HB, Aguiar FJC, Freitas EMP et al. (1983) Isolation and convulsivant effect of the insulin-like protein obtained from the exocrine pancreas of *Bradypus tridactylus* L. Comp Biochem Physiol 74A:951–954

Cowey BC, de la Higuera M, Adron JW (1977a) The effect of dietary composition and of insulin on glucogeogenesis in rainbow trout (*Salmo gairdneri*). Br J Nutr 38:385–395

Cowey BC, de la Higuera M, Adron JW (1977b) The regulation of glucogenesis by diet and insulin in rainbow trout (*Salmo gairdneri*). Br J Nutr 38:463–470

Crabbé J (1981) Stimulation by insulin of transepithelial sodium transport. Ann NY Acad Sci 372:220–234

Craighead JE (1985) Viral diabetes. In: Volk BW, Arquilla ER (eds) The diabetic pancreas, second edition. Plenum Press, New York, pp 439–466

Cramb G, Langslow DR, Phillips JH (1982a) Hormonal effects on cyclic nucleotides and carbohydrate and lipid metabolism in isolated chicken hepatocytes. Gen Comp Endocrinol 46:297–309

Cramb G, Langslow DR, Phillips JH (1982b) The binding of pancreatic hormones to isolated chicken hepatocytes. Gen Comp Endocrinol 46:297–309

Creutzfeldt W (1985) Endocrine tumors to the pancreas. In: Volk BW, Arquilla ER (eds) The diabetic pancreas, second edition. Plenum Press, New York, pp 543–586

Creutzfeldt W, Ebert R (1985) New developments in the incretin concept. Diabetologia 28:565–573

Cryer P (1983) Coordinated responses of glucogenic hormones to central glucopenia: the role of the sympathoadrenal system. Adv Metabol Disorders 10:469–483

Cryer P (1985) Glucose homeostasis and hypoglycemia. In: Wilson JD, Foster DW (eds) Textbook of endocrinology, seventh edition. WB Saunders, Philadelphia, pp 989–1017

Cryer PE, Tse TF, Clutter WE, Shah SD (1984) Roles of glucagon and epinephrine in hypoglycemic and nonhypoglycemic glucose counterregulation in humans. Am J Physiol 247:E198–E205

Csaba G (1980) Phylogeny and ontogeny of hormone receptors: the selection theory of receptor formation and hormonal imprinting. Biol Rev Camb Phil Soc 55:47–63

Csaba G (1981) Ontogeny and phylogeny of hormone receptors. Karger, Basel

Curry DL (1984) Reflex inhibition of insulin secretion: vagus nerve involvement via CNS. Am J Physiol 247:E827–E832

Cutfield JF, Cutfield SM, Carne A, Emdin SO, Falkmer S (1986) Isolation, purification and amino acid sequence of insulin from the teleost fish *Cottus scorpius* (daddy sculpin). Europ J Biochem 158:117–123

Czech MP (ed) (1985) Molecular basis of insulin action. Plenum Press, New York

Czech MP, Davis RJ, Pessin JE et al. (1985) Mechanisms that regulate membrane growth factor receptor. In: Poste G, Crooke S (eds) Mechanisms of receptor regulation, Plenum Press, New York, pp 395–406

D'Agostino J, Field JB, Frazier ML (1985) Ontogeny of immunoreactive insulin in the fetal bovine pancreas. Endocrinology 116:1108–1116

Da Prada M, Zürcher G (1976) Simultaneous radioenzymatic determination of plasma and tissue adrenaline, noradrenaline and dopamine within the fentomole range. Life Sci 19:1161–1174

Dashow L, Epple A (1983) Effects of exogenous catecholamines on plasma catecholamines and glucose in the sea lamprey, *Petromyzon marinus*. J Comp Physiol 152:35–41

Dashow L, Epple A (1985) Plasma catecholamines in the lamprey: intrinsic cardiovascular messengers. Comp Biochem Physiol 82C:119–122

Dashow L, Epple A, Nibbio B (1982) Catecholamines in anadromous lampreys: baseline levels and stress-induced changes with a note on cardiac cannulation. Gen Comp Endocrinol 46:500–504

Dashow L, Nibbio B, Epple A (1983) Circulating catecholamines: evolutionary aspects. Am Zool 23:1011A

Daughaday WH, Kapadia M, Yanow CE et al. (1985) Insulin-like growth factors I and II of nonmammalian sera. Gen Comp Endocrinol 59:316–325

Dave G, Johansson-Sjöbeck M-L, Larsson A et al. (1979) Effects of insulin on the fatty acid composition of the total blood plasma lipids in the European eel, *Anguilla anguilla* L. Comp Biochem Physiol 62A:649–653

Davidson JK, Falkmer S, Mehrotra BK, Wilson S (1971) Insulin assays and light microscopical studies of digestive organs in protostomian and deuterostomian species and in coelenterates. Gen Comp Endocrinol 17:388–401

Davis MS, Shuttleworth TJ (1985) Peptidergic and adrenergic regulation of electrogenic ion transport in isolated gills of the flounder (*Platichthys flesus* L). J Comp Physiol B 155:471–478

Dayhoff MO (ed) (1978) Atlas of Protein Sequence and Structure, Supplement 3. Nat Biomed Res Found, Silver Spring, Maryland

De Fronzo RA (1981) The effect of insulin on renal sodium metabolism. Diabetologia 21:165–171

De Fronzo RA, Sherwin RS, Felig P, Bia M (1977) Nonuremic diabetic hyperkalemia. Possible role of insulin deficiency. Arch Int Med 137:842–843

De Leiva A, Tanenberg RJ, Anderson G et al. (1978) Serotonergic activation and inhibition: effects on carbohydrate tolerance and plasma insulin and glucagon. Metabolism 27:511–520

De Oya M, Prigge WF, Grande F (1971) Suppression by hepatectomy of glucagon-induced hypertriglyceridemia in geese. Proc Soc Exp Biol Med 136:107–110

De Pablo F, Roth J, Hernandez E, Pruss RM (1982) Insulin is present in chicken eggs and early chick embryos. Endocrinology 111:1909–1916

De Pablo F, Hernandez E, Collia F, Gomez JA (1985a) Untoward effects of pharmacological doses of insulin in early chick embryos: through which receptor are they mediated? Diabetologia 28:308–313

De Pablo F, Birbau M, Gomez JA et al. (1985b) Insulin antibodies retard and insulin accelerates growth and differentiation in early embryos. Diabetes 34:1063–1067

De Roos R, De Roos CC (1979) Severe insulin-induced hypoglycemia in spiny dogfish shark (*Squalus acanthias*). Gen Comp Endocrinol 37:186–191

De Roos R, De Roos CC, Werner CS, Werner H (1985) Plasma levels of glucose, alanine lactate, and β-hydroxybutyrate in the unfed spiny dogfish shark (*Squalus acanthius*) after surgery and following mammalian insulin infusion. Gen Comp Endocrinol 58:28–43

De Roos R, Parker AV (1982) Nondetectable plasma glucose levels after insulin administration in the American bullfrog (*Rana catesbeiana*). Gen Comp Endocrinol 46:505–510

Desbals P (1972) Effects de la pancreatectomie et de l'hypophysectomie sur la circulation des lipides chez le Canard. These Sci Nat, Toulouse, France, pp 242

Desbals P, Desbals S, Mialhe P (1967) Variations des lipides plasmatiques et hepatiques chez le Canard, totalement depancreaté. J Physiol 59:232

DeVlaming VL, Pardo RJ (1975) In vitro effects of insulin on liver lipid and carbohydrate metabolism in the teleost, *Notemigonus chrysoleucus*. Comp Biochem Physiol 51B:489–497

Dickson AJ, Langslow DR (1978) Hepatic gluconeogenesis in chickens. Mol Cell Biochem 22:167–181

Dieterlen-Lièvre F (1970) Tissus exocrine et endocrine du pancreas chez l'embryon de Poulet: origine et interactions tissulaires dans la differentiation. Dev Biol 22:138–156

DiScala-Guenot D, Strosser M-T, Mialhe P (1984) Characterization of somatostatin in peripheral and portal plasma in the duck: *in vivo* metabolism of somatostatin-28. J Endocrinol 100:329–335

Docherty K, Steiner DF (1982) Post-translational proteolysis in polypeptide hormone biosynthesis. Ann Rev Physiol 44:625–638

Dogterom AA (1980) The effect of growth hormone of the freshwater snail *Lymnaea stagnalis* on biochemical composition and nitrogenous waste. Comp Biochem Physiol 65B:163–167

Dolnik VR (1975) Migratory condition of birds. Nauka, Moscow, pp 400

Dorn A, Bernstein H-G, Hahn H-G et al. (1980a) Regional distribution of glucagon-like immunoreactive material in the brain of rats and sand rats. An immunohistochemical investigation. Acta Histochem 66:269–272

Dorn A, Bernstein H-G, Kostmann G et al. (1980b) An immunofluorescent reaction appears to insulin-antiserum in different CNS regions of two rat species. Acta Histochem 66:276–278

Dorn A, Rinne A, Bernstein H-G et al. (1983a) The glucagon/glucagon-like immunoreactivities in neurons of human brain. Exp Clin Endocrinol 81:24–32
Dorn A, Bernstein H-G, Rinne A et al. (1983b) Insulin- and glucagon-like peptides in the brain. Anat Rec 207:69–77
Dubois MP (1975) Immunoreactive somatostatin is present in discrete cells of the endocrine pancreas. Proc Nat Acad Sci USA 72:1340–1343
Dubois PM, Paulin C, Assan R, Dubois MP (1975) Evidence for immunoreactive somatostatin in the endocrine cells of the human foetal pancreas. Nature (London) 256:731–732
Dulin WE, Gerritsen GC, Chang AY (1983) Experimental and spontaneous diabetes in animals. In: Ellenberg M, Rifkin H (eds) Diabetes mellitus, theory and practice, Chap 18. Medical Examination Publishing Co, Inc, pp 361–408
Dunn JS, Sheckan HL, McLetchie NGM (1943) Necrosis of islets of Langerhans produced experimentally. Lancet i:484–487
Dunny GM, Craig RA, Carron RL, Clewell DB (1979) Plasmid transfer in *Streptococcus fecalis*: production of multiple sex pheromones by recipients. Plasmid 2:454–465
Dunwiddie TV (1985) The physiological role of adenosine in the central nervous system. Int Rev Neurobiol 27:63–139
Dupé-Godet M (1984) Characterization and measurement of plasma somatostatin-like immunoreactivity in a Sahelian lizard (*Varanus exanthematicus*) during starvation. Comp Biochem Physiol 78A:53–58
Dupé-Godet M, Adjovi Y (1981a) Seasonal variations of immunoreactive glucagon contents in pancreatic extracts of a sahelian lizard (*Varanus exanthematicus*). Comp Biochem Physiol 69A:31–42
Dupé-Godet M, Adjovi Y (1981b) Seasonal variations of immunoreactive insulin contents in pancreatic extracts of a sahelian lizard (*Varanus exanthematicus*). Comp Biochem Physiol 69A:717–729
Dupé-Godet M, Adjovi Y (1983) Somatostatin-like immunoreactivity in pancreatic extracts of a Sahelian lizard (*Varanus exanthematicus*) during starvation. Comp Biochem Physiol 75A:347–352
Duve H (1978) The presence of a hypoglycemic and hypotrehalocemic hormone in the neurosecretory system of the blowfly *Calliphora erythrocephala*. Gen Comp Endocrinol 36:102–110
Duve H, Thorpe A (1979) Immunofluorescent localization of insulin-like material in the median neurosecretory cells of the blowfly, *Calliphora vomitora* (Diptera). Cell Tissue Res 200:187–191
Duve H, Thorpe A (1980) Localisation of pancreatic polypeptide (PP)-like immunoreactive material in neurons of the brain of the blowfly, *Calliphora erythrocephala* (Diptera). Cell Tissue Res 210:101–109
Duve H, Thorpe A, Lazarus NR (1979) Isolation of material displaying insulin-like immunological and biological activity from the brain of the blowfly, *Calliphora vomitora*. Biochem J 184:221–227
Eddlestone GT, Rojas E (1980) Evidence of electrical coupling between mouse pancreatic B-cells. J Physiol (London) 303:76P–77P
Edwards AV, Bloom SR, Järhult J (1980) Neural influences on the endocrine pancreas. Front Horm Res 7:30–40
Efendić S, Wajngot A, Cerasi E, Luft R (1980) Insulin release, insulin sensitivity and glucose intolerance. Proc Nat Acad Sci USA 77:7425–7429
El-Denshary ESM, Gagerman E, Täljedal I-B (1981) Alpha, and alpha$_2$ adrenoreceptors and the regulation of insulin release from isolated islets in normal and ob/ob mice. Diabetologia 21:267A
Eldridge RK, Fields PA (1985) Rabbit placental relaxin: purification and immunohistochemical localization. Endocrinology 117:2512–2519
El-Etr M, Schorderet-Slatkine S, Baulieu E-E (1979) Meiotic maturation in *Xenopus laevis* oocytes initiated by insulin. Science 205:1397–1399
El Hakeem OH, Babiker MM (1983) Bioassay of insulin activity of pancreatic tissue in active and aestivated lungfish, *Protopterus annectens* (Owen). J Fish Biol 23:277–282
Elliott WM, Youson JH (1986) Immunocytochemical localization of insulin and somatostatin in the endocrine pancreas of the sea lamprey, *Petromyzon marinus* L., at various stages of its life cycle. Cell Tissue Res 243:629–634
El-Salhy M (1984) Immunocytochemical investigation of the gastro-entero-pancreatic (GEP) neurohormonal peptides in the pancreas and gastrointestinal tract of the dogfish *Squalus acanthias*. Histochemistry 80:193–205

El-Salhy M, Grimelius L (1981) Histochemical and immunohistochemical studies of the endocrine pancreas of lizards. Histochemistry 72:237–247

El-Salhy M, Grimelius L, Wilander E et al. (1981) Histological and immunohistochemical studies of the endocrine cells of the gastrointestinal mucosa of the toad (*Bufo regularis*). Histochemistry 71:53–65

El-Salhy M, Wilander E, Abu-Sinna G (1982) The endocrine pancreas of anuran amphibians: a histological and immunocytochemical study. Biomed Res 3:579–589

El-Salhy M, Abu-Sinna G, Wilander E (1983) The endocrine pancreas of a squamate reptile, the desert lizard (*Chalcides ocellatus*). Histochemistry 78:391–397

El-Salhy M, Falkmer S, Kramer KJ, Speirs RD (1984) Immunocytochemical evidence for the occurrence of insulin in the frontal ganglion of a lepidopteran insect, the tobacco hornworm moth, *Nanduca sexta* L. Gen Comp Endocrinol 54:85–88

El-Tayeb KMA, Brubaker PL, Vranic M, Lickley HLA (1985) Beta-endorphin modulation of the glucoregulatory effects of repeated epinephrine infusion in normal dogs. Diabetes 34:1293–1300

Emdin SO (1982a) Insulin release in the Atlantic hagfish *Myxine glutinosa in vitro*. Gen Comp Endocrinol 48:333–341

Emdin SO (1982b) Effects of hagfish insulin in the Atlantic hagfish, *Myxine glutinosa*. The *in vivo* metabolism of [^{14}C] glucose and [^{14}C] leucine and studies on starvation and glucose loading. Gen Comp Endocrinol 47:414–425

Emdin SO, Falkmer S (1977) Phylogeny of insulin. Acta Paediatr Scand Suppl 270:15–23

Emdin SO, Steiner DF (1980) A specific antiserum against insulin from the Altantic hagfish, *Myxine glutinosa*: characterization of the antiserum, its use in homologous radioimmunoassay, and immunofluorescent microscopy. Gen Comp Endocrinol 42:251–258

Emdin SO, Steiner DF, Chan SJ, Falkmer S (1985) Hagfish insulin: evolution of insulin. In: Foreman RE, Gorbman A, Dodd JM, Olsson R (eds) Evolutionary biology of primitive fishes. Plenum Press, New York, pp 363–378

Emson PC, de Quidt ME (1984) NPY – a new member of the pancreatic polypeptide family. Trends Neurosci 7:31–34

Engberg G, Carlquist M, Jörnball H, Hall K (1984) The characterization of somatomedin A, isolated by microcomputer-controlled chromatography, reveals an apparent identity to insulin-like growth factor I. Eur J Biochem 143:117–124

Epple A (1963) Zur vergleichenden Zytologie des Inselorgans. Verh Dtsch Zool Ges 27:461–470

Epple A (1965) Weitere Untersuchungen über ein drittes Pankreashormon. Verh Dtsch Zool Ges 29:459–470

Epple A (1966a) Cytology of pancreatic islet tissue in the toad, *Bufo bufo* (L). Gen Comp Endocrinol 7:191–196

Epple A (1966b) Islet cytology in urodele amphibians. Gen Comp Endocrinol 7:207–214

Epple A (1967) Further observations on amphiphil cells in the pancreatic islets. Gen Comp Endocrinol 9:137–142

Epple A (1968a) Körpergewicht und Pankreas von Küken bei einseitiger Diät. Zool Anz 181:190–195

Epple A (1968b) Comparative studies on the pancreatic islets. Endocrinol Jap 15:107–122

Epple A (1969) The endocrine pancreas. In: Hoar W, Randall DJ (eds) Fish physiology, vol 2, The endocrine system. Academic Press, New York, London, pp 271–319

Epple A (1982) Functional principles of vertebrate endocrine systems. Verh Dtsch Zool Ges 117–126

Epple A, Brinn JE (1975) Islet histophysiology: evolutionary correlations. Gen Comp Endocrinol 27:320–349

Epple A, Brinn JE (1976) New perspectives in comparative islet research. In: Grillo TAI, Leibson L, Epple A (eds) The evolution of pancreatic islets. Pergamon Press, Oxford, pp 83–95

Epple A, Brinn JE (1980) Morphology of the islet organ. In: Ishii S, Hirano T, Wada M (eds) Hormones, adaptations and evolution. Jap Sci Soc Press, Tokyo, pp 213–220

Epple A, Brinn JE (1986) Pancreatic islets. In: Pang PKT, Schreibman MT, Gorbman A (eds) Vertebrate endocrinology, vol 1, Fundamentals and biomedical implications, Chap 10. Academic Press, New York, pp 279–317

Epple A, Koscis JJ (1980) The effects of pancreatectomy on tissue taurine of the American eel under varying osmotic and nutritional conditions. Comp Biochem Physiol 65A:139–142

Epple A, Lewis TL (1973) Comparative histophysiology of the pancreatic islets. Am Zool 13:567–590

Epple A, Lewis TL (1975) The effect of pancreatectomy on the survival of *Anguilla rostrata* in different salinities. J Exp Zool 192:457–462
Epple A, Lewis TL (1977) Metabolic effects of pancreatectomy and hypophysectomy in the yellow American eel, *Anguilla rostrata* LeSuer. Gen Comp Endocrinol 32:294–315
Epple A, Miller SB (1981) Pancreatectomy in the eel: osmoregulatory effects. Gen Comp Endocrinol 45:453–457
Epple A, Nibbio B (1985) Catecholaminotropic effects of catecholamines in a teleost fish, *Anguilla rostrata*. J Comp Physiol B155:285–290
Epple A, Brinn JE, Young JB (1980) Evolution of pancreatic islet functions. In: Pang PKT, Epple A (eds) Evolution of vertebrate endocrine systems. Texas Tech Press, Lubbock, pp 269–321
Epple A, Nibbio B, Trachtman MS (1983) Effects of somatostatin and antisomatostatin on serum parameters of the American eel. Comp Biochem Physiol 74A:671–675
Erlandsen SL (1980) Types of pancreatic islet cells and their immunocytochemical identification. In: Fitzgerald PJ, Morrison AM (eds) The pancreas, Chap 8. Williams and Wilkins, Baltimore, London, pp 140–155
Erlandsen SL, Hegre OD, Parsons JA et al. (1976) Pancreatic islet cell hormones. Distribution of cell types in the islet and evidence for the presence of somatostatin and gastrin within the D cell. J Histochem Cytochem 24:883–897
Ermisch A (1965) Zum physiologischen und immunologischen Insulin-Nachweis bei Neunaugen. Acta Biol Med Ger 15:193–196
Ermisch A (1966) Beiträge zur Histologie und Topochemie des Inselsystems der Neunaugen unter natürlichen und experimentellen Bedingungen. Zool Jahrb 83:52–106
Esteve JP, Susini C, Vaysse N et al. (1984) Binding of somatostatin to pancreatic acinar cells. Am J Physiol 247:G62–G69
Eusebi V, Capella C, Bondi A et al. (1981) Endocrine-paracrine cells in pancreatic exocrine carcinomas. Histopathology 5:599–613
Falkmer S (1961) Experimental diabetes research in fish. Acta Endocrinol 37(Suppl 59):1–122
Falkmer S, Matty AJ (1966) Blood sugar regulation in the hagfish, *Myxine glutinosa*. Gen Comp Endocrinol 6:334–346
Falkmer S, Östberg Y (1977) Comparative morphology of pancreatic islets in animals. In: Volk BW, Wellman KF (eds) The diabetic pancreas. Plenum Press, New York, pp 15–58
Falkmer S, Patent GJ (1972) Comparative and embryological aspects of the pancreatic islets. In: Steiner DF, Freinkel N (eds) Endocrine pancreas. Williams and Wilkins, Baltimore, Handbook of physiology, sect 7, vol 1, pp 1–23
Falkmer S, van Noorden S (1983) Ontogeny and phylogeny of the glucagon cell. In: Lefébvre PJ (ed) Glucagon I, Chap 5. Springer, Berlin Heidelberg New York Tokyo, pp 81–119
Falkmer S, Wilson S (1967) Comparative aspects of the immunology and biology of insulin. Diabetologia 3:519–528
Falkmer S, Östberg Y, van Noorden S (1978) Entero-insular systems of cyclostomes. In: Bloom SR (ed) Gut hormones, Chap 7. Churchill Livingstone, Edinburgh, London, New York, pp 57–63
Falkmer S, El-Salhy M, Titlbach M (1984) Evolution of the neuroendocrine system in vertebrates. In: Falkmer S, Håkanson R, Sundler F (eds) Evolution and tumour pathology of the neuroendocrine system, Chap 6. Elsevier, Amsterdam, New York, pp 59–87
Falkmer S, Ebert R, Arnold R, Creutzfeldt W (1980) Some phylogenetic aspects on the enteroinsular axis with particular regard to the appearance of the gastric inhibitory polypeptide. Front Horm Res 7:1–6
Falkmer S, Carraway RE, El-Salhy M et al. (1981) Phylogeny of the gastroenteropancreatic neuroendocrine system: a review. In: Grossman MI, Brazier MA, Lechago J (eds) Cellular basis of chemical messengers in the digestive system. Academic Press, New York, pp 21–42
Farah AE (1983) Glucagon and the circulation. Pharmacol Rev 35:181–207
Farina J, Pinto J, Basabe JC, Chieri RA (1975) Aspects of intermediary metabolism and insulin level in the penguin (*Pygocellis papua*). Gen Comp Endocrinol 27:209–213
Farkas T (1969) Studies of the mobilization of fats in lower vertebrates. Acta Biochem Biophys Acad Sci Hungaricae 4:237–244
Farrar ES, Frye BE (1979) A comparison of adrenalin and glucagon effects on carbohydrate levels of larval and adult *Rana pipiens*. Gen Comp Endocrinol 39:372–380
Feldman JM (1979) Species variation in the islets of Langerhans. Diabetologia 16:1–4

Feldman JM, Chapman B (1975) Characterization of pancreatic islet monoamine oxidase. Metabolism 24:581–588
Felig P (1983) Physiologic action of insulin, In: Ellenberg M, Rifkin H (eds) Diabetes Mellitus, theory and practice, Chap 4. Medical Examination Publishing Co, Inc, pp 77–88
Ferner H (1952) Das Inselsystem des Pankreas. Georg Thieme, Stuttgart
Ferner H (1957) Die Dissemination der Hodenzwischenzellen und der Langerhansschen Inseln als funktionelles Prinzip für die Samenkänalchen und das exokrine Pankreas. Z Mikroskop Anat Forsch 63:35–52
Feyrter F (1953) Über die peripheren endokrinen (parakrinen) Drüsen des Menschen. 2. Aufl, Wilhelm Maudrich, Wien, 83 pp
Figlewicz DP, Ikeda H, Hunt TR et al. (1986) Brain insulin binding is decreased in Wistar Kyoto rats carrying the 'fa' gene. Peptides 7:61–65
Findlay I, Petersen OH (1983) The extent of dye-coupling between exocrine acinar cells of the mouse pancreas. Cell Tissue Res 232:121–127
Fiocca R, Sessa F, Tenti P et al. (1983) Pancreatic polypeptide (PP) cells in the PP-rich lobe of the human pancreas are identified ultrastructurally and immunocytochemically as F cells. Histochemistry 77:511–523
Fister P, Eigenbrodt E, Schoner W (1983) Glucagon induced inactivation of phosphofructokinase and its counteraction by insulin in isolated hepatocytes from the domestic fowl (*Gallus domesticus*). Comp Biochem Physiol 75B:341–345
Flaten O (1983) Gastric inhibitory polypeptide: physiology and novel aspects. Scand J Gastroenterol 18:1–4
Fletcher DJ, Noe BD, Hunt EL (1978) Studies on insulin biosynthesis in the channel catfish (*Ictalurus punctatus*). Gen Comp Endocrinol 35:127–132
Fletcher DJ, Trent DF, Weir GC (1983) Catfish somatostatin is unique to piscine tissues. Regul Pept 5:181–187
Floyd JC, Vinik AI (1981) Pancreatic polypeptide. In: Bloom SR, Polak JM (eds) Gut hormones, second edition. Churchill Livingston, Edinburgh, London, New York, pp 195–201
Floyd JC, Fajans SS, Pek S, Chance RE (1977) A newly recognized pancreatic polypeptide; plasma levels in health and disease. Rec Prog Horm Res 33:519–570
Foà PP (1968) Glucagon. Ergeb Physiol 60:142–219
Foà PP (1973) Glucagon: an incomplete and biased review with selected references. Am Zool 13:613–623
Foglia VG, Wagner EM, de Barros M, Marques M (1955) La diabetes por pancreatectomia en la tortuga normal e hipofisorpriva. Rev Soc Argent Biol 31:87–95
Foltzer C, Haffen K, Kedinger M, Mialhe P (1982) Stimulation of insulin and glucagon secretion in organ culture of chick endocrine pancreas during embryonic life and after hatching. Gen Comp Endocrinol 47:213–230
Foltzer C, Haffen K, Kedinger M, Mialhe P (1983) Insulin secretion in organ culture of chick endocrine pancreas: effect of somatostatin, growth hormone and dexamethasone. Horm Metab Res 15:513–514
Foltzer C, Harvey S, Mialhe P (1984) Ontogenetic variations of tissue somatostatin-like immunoreactivity (SLI) and growth hormone (GH) in ducks. J Steroid Biochem 20:1544A
Forey PL (1980) Latimeria: a paradoxical fish. Proc R Soc Lond B 208:369–384
Forssmann A (1976) The ultrastructure of the cell types in the endocrine pancreas of the horse. Cell Tissue Res 167:179–195
Forssmann WG, Greenberg J (1978) Innervation of the endocrine pancreas in primates. In: Coupland RE, Forssmann WG (eds) Peripheral neuroendocrine interaction. Springer, Berlin, pp 124–133
Forssmann WG, Helmstaedter V, Metz J, Greenberg J, Chance R (1977) The identification of the F-cell in the dog pancreas as the pancreatic polypeptide producing cell. Histochemistry 50:281–290
Foskett JK, Hubbard GM, Machen TE, Bern HA (1982) Effects of epinephrine, glucagon and vasoactive intestinal polypeptide on chloride secretion by teleost operculum membrane. J Comp Physiol 146:27–34
Foskett JK, Bern HA, Machen TE, Conner M (1983) Chloride cells and the hormonal control of teleost fish osmoregulation. J Exp Biol 106:255–281
Frank HJL, Jankovic-Vokes T, Pardridge WM, Morris WL (1985) Enhanced insulin binding to blood barrier *in vivo* and to brain microvessels *in vitro* in newborn rabbits. Diabetes 34:728–733

Frazier WA, Angelitti RH, Bradshaw RA (1972) Nerve growth factor and insulin. Science 176:482–488

Fritsch HAR, van Noorden S, Pearse AGE (1976) Cytochemical and immunofluorescence investigations on insulin-like producing cells in the intestine of *Mytilus edulis* L. (Bivalvia). Cell Tissue Res 165:365–369

Fritsch HAR, van Noorden S, Pearse AGE (1982) Gastro-intestinal and neurohormonal peptides in the alimentary tract and cerebral complex of *Ciona intestinalis* (Ascidiaceae). Cell Tissue Res 223:369–402

Froesch ER, Zapf J (1985) Insulin-like growth factors and insulin: comparative effects. Diabetologia 28:485–493

Froesch ER, Zapf J, Humbel RE (1983) Insulin-like activity, IGF I and II and the somatomedins. In: Ellenberg M, Rifkin H (eds) Diabetes mellitus, theory and practice, Chap 9. Medical Examination Publishing Co, Inc, pp 179–201

Frye BE (1958) Development of the pancreas in *Amblystoma opacum*. Am J Anat 102:117–139

Frye BE (1964) Metamorphic changes in the blood sugar and the pancreatic islets of the frog, *Rana clamitans*. J Exp Zool 155:215–224

Frye BE (1962) Extirpation and transplantation of the pancreatic rudiments of the salamanders, *Amblystoma punctatum* and *Eurycea bislineata*. Anat Rec 144:97–107

Fujii S, Kobayashi S, Fujita T, Yanaihara N (1980) VIP-immunoreactive nerves in the pancreas of the snake, *Elaphe quadrivirgata* (Boie): another model for insular neurosecretion. Biomed Res 1:180–187

Fujii S, Baba S, Fujita T (1982) Pancreatic polypeptide immunoreactive cells and nerves in the canine pituitary. Biomed Res 3:525–533

Fujimoto WY, Kawazii S, Cheuchi M, Kanazawa Y (1983) *In vitro* paracrine regulation of islet B-cell function by A and D cells. Life Sci 32:1873–2188

Fujita T (1962) Über das Inselsystem des Pankreas von *Chimaera monstrosa*. Z Zellforsch 57:487–494

Fujita T (1968) D cell, the third endocrine element of the pancreatic islet. Arch Histol Jap 29:1–40

Fujita T (1973) Insulo-acinar portal system in the horse pancreas. Arch Histol Japon 35:161–171

Fujita T (1977) Concept of paraneurons. Arch Histol Jap 40(Suppl):1–12

Fujita T (1980) Paraneuron, its current implications. Biomed Res 1(Suppl):3–9

Fujita T (1983) New aspects of cells secreting neuropeptides. In: Shizume K, Imura H, Shimizu N (eds) Endocrinology. Excerpta Medica, Amsterdam, pp 35–43

Fujita T, Kobayashi S (1973) The cells and hormones of the GEP endocrine system. The current of studies. In: Fujita T (ed) Gastro-enteropancreatic endocrine system – a cell-biological approach. Igaku Shoin, Tokyo, pp 1–16

Fujita T, Kobayashi S, Serizawa Y (1979) The insulin producing B-cell as a paraneuron. In: Baba S, Kaneko T, Yanaihara N (eds) Proinsulin, insulin, c-peptide. Excerpta Medica, Amsterdam, pp 327–334

Fujita T, Kobayashi S, Yui R, Iwanaga T (1980) Evolution of neurons and paraneurons. In: Ishii S, Hirano T, Wada M (eds) Hormones, adaptation and evolution. Japan Sci Soc Press, Tokyo. Springer, Berlin, pp 35–43

Fujita T, Ryogo Y, Iwanaga T et al. (1981a) Evolutionary aspects of "brain-gut peptides": an immunohistochemical study. Peptides 2:123–131

Fujita T, Kobayashi S, Fujii S et al. (1981b) Langerhans islets as the neuro-paraneuronal control center of the exocrine pancreas. In: Grossman MI, Brazier MAB, Lechago J (eds) Cellular basis of chemical messengers in the digestive system. Academic Press, New York, pp 231–242

Fujita T, Kobayashi S, Serizawa Y (1981c) Intercellular canalicule system in pancreatic islet. Biomed Res 2(Suppl):115–118

Fujita T, Iwanaga T, Kusumoto Y, Yoshie S (1982) Paraneurons and neurosecretion. In: Farner DS, Lederis K (eds) Neurosecretion: molecules, cells, systems. Plenum Press, New York, pp 2–13

Fukagawa NK, Minaker KL, Young VR, Rowe JW (1986) Insulin dose-dependent reductions in plasma amino acids in man. Am J Physiol 250:E13–E17

Furuichi M, Nakamura Y, Yone Y (1980) A radioimmunoassay method for determination of fish plasma insulin. Bull Jap Soc Sci Fish 46:1177–1181

Gabe M (1970) Pancreas endocrine. In: Grassé PP (ed) Traite de Zoologie, vol 14. Masson, Paris, pp 1333

Gagliardino TT (1983) Concurrent factors in the regulation of insulin secretion. In: Mngola EN (ed) Diabetes 1982. Excerpta Medica, Amsterdam, pp 19–27
Ganda OP, Soeldner JS (1983) Diabetes secondary to endocrinopathies. In: Ellenberg M, Rifkin H (eds) Diabetes mellitus, theory and practice, Chap 51. Medical Examination Publishing Co, Inc, pp 1005–1020
Gamse R, Vaccaro ED, Gamse G et al. (1980) Release of immunoreactive somatostatin from hypothalamic cells in culture: inhibition by γ-aminobutyric acid. Proc Nat Acad Sci USA 77:5552–5556
Gapp DA, Polak JM (1983) The endocrine pancreas of the turtle, *Chrysemys picta*. Am Zool 23:910
Gapp DA, Kenny MP, Polak JM (1986) The gastro-entero-pancreatic system of the turtle, *Chrysemys picta*. Peptides 6(Suppl 1):347–352
Gater S, Balls M (1977) Amphibian pancreas function in long-term organ culture. Control of insulin release. Gen Comp Endocrinol 31:249–256
George DT, Bailey PT (1978) The effect of adrenergic and ganglionic blockers upon the L-dopa-stimulated release of glucagon in the rat. Proc Soc Exp Biol Med 157:1–4
Geraerts WPM (1976) Control of growth by the neurosecretory hormone of the light green cells in the freshwater snail *Lymnaea stagnalis*. Gen Comp Endocrinol 29:61–71
Gerber JC, Hare TA (1979) Gamma-amino butyric acid in peripheral tissue, with emphasis on the endocrine pancreas. Diabetes 28:1073–1076
Gerber JC, Hare TA (1980) GABA in peripheral tissues: presence and actions in endocrine pancreatic function. Brain Res Bull 5(Suppl 2):341–346
Gerich JE (1981) Somatostatin. In: Brownlee M (ed) Diabetes mellitus, vol I, Chap 6. Garland STPM Press, New York, London, pp 297–354
Gerich JE (1983a) Somatostatin and analogues. In: Ellenberg M, Rifkin H (eds) Diabetes mellitus, theory and practice, Chap 11. Medical Examination Publishing Co, Inc, pp 225–254
Gerich JE (1983b) Glucose in the control of glucagon secretion. In: Lefébvre PJ (ed) Glucagon II, Chap 23. Springer, Berlin Heidelberg New York Tokyo, pp 3–18
Gerich JE, Lorenzi M (1978) The role of the autonomic nervous system and somatostatin in the control of insulin and glucagon secretion. In: Ganong WF, Martini L (eds) Frontiers in neuroendocrinology, vol 5, Chap 10. Raven Press, New York, pp 265–288
Gerich JE, Charles MA, Grodsky JM (1976) Regulation of pancreatic insulin and glucagon secretion. Ann Rev Physiol 38:353–388
Gerlovin ESH (1976) Some principles of cytodifferentiation of pancreatic islets in vertebrata during onto- and phylogeny from the standpoint of molecular biology and genetics. In: Grillo TAI, Leibson L, Epple A (eds) The evolution of pancreatic islets. Pergamon Press, Oxford, New York, pp 97–112
Gershon MD (1981) The enteric nervous system. Ann Rev Neurosci 4:227–272
Gershon MD, Payette RF, Rothman TP (1983) Development of the enteric nervous system. Fed Proc 42:1620–1625
Gertner JM, Tamborlane WV, Horst RL et al. (1980) Mineral metabolism in diabetes mellitus: changes accompanying treatment with a portable subcutaneous insulin infusion system. J Clin Endocrinol Metab 50:862–866
Gepts W, LeCompte PM (1985) The pathology of type I (juvenile) diabetes. In: Volk BW, Arquilla ER (eds) The diabetic pancreas, second edition. Plenum Press, New York, pp 337–365
Gibson R, Barker PL (1979) The decapod hepatopancreas. In: Barnes M (ed) Oceanogr Mar Biol Ann Rev. Aberdeen University Press, Aberdeen, pp 285–346
Gingerich RL, Kramer JL (1983) Identification of pancreatic polypeptide secretagogous in canine duodenal mucosa. Endocrinology 112:696–700
Girard J, Sperling M (1983) Glucagon in the fetus and newborn. In: Lefébvre PJ (ed) Glucagon II, Chap 36. Springer, Berlin Heidelberg New York Tokyo, pp 251–274
Girone E (1928) Il tessuto insulare nel pancreas dei cheloni. Monit zool ital 39:38–44
Glaser B, Floyd JC Jr, Vinik AI (1983) Secretion of pancreatic polypeptide in man in response to beef ingestion is mediated in part by an extravagal cholinergic mechanism. Metabolism 32:57–61
Gluschankof P, Morel A, Gomez S et al. (1984) Enzymes processing somatostatin precursors: An arg-lys enteropeptidase from rat brain cortex converting somatostatin-28 into somatostatin-14. Proc Nat Acad Sci USA 81:6662–6666
Go VLW, Brooks FP, Dimagno EP et al. (1986) The exocrine pancreas. Raven Press, New York, 904 pp

Gomih YK, Grillo TAI (1976) Insulin-like activity of the extract of the digestive gland and the pylorus of the giant African snail, *Achatina fulica*. A preliminary report. In: Grillo TAI, Leibson L, Epple A (eds) The evolution of pancreatic islets. Pergamon Press, Oxford, pp 153–162

Goodman RH, Lund PK, Jacobs JW, Habener JF et al. (1980a) Pre-prosomatostatin. Products of cell-free translations of messenger RNA's from anglerfish islets. J Biol Chem 255:6549–6552

Goodman RH, Jacobs JW, Chin WW et al. (1980b) Nucleotide sequence of a cloned structural gene coding for a precursor of pancreatic somatostatin. Proc Nat Acad Sci USA 77:5869–5873

Goodridge AG (1968) Lipolysis in vitro in adipose tissue from embryonic and growing chicks. Am J Physiol 214:897

Goodridge AG (1973) Regulation of fatty acid synthesis in isolated hepatocytes prepared from livers of neonatal chicks. J Biol Chem 248:1924–1931

Gorelick FS, Jamieson JD (1981) Structure-function relationships of the pancreas. In: Johnson LR (ed) Physiology of the gastrointestinal tract. Raven Press, New York, pp 773–794

Gourdoux L, Lequellec Y, Moreau R, Dutrieu J (1983) Gluconeogenesis from some amino acids and its endocrine modification in *Tenebrio molitor* L. (Coleoptera). Comp Biochem Physiol 74B:273–276

Gourley DRH, Kyu ST, Brunton LL (1969) Seasonal differences and the effect of insulin on pyruvate uptake, oxidation and synthesis to glycogen by frog skeletal muscle. Comp Biochem Physiol 29:509–524

Graf R (1981) Immunocytochemical detection of anti-ACTH reactivity in pancreatic islet cells of normal and steroid diabetic rats. Histochemistry 73:233–238

Grande F (1969) Lack of insulin effects on free fatty acid mobilization produced by glucagon in birds. Proc Soc Exp Biol Med 130:711–713

Grande F (1970) Effects of glucagon and insulin on plasma free fatty acids and blood sugar in owls. Proc Soc Exp Biol Med 133:540–543

Grant WC Jr, Hendler FJ, Banks PM (1969) Studies on blood sugar regulation in the little skate, *Raja erinacea*. Physiol Zool 42:231–247

Gray DE, Lickey HLA, Vranic M (1980) Physiological effects of epinephrine on glucose turnover and plasma free fatty acid concentrations mediated independently of glucagon. Diabetes 29:600–608

Greeley Jr GH, Trowbridge J, Burdett J et al. (1984) Radioimmunoassay of pancreatic polypeptide in mammalian and submammalian vertebrates using a carboxy-terminal hexapeptide antiserum. Regul Pept 8:177–187

Greider MH, Gersell DJ, Gingerich RL (1978) Ultrastructural localization of pancreatic polypeptide in the F cell of the dog pancreas. J Histochem Cytochem 26:1103–1108

Greenwald L, Munford J (1976) Insulin transport across intact frog skin. Gen Comp Endocrinol 29:426–429

Grimmelikhuijzen CJP (1984) Peptides in the nervous system of coelenterates. In: Falkmer S, Håkanson R, Sundler F (eds) Evolution and tumour pathology of the neuroendocrine system, Chap 3. Elsevier, Amsterdam, New York, pp 39–58

Groscolas R, Bezard J (1977) Effect of glucagon and insulin on plasma free fatty acids and glucose in the emperor penguin, *Aptenodytes forsteri*. Gen Comp Endocrinol 32:230–235

Gross R, Mialhe P (1982) Beta adrenergic receptors mediated effects, insulin and plasma free fatty acids in the duck. Horm Metab Res 14:18–25

Gross R, Mialhe P (1986) Free fatty acids and pancreatic function in the duck. Acta Endocrinol 112:100–104

Grube D, Aebert H (1981) Immunocytochemical investigations of gastroenteropancreatic endocrine cells using semithin and thin serial section. In: Grossman MI, Brazier MAB, Lechago J (eds) Cellular basis of chemical messengers in the digestive system. Academic Press, New York, pp 83–95

Grube D, Maier V, Raptis S, Schlegel W (1978a) Immunoreactivity of the endocrine pancreas. Evidence for the presence of cholecystokinin-pancreozymin within the A-cell. Histochemistry 56:13–35

Grube D, Voigt KH, Weber E (1978b) Pancreatic glucagon cells contain endorphin-like immunoreactivity. Histochemistry 59:75–79

Grunstein HS, James DE, Storlien LH et al. (1985) Hyperinsulinemia suppresses glucose utilization in specific brain regions: in vivo studies using the euglycemic clamp in the rat. Endocrinology 116:604–610

Guha B, Ghosh A (1978) A cytomorphological study of the endocrine pancreas of some Indian birds. Gen Comp Endocrinol 34:38–44
Guidotti GG, Luneburg B, Borghetti AF (1969) Amino acid uptake in isolated chick embryo heart cells. Biochem J 114:97
Guilleman R, Brazeau P, Böhlen P et al. (1984) Somatostatin, the growth hormone releasing factor. Rec Prog Horm Res 40:233–299
Gullo L, Labo G (1981) Thyrotropin-releasing hormone inhibits pancreatic enzyme secretion in humans. Gastroenterology 80:735–739
Gunesch VK-D (1974) Beeinflussung des Fettstoffwechsels beim Krallenfrosch (*Xenopus laevis* Daudin) durch ACTH, Corticosteroide und Insulin. Zool Jahrb 78:108–127
Guntupalli J, Rogers A, Bourke E (1985) Effect of insulin on renal phosphorus handling in the rat: interaction with PTH and nicotinamide. Am J Physiol 2:F610–F618
Gutierrez J, Carrillo M, Lancey S, Planas J (1984) Daily rhythms of insulin and glucose levels in the plasma of sea bass *Dicentrarchus labrax* after experimental feeding. Gen Comp Endocrinol 55:393–397
Gutierrez J, Fernandez J, Blasco J et al. (1986) Plasma glucagon levels in different species of fish. Gen Comp Endocrinol 63:328–333
Hagedorn HC, Jensen BN, Krarup NN, Wodstrup I (1936) Protamine insulinate. JAMA 106:177A
Hager SR, Bittar EE (1985) Hormones and the barnacle muscle fiber as a preparation. Comp Biochem Physiol 81C(No. 2):247–252
Hahn von Dorsche H, Drause R, Fehrmann P, Sulzmann R (1976) The verification of neurons in the pancreas of the spiny mice (*Acomys cahirinus*). Endokrinologie 67:115–118
Håkanson R, Lundquist I (1971) Occurrence of insulin in rat duodenum and its depletion with alloxan. Experientia (Basel) 27:1220–1221
Halter JB, Beard JC, Porte D Jr (1984) Islet function and stress hyperglycemia: plasma glucose and epinephrine interaction. Am J Physiol 247:E47–E52
Halushka PV, Colwell JA (1983) Prostaglandins and diabetes mellitus. In: Ellenberg M, Rifkin H (eds) Diabetes mellitus, theory and practice, Chap 14. Medical Examination Publishing Co, Inc, pp 295–308
Hammerman MR (1985) Interaction of insulin with the renal proximal tubular cell. Am J Physiol 249:F1–F11
Hammond JM, Baranao JLS, Shaleris D et al. (1985) Production of insulin-like growth factors by ovarian granulosa cells. Endocrinology 117:2553–2555
Hanke W (1974a) Endocrinology of amphibia. In: Florkin M, Scheer BT (eds) Chemical zoology. Academic Press, New York, pp 123–159
Hanke W (1974b) Hormonal regulation of metabolism in amphibia. Fortschr Zool 22:431–455
Hanke W (1974c) Amphibienentwicklung. Fortschr Zool 22:468–470
Hanke W, Neumann U (1972) Carbohydrate metabolism in amphibia. Gen Comp Endocrinol 3:198–208
Hanke W, Janssens PA (1983) The role of hormones in regulation of carbohydrate metabolism in the Australian lungfish *Neoceratodus forsteri*. Gen Comp Endocrinol 51:364–369
Hansen SE, Hedeskov CJ (1977) Simultaneous determination of the content of serotonin, dopamine, noradrenaline and adrenaline in pancreatic islets isolated from fed and starved mice. Acta Endocrinol 86:820–832
Hardisty MW (1979) Biology of the cyclostomes. Chapman and Hall, London
Hardisty MW (1982) Lampreys and hagfishes: analysis of cyclostome relationships. In: Hardisty MW, Potter IC (eds) The biology of lampreys, vol 4B. Academic Press, London, New York, pp 166–259
Hardisty MW, Baker BI (1982) Endocrinology of lampreys. In: Hardisty MW, Potter IC (eds) The biology of lampreys, vol 4B. Academic Press, London, New York, pp 1–115
Hardisty MW, Zelnik R, Moore IA (1975) The effects of subtotal and total isletectomy in the river lamprey, *Lampetra fluviatilis*. Gen Comp Endocrinol 27:179–192
Harri MNE, Puuska M (1973) Hormonal control of fat metabolism in the frog, *Rana temporaria*. Gen Comp Endocrinol 21:129–137
Harvey S, Klandorf H, Foltzer C et al. (1982) Endocrine responses of ducks (*Anas platyrhynchus*) to treadmill exercise. Gen Comp Endocrinol 48:415–420
Haskell JF, Meezan E, Pillion DJ (1985) Neonatal porcine cerebral microvessels have a greater number of insulin receptors/unit protein than the adult. Am J Physiol 248:E115–E125

Hatfield JS, Pansky B, Waller HJ, Budd GC (1981) Insulin-like immunoreactivity in cells of the mouse anterior pituitary. Micron 12:205–206

Hatton TW, Yip CC, Vranic M (1985) Biosynthesis of glucagon (IRG^{3500}) in canine gastric mucosa. Diabetes 34:38–46

Havrankova J, Brownstein M, Roth JJ (1979) Concentrations of insulin and or insulin receptors in the brain are independent of peripheral insulin levels: studies of obese and streptozotocin-treated rodents. J Clin Invest 64:636–642

Havrankova J, Roth J, Brownstein MJ (1983) Insulin receptors in brain. Adv Metab Disorders 10:259–267

Hayford JT, Danney MM, Hendrix JA, Thompson RG (1980) Integrated concentration of growth hormone in juvenile-onset diabetes. Diabetes 29:391–398

Hazelwood RL (1965) Carbohydrate metabolism. In: Sturkie PD (ed) Avian physiology, second edition. Cornell University Press, New York, pp 313–317

Hazelwood RL (1980) The avian gastro-enteric-pancreatic system: structure and function. In: Epple A, Stetson M (eds) Avian endocrinology. Academic Press, New York, London, Toronto, Sydney, San Francisco, pp 231–250

Hazelwood RL (1981) Synthesis, storage, secretion, and significance of pancreatic polypeptide in vertebrates. In: Cooperstein SJ, Watkins D (eds) The islets of Langerhans. Biochemistry, physiology and pathology, Chap 12. Academic Press, London, New York, pp 276–318

Hazelwood RL (1984) Pancreatic hormones, insulin/glucagon molar ratios, and somatostatin as determinants of avian carbohydrate metabolism. J Exp Zool 232:647–652

Hazelwood RL (1986) Carbohydrate metabolism. In: Sturkie PD (ed) Avian physiology, fourth edition, Chap 13. Springer, New York, pp 303–325

Hazelwood RL, Lorenz FW (1959) Effects of fasting and insulin on carbohydrate metabolism of the domestic fowl. Am J Physiol 197:47–51

Hedeskov CJ (1980) Mechanism of glucose-induced insulin secretion. Physiol Rev 60:442–509

Heding L (1983) The immunogenicity of glucagon. In: Lefébvre PJ (ed) Glucagon I, Chap 9. Springer, Berlin Heidelberg New York Tokyo, pp 189–201

Heidenreich K, Zahniser NR, Berhanu P et al. (1983) Structural differences between insulin receptors in the brain and peripheral target tissues. J Biol Chem 258:8527–8530

Helgeson AS, Lawson T, Pour P (1984) Exocrine pancreatic secretion in the Syrian golden hamster *Mesocricetus auratus* – III. Effects of carcinogen administration and development of pancreas cancer. Comp Biochem Physiol 77C:191–197

Hellerström CL, Asplund K (1966) The two types of A-cells in the pancreatic islets of snakes. Z Zellforsch 70:68–80

Hellmann B, Lernmark A (1969) Inhibition of the in vitro secretion of insulin by an extract of pancreatic a_1 cells. Endocrinology 84:1484–1487

Hemminga VA (1984) Regulation of glycogen metabolism in the freshwater snail *Limnea stagnalis*. Ph.D. Thesis, Academic Press, Amsterdam

Henderson JR (1969) Why are the islets of Langerhans? Lancet ii:469–470

Henderson JR (1974) Insulin in body fluids other than blood. Physiol Rev 54:1–22

Henderson JR, Daniel PM, Fraser PA (1981) The pancreas as a single organ: the influence of the endocrine upon the exocrine part of the gland. Gut 22:158–167

Hermansen K, Schwartz TW (1979) Differential sensitivity to somatostatin of pancreatic polypeptide, glucagon and insulin secretion from the isolated, perfused canine pancreas. Metabolism 28:1229–1233

Hernandez T, Coulson RA (1968) Effect of insulin on free amino acids in caiman tissue and plasma. Comp Biochem Physiol 26:991–996

Heumann R, Schwab ME (1985) The production of NGF mRNA in peripheral organs. Trends Neurosci 8:373–374

Hill DE, Mayes S, DiBattista D et al. (1977) Hypothalamic regulation of insulin release in rhesus monkeys. Diabetes 26:727–731

Hilliard RW, Epple A, Potter IC (1985) The morphology and histology of the endocrine pancreas of the Southern Hemisphere lamprey, *Geotria australis* Gray. J Morphol 184:253–261

Hisatomi A, Maruyama H, Orci L et al. (1985) Adrenergically mediated intrapancreatic control of the glucagon response to glucopenia in the isolated rat pancreas. J Clin Invest 75:420–426

Hobart P, Crawford R, Shen L-P et al. (1980) Cloning and sequence analysis of cDNAs encoding two distinct somatostatin precursors found in the endocrine pancreas of anglerfish. Nature (Lond) 288:137–141

Hodgkin DC, Mercola D (1972) The secondary and tertiary structure of insulin. In: Freinkel N, Steiner DF (eds) Endocrine pancreas. Williams and Wilkins, Baltimore, Handbook of physiology, sect 7, vol 1, pp 139–157

Hökfelt T, Elfvin LG, Elde R et al. (1977) Occurrence of somatostatin-like immunoreactivity in some peripheral sympathetic noradrenergic neurons. Proc Nat Acad Sci USA 74:3587–3591

Holmgren S, Vaillant C, Dimaline R (1982) VIP-, substance P-, gastrin/CCK-, bombesin-, somatostatin- and glucagon-like immunoreactivities in the gut of the rainbow trout, *Salmo gairdneri*. Cell Tissue Res 223:141–153

Holst JJ (1977) Extraction, gel filtration pattern, and receptor binding of porcine gastrointestinal glucagon-like immunoreactivity. Diabetologia 13:159–169

Holst JJ, Grønholt, Schaffalitzky de Muckadell OB, Fahrenkrug J (1981a) Nervous control of pancreatic endocrine secretion in pigs II. The effect of pharmacological blocking agents on the response to vagal stimulation. Acta Physiol Scand 111:9–14

Holst JJ, Grønholt R, Schaffalitzky de Muckadell OB, Fahrenkrug J (1981b) Nervous control of pancreatic endocrine secretion in pigs V. Influence of the sympathetic nervous system on the pancreatic secretion of insulin and glucagon, and on the insulin and glucagon response to vagal stimulation. Acta Physiol Scand 113:279–283

Holst JJ, Pedersen JH, Baldissera F, Stadil F (1983) Circulating glucagon after total pancreatectomy in man. Diabetologia 25:396–399

Honey RN, Weir GC (1980) Acetylcholine stimulates insulin, glucagon, and somatostatin release in the perfused chicken pancreas. Endocrinology 107:1065–1068

Honey RN, Fallon MB, Weir GC (1980) Effects of exogenous insulin, glucagon, and somatostatin on islet hormone secretion in the perfused chicken pancreas. Metabolism 29:1242–1246

Honey RN, Arimura A, Weir GC (1981) Somatostatin neutralization stimulates glucagon and insulin secretion from the avian pancreas. Endocrinology 109:1971–1974

Honjin R (1956) The innervation of the pancreas of mouse with special reference to the structure of the peripheral extension of the vegetative nervous system. J Comp Neurol 104:331–370

Hoo-Paris R, Hamsany M, Sutter BChJ et al. (1982) Plasma glucose and glucagon concentrations in the hibernating hedgehog. Gen Comp Endocrinol 46:246–254

Hoosein NM, Gurd RS (1984) Human glucagon-like peptides 1 and 2 activate rat brain adenylate cyclase. FEBS Lett 178:83–86

Hopcroft DW, Mason DR, Scott RS (1985) Structure-function relationships in pancreatic islets: support for intraislet modulation of insulin secretion. Endocrinology 117:2073–2080

Horuk R, Goodwin P, O'Conner K et al. (1979) Evolutionary change in the insulin receptor of hystricomorph rodents. Nature 279:439–440

Huang CG, Eng J, Pan YCE et al. (1986) Guinea pig glucagon differs from other mammalian glucagons. Diabetes 35:508–512

Hulsebus J, Farrar ES (1985) Insulin-like immunoreactivity in serum and pancreas of metamorphosing tadpoles. Gen Comp Endocrinol 58:114–119

Humbel RE, Bosshard HR, Zahn H (1972) Chemistry of insulin. In: Freinkel N, Steiner DF (eds) Endocrine pancreas. Williams and Wilkins, Baltimore, Handbook of physiology, sect. 7, vol 1, pp 111–132

Hunt LT, Dayhoff MO (1979) Structural and functional similarities among hormones and active peptides from distantly related eukaryotes. In: Gross E, Meienhofer J (eds) Peptides: structure and biological function. Proceedings of the Sixth American Peptide Symposium. Rockford, IL. Pierce Chem Co, pp 757–760

Huth A, Rapoport TA (1982) Regulation of the biosynthesis of insulin isolated Brockmann bodies of the carp (*Cyprinus carpio*). Gen Comp Endocrinol 46:158–167

Ichikawa T, Lederis K, Kobayashi H (1984) Primary structure of multiple forms of urotensin II in the urophysis of the carp, *Cyprinus carpio*. Gen Comp Endocrinol 55:133–141

Ince BW (1979) Insulin secretion from the in situ perfused pancreas of the European silver eel, *Anguilla anguilla* L. Gen Comp Endocrinol 37:533–540

Ince BW (1980) Amino acid stimulation of insulin secretion from the in situ perfused eel pancreas, modification by somatostatin, adrenaline and theophylline. Gen Comp Endocrinol 40:275–282

Ince BW (1982) Plasma clearance kinetics of unlabelled bovine insulin in rainbow trout (*Salmo gairdneri*). Gen Comp Endocrinol 46:463–472

Ince BW (1983) Secretin-stimulated insulin release in vivo in European eels, *Anguilla anguilla* L. J Fish Biol 22:259–263

Ince BW, So STC (1984) Differential secretion of glucagon-like and somatostatin-like immunoreactivity from the perfused eel pancreas in response to D-glucose. Gen Comp Endocrinol 53:389–397

Ince BW, Thorpe A (1975) Hormonal and metabolite effects on plasma free fatty acids in the northern pike, *Esox lucius* L. Gen Comp Endocrinol 27:144–152

Ince BW, Thorpe A (1977) Glucose and amino acid-stimulated insulin release in vivo in the European silver eel (*Anguilla anguilla* L.). Gen Comp Endocrinol 31:249–256

Ince BW, Thorpe A (1978) Insulin kinetics and distribution in rainbow trout (*Salmo gairdneri*). Gen Comp Endocrinol 35:1–9

Inokuchi A, Tomida Y, Yanaihara C et al. (1986) Glucagon-related peptides in the rat hypothalamus. Cell Tissue Res 246:71–75

In't Veld PA, Pipeleers DG, Gepts W (1984) Evidence against the presence of tight junctions in normal endocrine pancreas. Diabetes 33:101–104

Inui A, Mizuno N, Oya M et al. (1986) Effects of amino acids on pancreatic polypeptide before and after vagotomy in the dog. Diabetologia 29:262–264

Inui Y (1969) Hepatectomy in eels. Its operation technique and effects on blood glucose. Bull Jap Soc Sci Fish 35:975–978

Inui Y, Gorbman A (1977) Sensitivity of Pacific hagfish, *Eptatretus stouti*, to mammalian insulin. Gen Comp Endocrinol 33:423–427

Inui Y, Gorbman A (1978) Role of the liver in regulation of carbohydrate metabolism in hagfish, *Eptatretus stouti*. Comp Biochem Physiol 60A:181–183

Inui Y, Ishioka H (1983a) Effects of insulin and glucagon on the incorporation of [^{14}C] glycine into the protein of the liver and opercular muscle of the eel in vitro. Gen Comp Endocrinol 51:208–212

Inui Y, Ishioka H (1983b) Effects of insulin and glucagon on amino acid transport into the liver and opercular muscle of the eel in vitro. Gen Comp Endocrinol 51:213–218

Inui Y, Yokote M (1975) Gluconeogenesis in the eel – III. Effects of mammalian insulin on the carbohydrate metabolism of the eel. Bull Jap Soc Sci Fish 41:965–972

Inui Y, Yokote M (1977) Effects of glucagon on amino acid metabolism in Japanese eels, *Anguilla japonica*. Gen Comp Endocrinol 33:167–173

Inui Y, Arai S, Yokote M (1975) Gluconeogenesis in the eel. VI. Effects of hepatectomy, alloxan, and mammalian insulin on the behavior of plasma amino acids. Jap Soc Sci Fish Bull 41:1105–1111

Inui Y, Yu JY-L, Gorbman A (1978) Effect of bovine insulin on the incorporation of [^{14}C] glycine into protein and carbohydrate in liver and muscle of hagfish, *Eptatretus stouti*. Gen Comp Endocrinol 36:133–141

Ipp E, Rivier J, Dobbs RE et al. (1979) Somatostatin analogs inhibit somatostatin release. Endocrinology 104:1270–1273

Ishida T, Rojdmark S, Bloom G et al. (1980) The effect of somatostatin on the hepatic extraction of insulin and glucagon in the anesthetized dog. Endocrinology 106:220–230

Itoh M, Reach G, Furman B, Gerich JE (1981) Secretion of glucagon. In: Cooperstein SJ, Watkins D (eds) The islets of Langerhans. Biochemistry, physiology and pathology, Chap 10. Academic Press, London, New York, pp 225–255

Iversen J, Miles DW (1971) Evidence for feedback inhibition of insulin on insulin secretion in the isolated, perfused canine pancreas. Diabetes 20:1–8

Iwanaga T, Yui R, Fujita T (1983) The pancreatic islets of the chicken. In: Mikami S, Hamna K, Wada M (eds) Avian endocrinology. Springer, Berlin Heidelberg New York, pp 81–94

Jackson IMD (1978) Extrahypothalamic and phylogenetic distribution of hypothalamic peptides. In: Reichlin S, Baldessarini RJ, Martin JB (eds) The hypothalamus. Raven Press, New York, pp 217–231

Janssens PA, Maher F (1986) Glucagon and insulin regulate in vitro hepatic glycogenolysis in the axolotl *Ambystoma mexicanum* via changes in tissue cyclic AMP concentration. Gen Comp Endocrinol 61:64–70

Jansson L, Hellerström C (1983) Stimulation by glucose of the blood flow to the pancreatic islets of the rat. Diabetologia 25:45–50

Järhult J, Farnebo L-O, Hambergh B, Holst JJ, Schwartz TW (1981) The relation between catecholamines, glucagon and pancreatic polypeptide during hypoglycaemia in man. Acta Endocrinologica 98:402–406

Jarry H, Düker E-M, Wuttke W (1985) Adrenal release of catecholamines and met-enkephalin before and after stress as measured by a novel in vivo dialysis method in the rat. Neurosci Lett 60:273–278

Jarotzky AJ (1899) Über die Veränderungen in der Größe und im Bau der Pankreaszellen mit einigen Arten der Inanition. Virchows Arch (Pathol Anat) 156:409–429

Jaspan J, Polonsky K, Lewis M, Moosa A (1979) Reduction in portal vein blood flow by somatostatin. Diabetes 28:888–892

Jeanrenaud B (1985) An hypothesis on the aetiology of obesity: dysfunction of the central nervous system as a primary cause. Diabetologia 28:502–513

Jefferson LS (1980) Role of insulin in the regulation of protein synthesis. Diabetes 29:487–496

Jefferson LS, Rannels DE, Munger BL, Morgan HE (1974) Insulin in the regulation of protein turnover in heart and skeletal muscle. Fed Proc 33:1098A

Jessen KR, Mirsky R, Dennison ME, Burnstock G (1979) GABA may be a neurotransmitter in the vertebrate peripheral nervous system. Nature (London) 281:71–74

Johnson DE, Torrence JL, Elde RP et al. (1976) Immunohistochemical localization of somatostatin, insulin and glucagon in the principal islets of the anglerfish (*Lophius americanus*) and the channel catfish (*Ictalurus punctatus*). Am J Anat 147:119–124

Johnson DE, Noe BD, Bauer GE (1982) Pancreatic polypeptide (PP)-like immunoreactivity in the pancreatic islets of the anglerfish (*Lophius americanus*) and the channel catfish (*Ictalurus punctatus*). Anat Rec 204:61–67

Jörnvall H, Carlston A, Pettersson T et al. (1981) Structural homologies between prealbumin, gastrointestinal hormones and other proteins. Nature (London) 291:261–263

Josefsson JO, Johansson P (1979) Naloxone – reversible effects of opioids on pinocytosis in *Amoeba proteus*. Nature (London) 282:78–80

Kadowaki S, Norman AW (1985a) Time course study of insulin secretion after 1, 25-dihydroxyvitamin D_3 administration. Endocrinology 117:1765–1771

Kadowaki S, Norman AW (1985b) Demonstration that the vitamin D metabolite 1, $25(OH)_2$-vitamin D_3 and not 24R, $25(OH)_2$-vitamin D_3 is essential for normal insulin secretion in the perfused rat pancreas. Diabetes 34:315–320

Kaiser D, Manoil C, Dworkin M (1979) Myxobacteria: cell interactions, genetics, and development. Ann Rev Microbiol 33:595–639

Kameda Y, Oyama H, Horino M (1984) Ontogeny of immunoreactive somatostatin in thyroid C cells from dogs and guinea pigs. Anat Rec 208:89–101

Kallen JL, Reijntjens FMJ, Peters KJM, van Herp F (1986) Biochemical analyses of the crustacean hyperglycemic hormone of the crayfish *Astacus leptodactylus*. Gen Comp Endocrinol 61:248–259

Katzeff HL, Savage PJ, Barclay-White B et al. (1985) C-peptide measurement in the differentiation of type 1 (insulin-dependent) and type 2 (non-insulin-dependent) diabetes mellitus. Diabetologia 28:264–268

Kaung H-LC (1981) Immunocytochemical localization of pancreatic endocrine cells in frog embryos and young larvae. Gen Comp Endocrinol 45:204–211

Kaung H-LC (1983) Changes of pancreatic beta cell population during larval development of *Rana pipiens*. Gen Comp Endocrinol 49:50–56

Kaung H-LC, Elde R (1980) Distribution and morphometric quantitation of pancreatic endocrine cell types in the frog, *Rana pipiens*. Anat Rec 196:173–181

Kawai K, Rouiller D (1981) Evidence that the islet interstitium contains functionally separate 'arterial' and 'venous' compartments. Diabetes 30(Suppl 1):14A

Kawai K, Orci L, Unger RH (1982a) High somatostatin uptake by the isolated perfused dog pancreas consistent with an "insulo-acinar" axis. Endocrinology 110:660–662

Kawai K, Ipp E, Orci L, Perrelet A, Unger RH (1982b) Circulating somatostatin acts on the islets of Langerhans by way of a somatostatin-poor compartment. Science 218:477–478

Kawano H, Daikoku S, Saito S (1983) Location of thyrotropin-releasing hormone-like immunoreactivity in rat pancreas. Endocrinology 112:951–955

Keller R, Jaros PP, Kegel G (1985) Crustacean hyperglycemic neuropeptides. Am Zool 25:207–221

Khanna SS, Gill TS (1972) Further observations on the blood glucose level in *Channa punctatus* (Bloch). Acta Zool 53:127–133

Khemiss F, Sitbon G (1982) Effects of amino acids on pancreatic hormones and gut glucagon-like immunoreactivity in the goose. Horm Metab Res 14:122–127
Kimmel JR, Pollock HG (1981) Target organs for avian pancreatic polypeptide. Endocrinology 109:1693–1699
Kimmel JR, Pollock HG, Hazelwood RL (1968) Isolation and characterization of chicken insulin. Endocrinology 83:1323–1330
Kimmel JR, Pollock HG, Hazelwood RL (1971) A new pancreatic polypeptide hormone. Fed Proc 30:1318A
Kimmel JR, Maher MJ, Pollock HG, Vensel WH (1976) Isolation and characterization of reptilian insulin: partial amino acid sequence of rattlesnake (*Crotalus atrox*) insulin. Gen Comp Endocrinol 28:320–333
Kimmel JR, Pollock HG, Hayden LJ (1978) Biological activity of avian PP in the chicken. In: Bloom SR (ed) Gut Hormones, Churchill Livingstone, Edinburgh, London, New York, pp 234–241
Kimmel JR, Pollock HG, Chance RE et al. (1984) Pancreatic polypeptide from rat pancreas. Endocrinology 114:1725–1731
King DL, Hazelwood RL (1976) Regulation of avian insulin secretion by isolated perfused chicken pancreas. Am J Physiol 231:1830–1839
King GL, Kahn CR (1981) Non-parallel evolution of metabolic and growth-promoting functions of insulin. Nature 292:644–646
King JA, Millar RP (1979) Phylogenetic and anatomical distribution of somatostatin in vertebrates. Endocrinology 105:1322–1329
King JR (1972) Adaptive periodic fat storage by birds. Proc XV. Int Ornithol Congress, The Hague, pp 200–217
Kitabchi AE (1977) Proinsulin and c-peptide: a review. Metabolism 26:547–587
Kitabchi AE (1983) Proinsulin and c-peptides. In: Ellenberg M, Rifkin H (eds) Diabetes mellitus, theory and practice, Chap 6. Medical Examination Publishing Co, Inc, pp 97–117
Klein C (1971) Innervation des cellules du pancreas endocrine du Poisson Teleosteen *Xiphophorus helleri*. Z Zellforsch 113:564–580
Klein C, van Noorden S (1978) Use of immunocytochemical staining of somatostatin for correlative light and electron microscopic investigation of D cells in the pancreatic islet of *Xiphophorus helleri* H. (Teleostei). Cell Tissue Res 194:399–404
Klein C, van Noorden S (1980) Pancreatic polypeptide (PP)- and glucagon cells in the pancreatic islet of *Xiphophorus helleri* H. (Teleostei). Cell Tissue Res 205:187–198
Knip M, Pentti L, Akerblom HK et al. (1983) Partial purification of an insulin-releasing activity in human serum. Life Sci 33:2311–2319
Knudtzon J (1984) Adrenergic effects on plasma levels of glucagon, insulin, glucose and free fatty acids in rabbits. Horm Metab Res 16:415–422
Kobayashi K (1969) Light and electron microscopic studies on the pancreatic acinar and islet cells in *Xenopus laevis*. Gunma J Med Sci 18:60–103
Kobayashi K, Syed Ali S (1981) Cell types of the endocrine pancreas in the shark *Scyliorhinus stellaris* as revealed by correlative light and electron microscopy. Cell Tissue Res 215:475–490
Kochert G (1978) Sex pheromones in algae and fungi. Ann Rev Plant Physiol 29:41–486
Kohen E, Kohen C, Thorell B et al. (1979) Intercellular communication in pancreatic islet monolayer cultures: a microfluorometric study. Science 204:862–865
Köhler E (1963) Versuche zur Genese und Therapie der Fettleber bei Fischen. Z Gesamte Inn Med Ihre Grenzgeb 18:936–942
Kolanowski J (1983) Influence of glucagon on water and electrolyte metabolism. In: Lefébvre PJ (ed) Glucagon II, Chap 50. Springer, Berlin New York Tokyo, pp 525–536
Kolata G (1984) Steroid hormone systems found in yeast. Science 225:913–914
Komiya I, Yamada T, Kanns Y et al. (1984) Inhibitory action of thyrotropin-releasing hormone on serum amylase activity and its mechanism. J Clin Endocrinol Metab 58:1059–1063
Kondo H, Yui R (1984) Co-existence of enkephalin and adrenalin in the frog adrenal gland. Histochemistry 80:243–246
Koranyi L, Peterfy F, Szabo T et al. (1981) Evidence for transformation of glucagon-like immunoreactivity of gut into pancreatic glucagon in vivo. Diabetes 30:792–794
Kramer KJ, Childs CN, Speirs RD, Jacobs RM (1982) Purification of insulin-like peptides from insect haemolymph and royal jelly. Insect Biochem 12:91–98

Kream BE, Smith MD, Canalis E, Raisy LG (1985) Characterization of the effect of insulin on collagen synthesis in fetal rat bone. Endocrinology 116:296–302
Krejs GJ (1985) Non-insulin-secreting tumors of the pancreatic islets. In: Wilson JD, Foster DW (eds) Textbook of endocrinology. WB Saunders Co, Philadelphia, pp 1301–1308
Krieger DT (1983) Brain peptides: what, where and why? Science 222:975–985
Krogsgaard-Larsen P, Falch E (1981) GABA agonists. Molec Cell Biochem 38:129–146
Krulich L, Dhariwal APS, McCann SM (1968) Stimulatory and inhibitory effects of purified hypothalamic extracts on growth hormone release from rat pituitary in vitro. Endocrinology 83:783–790
La Barre J, Stille EV (1930) Studies on the physiology of secretin III. Further studies on the effects of secretin on the blood sugar. Am J Physiol 91:649–653
Lacy PE (1977) The physiology of insulin release. In: Volk BW, Wellman KF (eds) The diabetic pancreas. Plenum Press, New York, p 211–230
Lacy PE, Greider MH (1972) Ultrastructural organization of mammalian pancreatic islets. In: Freinkel N, Steiner DF (eds) Endocrine pancreas. Williams and Wilkins, Baltimore, Handbook of physiology, sect 7, vol 1, pp 77–89
Lacy PE, Greider MH (1979) Anatomy and ultrastructural organization of pancreatic islets. In: DeGroot LJ, Cahill GF Jr, Odell WD, Martini L, Potts JT Jr, Nelson DH, Steinberger E, Winegrad AI (eds) Endocrinology, vol 2. Grune and Stratton, New York, London, Toronto, Sydney, San Francisco, pp 907–919
Laguesse ME (1893) Sur la formation des îlots de Langerhans dans le pancréas. CR Soc Biol 45:819–920
Laito M, Lev R, Orlic D (1974) The developing human fetal pancreas: an ultrastructural and histochemical study with special reference to exocrine cells. J Anat 117:619–634
Lamberton P, Wu P, Jackson IMD (1985) Thyrotropin-releasing hormone release from rat pancreas is stimulated by serotonin but inhibited by carbachol. Endocrinology 117:1834–1838
Lance V, Hamilton JW, Rouse JB et al. (1984) Isolation and characterization of reptilian insulin, glucagon, and pancreatic polypeptide: complete amino acid sequence of alligator (*Alligator mississippiensis*) insulin and pancreatic polypeptide. Gen Comp Endocrinol 55:112–124
Landau BR, Tahaoha Y, Abrams MA et al. (1983) Binding of insulin by monkey and pig hypothalamus. Diabetes 32:284–292
Landauer W (1945) Rumplessness of chicken embryos produced by the injection of insulin and other chemicals. J Exp Zool 98:65–77
Landsberg L, Young JB (1985) Catecholamines and the adrenal medulla. In: Wilson JD, Foster DW (eds) Textbook of endocrinology. WB Saunders Co, Philadelphia, pp 891–965
Lange RH (1973) Histochemistry of the islets of Langerhans. In: Graumann W, Neumann K (eds) Handbuch der Histochemie, vol 8/1. Gustav Fischer, Stuttgart, pp 1–141
Lange RH (1984) The vascular system of principal islets: semithin-section studies in teleosts fixed by perfusion. Gen Comp Endocrinol 54:270–276
Langer M, van Noorden S, Polak JM, Pearse AGE (1979) Peptide hormone-like immunoreactivity in the gastrointestinal tract and endocrine pancreas of eleven teleost species. Cell Tissue Res 199:493–508
Langerhans P (1869) Beiträge zur mikroskopischen Anatomie der Bauchspeicheldrüse. (Thesis) Friedrich-Wilhelms-Universität, Berlin
Langley JN (1921) The autonomic nervous system, part I. Heffer, London
Langslow DR, Hales CN (1969) Lipolysis in chicken adipose tissue in vitro. J Endocrinol 43:285–294
Langslow DR, Hales CN (1971) The role of the endocrine pancreas and catecholamines in the control of carbohydrate and lipid metabolism. In: Bell DJ, Freeman B (eds) Physiology and biochemistry of the domestic fowl. Academic Press, London, pp 521–547
Langslow DR, Kimmel JR, Pollock HG (1973) Studies of the distribution of a new avian pancreatic polypeptide and insulin among birds, reptiles, amphibians and mammals. Endocrinology 93:558–565
Langslow DR, Butler EJ, Hales CN, Pearson AW (1970) The response of plasma insulin, glucose and nonesterified fatty acids to various hormones, nutrients and drugs in the domestic fowl. J Endocrinol 46:243
Larsen LO (1976a) Blood glucose levels in intact and hypophysectomized river lampreys (*Lampetra fluviatilis* L.) treated with insulin, stress or glucose before and after the period of sexual maturation. Gen Comp Endocrinol 29:1–13

Larsen LO (1976b) Regulation of blood glucose in the river lamprey: the probable physiological role of insulin and hyperglycaemic hormone. In: Grillo TAI, Leibson L, Epple A (eds) The evolution of pancreatic islets. Pergamon Press, Oxford, pp 285–290
Larsen LO (1978) Subtotal hepatectomy in intact or hypophysectomized river lampreys (*Lampetra fluviatilis*): effects on regeneration, blood glucose regulation, and vitellogenesis. Gen Comp Endocrinol 35:197–204
Larsen LO (1980) Physiology of adult lampreys, with special regard to natural starvation, reproduction, and death after spawning. Can J Fish Aquat Sci 3:1762–1779
Larsson AL, Lewander K (1972) Effects of glucagon administration to eels (*Anguilla anguilla* L.). Comp Biochem Physiol 43A:831–836
Larsson L-I (1977a) Ontogeny of peptide-producing nerves and endocrine cells of the gastroduodeno-pancreatic region. Histochemistry 54:133–142
Larsson L-I (1977b) Corticotropin-like peptides in central nerves and in endocrine cells of gut and pancreas. Lancet ii:1321–1323
Larsson L-I (1984) Evidence for anterograde transport of secretory granules in processes of gastric paracrine (somatostatin) cells. Histochemistry 80:323–326
Larsson L-I, Rehfeld JF, Sundler F, Håkanson R (1976a) Pancreatic gastrin in foetal and neonatal rats. Nature 262:609–610
Larsson L-I, Sundler F, Håkanson R (1976b) Pancreatic polypeptide – a postulated new hormone: identification of its cellular storage site by light and electron microscopic immunocytochemistry. Diabetologia 12:211–226
Larsson L-I, Sundler F, Håkanson R et al. (1974) Localization of APP, a postulated new hormone, to a pancreatic endocrine cell type. Histochemistry 42:377–382
Larsson L-I, Fahrenkrug J, Hølst J, Schaffalitzky de Muckadell OB (1978) Innervation of the pancreas by vasoactive intestinal polypeptide (VIP) nerves. Life Sci 22:773–780
Larsson L-I, Goltermann NR, de Magistris L et al. (1979) Somatostatin cell processes as pathways for paracrine secretion. Science 205:1393–1394
Laughton W, Powley TL (1979) Four central nervous system sites project to the pancreas. Neuroscience 5:46(A)
Laurent R, Gross R, Lahili M, Mialhe P (1981) Effect of insulin on glucagon secretion mediated via glucose metabolism of pancreatic A cells in ducks. Diabetologia 20:72–77
Laurent F, Mialhe A, Boulanger Y, Mialhe P (1985) Amino acids in normal and diabetic ducks. Horm Metab Res 17:223–225
Lazarus SS, Volk BW (1962) The pancreas in human and experimental diabetes. Grune and Stratton, New York
Leboulenger F, Leroux P, Delarue C et al. (1983c) Co-localization of vasoactive intestinal peptide (VIP) and enkephalins in chromaffin cells of the adrenal gland of amphibia: stimulation of corticosteroid production by VIP. Life Sci 32:375–383
Leboulenger F, Leroux P, Delaru C et al. (1983b) Co-existence of vasoactive intestinal peptide and enkephalins in the adrenal chromaffin granules of the frog. Neurosci Lett 37:221–225
Lebovitz HE, Feldman JM (1973) Pancreatic biogenic amines and insulin secretion in health and disease. Fed Proc 32:1797–1802
Leclercq-Meyer V, Malaisse WJ (1983) Ions in the control of glucagon release. In: Lefébvre PJ (ed) Glucagon II, Chap 26. Springer, Berlin New York Tokyo, pp 59–74
Leclercq-Meyer V, Marchand J, Woussen-Colle MC et al. (1985a) Multiple effects on glucagon, insulin, and somatostatin secretion from the perfused rat pancreas. Endocrinology 116:1168–1174
Leclercq-Meyer V, Marchand J, Malaisse WJ (1985b) Insulin and glucagon release from the ventral and dorsal parts of the perfused pancreas of the rat. Horm Res 21:19–32
Leduque P, Moody AJ, Dubois PM (1982) Ontogeny of immunoreactive glicentin in the human gastrointestinal tract and endocrine pancreas. Regul Pept 4:261–274
Leduque P, Paulin C, Dubois PM (1983) Immunocytochemical evidence for a substance related to the bovine pancreatic polypeptide-peptide YY group of peptides in the human fetal gastrointestinal tract. Regul Pept 6:219–230
Leech AR, Goldstein L, Cha CJ, Goldstein JM (1979) Alanine biosynthesis during starvation in skeletal muscle of the spiny dogfish, *Squalus acanthias*. J Exp Zool 207:73–80
Lefébvre PJ (1983a) Glucagon I. Springer, Berlin Heidelberg New York Tokyo, 535 pp
Lefébvre PJ (1983b) Glucagon II. Springer, Berlin Heidelberg New York Tokyo, 700 pp

Lefébvre PJ, Luyckx AS (1983) The renal handling of glucagon. In: Lefébvre PJ (ed) Glucagon II, Chap 41. Springer, Berlin Heidelberg New York Tokyo, pp 389–396

Leibson LG (1981) Progress and problems of evolutionary endocrinology. (In Russian) Evol Biokh Fiziol 17:116–126

Leibson L, Plisetskaya EM (1968) Effect of insulin on blood sugar level and glycogen content in organs of some cyclostomes and fish. Gen Comp Endocrinol 11:381–392

Leibson L, Plisetskaya EM (1969) Hormonal control of blood sugar levels in cyclostomes. Gen Comp Endocrinol 2:528–534

Leibson L, Plisetskaya EM (1973) Effects of hormones in poikilothermic vertebrates. In: Michelson MJ (ed) International encyclopedia of pharmacological therapy. Pergamon Press, London, pp 625–684

Leibson L, Bondareva V, Soltitshaza L (1976a) The secretion and the role of insulin in chick embryos and chickens. In: Grillo TAI, Leibson L, Epple A (eds) The evolution of pancreatic islets. Pergamon Press, Oxford, pp 69–79

Leibson L, Plisetskaya EM, Leibush B (1976b) The comparative study of mechanisms of insulin action on muscle carbohydrate metabolism. In: Grillo TAI, Leibson L, Epple A (eds) The evolution of pancreatic islets. Pergamon Press, Oxford, pp 345–362

Leibush BN (1983) Insulin receptors of the brain in evolution of vertebrates. (In Russian) Zh Evol Biochem Physiol 19:407–413

Leibush BN, Bondareva VM (1981) Insulin receptor of liver plasma membranes from the scorpion-fish *Scorpaena porcus* in comparison with mammalian receptor. (In Russian) Zh Evol Biochem Physiol 17:141–147

Lernmark A (1985) Molecular biology of type 1 (insulin-dependent) diabetes mellitus. Diabetologia 28:195–203

LeRoith D, Roth JJ (1984) Evolutionary origins of messenger peptides: material in microbes that resemble vertebrate hormones. In: Falkmer S, Håkanson R, Sundler F (eds) Evolution and tumour pathology of the neuroendocrine system, Chap 9. Elsevier, Amsterdam, New York, pp 147–164

LeRoith D, Lesniak MA, Roth JJ (1981) Insulin in insects and annelids. Diabetes 30:70–76

LeRoith D, Hendricks SA, Lesniak MA et al. (1983a) Insulin in brain and other extrapancreatic tissues of vertebrates and nonvertebrates. Adv Metabol Disorders 10:303–340

LeRoith D, Shiloach J, Berelowitz M et al. (1983b) Are messenger molecules in microbes the ancestors of the vertebrate hormones and tissue factors? Fed Proc 42:2602–2607

LeRoith D, Pickens W, Crosby LK et al. (1985a) Evidence for multiple molecular forms of somatostatin-like material in *E. coli*. Biochim Biophys Acta 838:335–342

LeRoith D, Pickens W, Vinik AI, Shiloach J (1985b) *Bacillus subtilis* contains multiple forms of somatostatin-like material. Biochem Biophys Res Comm 127:713–719

LeRoith D, Pickens W, Wilson GL et al. (1985c) Somatostatin-like material is present in flowering plants. Endocrinology 117:2093–2097

Leshin M (1985) Multiple endocrine neoplasia. In: Wilson JD, Foster DW (eds) Textbook of endocrinology. WB Saunders Co, Philadelphia, pp 1274–1289

Lewis TL, Epple A (1972) Pancreatectomy in the eel: effects on serum glucose and cholesterol. Science 178:1286–1288

Lewis TL, Epple A (1984) Effects of fasting, pancreatectomy and hypophysectomy in the yellow eel, *Anguilla rostrata*. Gen Comp Endocrinol 55:182–194

Lewis TL, Parke WW, Epple A (1977) Pancreatectomy in a teleost fish, *Anguilla rostrata*. Lab Anim Sci 27:102–109

Lifson N, Lassa CV, Dixit PK (1985) Relation between blood flow and morphology in islet organ of rat pancreas. Am J Physiol 249:E43–E48

Lifson N, Kramlinger KG, Mayrand RR, Lender EJ (1980) Blood flow to the rabbit pancreas with special reference to the islets of Langerhans. Gastroenterology 79:466–473

Like AA, Orci L (1972) Embryogenesis of the human pancreatic islets: A light and electron microscopic study. Diabetes 21:511–534

Liljenquist JE, Horwitz DL, Jennings AS et al. (1978) Inhibition of insulin secretion in normal man as demonstrated by C-peptide assay. Diabetes 27:563–570

Lin TM, Chance RE (1972) Spectrum gastrointestinal actions of a new bovine pancreas polypeptide (BPP). Gastroenterology 62:852A

Lindström P, Sehlin J (1983) Opposite effects of 5-hydroxytryptophan and 5-hydroxytryptamine on the function of microdissected ob/ob mouse pancreatic islets. Diabetologia 24:52–57
Linnestad P, Guldvog I, Schrumpf E (1983) The effect of alpha- and beta-adrenergic agonists and blockers on postprandial pancreatic polypeptide release in dogs. J Gastroenterol 18:87–90
Lloyd RV, Wilson BS (1983) Specific endocrine tissue marker defined by a monoclonal antibody. Science 222:628
Locket NA (1980) Some advances in coelacanth biology. Proc R Soc Lond B 208:265–307
Lockhart-Ewart RB, Mok C, Martin JM (1976) Neuroendocrine control of insulin secretion. Diabetes 25:96–100
Long RG (1983) Recent advances in pancreatic hormone research. Postgrad Med J 59:277–282
Lopez LC, Frazier ML, Su C-J et al. (1983) Mammalian pancreatic preproglucagon contains three glucagon-related peptides. Proc Nat Acad Sci USA 80:5485–5489
Lorén I, Alumets J, Håkanson R, Sundler F (1979) Immunoreactive pancreatic polypeptide (PP) occurs in the central and peripheral nervous system: preliminary immunocytochemical observations. Cell Tissue Res 200:179–186
Loretz CA, Freel RW, Bern HA (1983) Specificity of response of intestinal ion transport systems to a pair of natural peptide hormone analogs: somatostatin urotensin II. Gen Comp Endocrinol 52:198–206
Loubatieres-Mariani MM, Chapal J, Ribes G, Loubatieres A (1977) Discrepancies in the response of the insulin secreting cells of the dog and rat to different adrenergic stimulating agents. Acta Diabetol Lat 14:144–155
Loubatieres-Mariani MM, Ribes G, Blayac JP, Campo P (1980) Insulin-secretory effect of a low dose of adrenalin in the dog. Horm Metab Res 12:126–127
Louie D, Williams JA, Owyang C (1985) Action of pancreatic polypeptide on rat pancreatic secretion: in vivo and in vitro. Am J Physiol 249:G489–G495
Louis-Sylvestre J (1978) Relationship between two stages of prandial insulin release in rats. Am J Physiol 235:E103–E111
Luiten PGM, Horst GJ, Koopmans SJ et al. (1984) Preganglionic innervation of the pancreas islet cells in the rat. J Auton Nerv Sys 10:27–42
Lund PK, Goodman RH, Montminy MR et al. (1983) Anglerfish islet preproglucagon II. J Biol Chem 258:3280–3284
Lundberg JM, Hökfelt T (1983) Coexistence of peptides and classical neurotransmitters. Trends Neurosci 6:325–332
Lundberg JM, Hökfelt T, Anggard A et al. (1982) Organizational principles in the peripheral sympathetic nervous system: subdivision by coexisting peptides (somatostatin-, avian pancreatic polypeptide-), and vasoactive intestinal polypeptide-like immunoreactive materials. Proc Nat Acad Sci USA 79:1303–1307
Luyckx AS, Lefébvre PJ (1983) Prostaglandins and glucagon secretion. In: Lefébvre PJ (ed) Glucagon II, Chap 28. Springer, Berlin Heidelberg New York, pp 83–98
MacCallum WG (1909) On the relations of the islands of Langerhans to glycosuria. Bull Hopkins Hosp 20:265–268
Macerollo PC (1977) Pancreatic islet cells in very young chick embryos with special reference to D cells. Gen Comp Endocrinol 33:139–146
Macey DJ, Epple A, Potter IC, Hilliard RW (1984) The effect of catecholamines on branchial and cardiac electrical recording in adult lampreys (*Geotria australis*). Comp Biochem Physiol 79C:295–300
Maclean JL (1965) Studies on the intestinal diverticulum of the East Australian lamprey, *Mordacia mordax*. Bachelor of Science Thesis, University of New South Wales, Australia
Macleod JJR (1922) The source of insulin. A study of the effect produced on blood sugar by extracts of the pancreas and the principal islets of fishes. J Metabol Res 2:149–172
Mahony C, Feldman JM (1977) Species variation in pancreatic islet monoamine uptake and action. Diabetes 26:257–261
Maier V, Kroder A, Groner E et al. (1975) Glucagon-like activity (GLI) in the intestine of *Porcus domesticus* and *Astacus fluviatilis*. Acta Endocrinol (Suppl) 193:41
Malaisse WJ, Malaisse-Lagae F, Senar A (1983) Anomeric specificity of hexose metabolism in pancreatic islets. Physiol Rev 63:773–786

Malaisse-Lagae F, Ravazzola M, Robberecht P et al. (1976) Hydrolases content in peri-insular and tele-insular exocrine pancreas. In: Fujita T (ed) Endocrine gut and pancreas. Elsevier, Amsterdam, pp 313–320
Malaisse-Lagae F, Stefan Y, Cox J et al. (1979) Identification of a lobe in the adult human pancreas rich in pancreatic polypeptide. Diabetologia 17:361–365
Mallatt J (1985) Reconstructing the life cycle and feeding of ancestral vertebrates. In: Foreman RE, Gorbman A, Dodd JM, Olsson R (eds) Evolutionary biology of primitive fishes. Plenum Press, New York London, pp 59–68
Mann GV, Croffard OB (1970) Insulin levels in primates by immunoassay. Science 169:1312–1313
Marchant TA, Peter RE (1985) The influence of various forms of somatostatin on the in vitro release of growth hormone from the pituitary gland of the goldfish, *Carassius auratus* L. Abstract | B-7.19, Seventh Conference of the European Society of Comparative Physiology and Biochemistry. Promociones y Publicaciones Universitarias Barcelona
Marco J, Correas I, Zulueta MA et al. (1983) Inhibitory effect of somatostatin-28 on pancreatic polypeptide, glucagon and insulin secretion in normal man. Horm Metab Res 15:363–366
Markussen J, Frandsen EK, Heding LG, Sundby F (1972) Turkey glucagon: crystallization, amino acid composition and immunology. Horm Metab Res 4:360–363
Marques M (1967) Effects of prolonged glucagon administration to turtles *Chrysemys dorbigni*. Gen Comp Endocrinol 9:102–109
Marques M, Kraemer A (1968) Extractable insulin and glucagon from turtle's (*Chrysemys d'orbignyi*) pancreas. Comp Biochem Physiol 27:439–446
Marques M, Bells AA, Machado VLA et al. (1982) In vivo specific uptake of labeled insulin by liver, adipose tissue, pituitary, and adrenals in the turtle *Chrysemys dorbigni*. Gen Comp Endocrinol 48:89–97
Marques M, da Silva RSM, Turyn D, Dellacha JM (1985) Specific uptake, dissociation, and degradation of ^{125}I-labeled insulin in isolated turtle (*Chrysemys dorbigni*) thyroid glands. Gen Comp Endocrinol 60:306–314
Martin G, Dubois MP (1981) A somatostatin-like antigen in the nervous system of an isopod *Porcellis ditatatus* Brandt. Gen Comp Endocrinol 45:125–130
Martin SM, Spencer AN (1983) Neurotransmitters in coelenterates. Comp Biochem Physiol 74C:1–14
Maruyama H, Hisatomi A, Orci L et al. (1984) Insulin within islets is a physiologic glucagon release inhibitor. J Clin Invest 74:2296–2299
Matsuyama T, Foà PP (1974) Effects of pancreatectomy and of enteral administration of glucose on plasma insulin, total and pancreatic glucagon. Diabetes 23:344A
Matty S, Gorbman A (1978) The effects of isletectomy and hypophysectomy on some blood plasma constituents of the hagfish, *Eptatretus stouti*. Gen Comp Endocrinol 34:94A
Maximovich AA, Barthova AD, Pliusnin BB et al. (1978) The ultrastructure and function of pancreatic B-cells of Pacific salmon during spawning migration. Biol Moria 6:61–69
Maxwell GD, Sietz PD, Jean S (1984) Somatostatin-like immunoreactivity is expressed in neural crest cultures. Dev Biol 101:357–366
Mazina TI, Vizek K (1969) The effect of insulin and glucagon on the NEFA content in chick embryo plasma. (In Russian) Dokl Akad Nauk SSSR 189:1154
Mazzi V (1976) Note sul pancreas endocrino del polipteriforme *Calamoichthys calabaricus*. Atti Acad Sci Torino Classe Sci Fis Math Nat 110:387–392
McCann SM, Krulich L, Negro-Vilar A et al. (1980) Regulation and function of panhibin (somatostatin). Adv Biochem Psychopharmacol 22:131–143
McCormick NA, Macleod JJR (1925) The effect on blood sugar of fish in various conditions including removal of principal islets (isletectomy). Proc R Soc Ser B 98:1–29
McCormick K, Donlon E, Dziwis P (1985) Fetal rat hyperinsulinism and hyperglucagonism: effects of hepatic ketogenesis, lipogenesis and gluconeogenesis. Endocrinology 116:1281–1287
McDonald JK, Lumpkin MD, Samson WK, McCann SM (1985) Pancreatic polypeptides affect luteinizing and growth hormone secretion in rats. Peptides 6:79–84
McNeill D, Brinn JE, Fletcher DF (1984) An immunocytochemical study of the pancreatic islet system of the channel catfish. Anat Rec 209:381–384
Meda P, Amherdt M, Perrelet A, Orci L (1981) Metabolic coupling between cultured pancreatic B-cells. Exp Cell Res 133:421–430

Meda P, Kohen E, Kohen C et al. (1982) Direct communication of homologous and heterologous endocrine islet cells in culture. J Cell Biol 92:221–226
Meda P, Michaels RL, Halban PA et al. (1983) In vivo modulation of gap junctions and dye coupling between B-cells of the intact pancreatic islet. Diabetes 32:858–868
Meglasson MD, Hazelwood RL (1983) Adrenergic regulation of avian pancreatic polypeptide secretion in vitro. Am J Physiol 245:E408–E413
Meglasson MD, Hazelwood RL (1984) Adrenergic regulation of insulin secretion from the chicken pancreas in vitro. Gen Comp Endocrinol 56:82–89
Mehring J von, Minkowski O (1889) Diabetes mellitus nach Pankreasextirpation. Arch Exp Path Pharm 26:371–387
Melnyk RB, Martin JM (1985) Insulin and central regulation of spontaneous fattening and weight loss. Am J Physiol 249:R203–R208
Merimee TJ, Zapf J, Froesch ER (1982) Insulin-like growth factors (IGFs) in pygmies and subjects with the pygmy trait: characterization of the metabolic actions of IGF I and IGF II in man. J Clin Endocrinol Metab 55:1081–1088
Merklin RJ (1974) Growth and distribution of human fetal brown fat. Anat Rec 178:636A
Meuris S, Verloes A, Robyn C (1983) Immunocytochemical localization of prolactin-like immunoreactivity in rat pancreatic islets. Endocrinology 12:2221–2223
Meyers CA, Murphy WA, Redding TW et al. (1980) Synthesis and biological actions of prosomatostatin. Proc Nat Acad Sci USA 77:6171–6174
Mialhe P (1958) Glucagon, insulinê et régulation endocrine de la glycémie chez le canard. Acta Endocrinol 36(Suppl 36):1–134
Mialhe P (1976) The role of glucagon in birds and mammals. In: Grillo TAI, Leibson L, Epple A (eds) The evolution of pancreatic islets. Pergamon Press, Oxford, pp 291–300
Michaels R, Sheridan JD (1981) Islets of Langerhans: dye coupling among immunocytochemically distinct cell types. Science 214:801–803
Mihail N, Ionescu M, Dusa L (1963) Morphologische Auswirkungen der Pankreatektomie bei der Taube. Anat Anz 112:97–100
Mikami S, Ono K (1962) Glucagon deficiency induced by extirpation of alpha islets of the fowl pancreas. Endocrinology 71:464–474
Miller MR (1961) Carbohydrate metabolism in amphibians and reptiles. In: Martin AW (ed) Comparative physiology of carbohydrate metabolism of heterothermic animals. University of Washington Press, Seattle, pp 125–144
Miller MR, Lagios M (1970) The pancreas. In: Gans C (ed) Biology of the reptilia, vol 3. Academic Press, New York, p 319–346
Miller RE (1981) Pancreatic neuroendocrinology: peripheral neural mechanisms in the regulation of the islets of Langerhans. Endocrine Rev 2:471–494
Mirsky IA, Jinks R, Perisutti G (1964) Production of diabetes mellitus in the duck with insulin antibodies. Am J Physiol 206:133–135
Mitchenere P, Adrian TE, Chadwick VS et al. (1981) Effect of gut regulatory peptides on intestinal luminal fluid in the rat. Life Sci 29:1563–1570
Mizoguchi A, Fujiwara Y, Suzuki A (1984) Amino-terminal amino acid sequence of the silkworm prothoracictropic hormone: homology with insulin. Science 2:1344–1345
Moltz JH, Fawcett CP (1983) Purification of a glucagon releasing factor from the rat hypothalamus. Life Sci 32:1271–1278
Moltz JH, Fawcett CP (1985) Corticotropin-releasing factor inhibits insulin release from perfused pancreas. Am J Physiol 248:E741–E743
Moltz JH, Dobbs RE, McCann SM, Fawcett CP (1979) Preparation and properties of hypothalamic factors capable of altering pancreatic hormone release in vitro. Endocrinology 105:1262–1268
Mommsen TP (1986) Comparative gluconeogenesis in hepatocytes from salmonid fishes. Can J Zool 65:1110–1115
Moody AJ, Thim L (1983) Glucagon, glicentin, and related peptides. In: Lefébvre PJ (ed) Glucagon I, Chap 7. Springer, Berlin Heidelberg New York, pp 139–174
Moody AJ, Thim L, Valverde I (1984) The isolation and sequencing of human gastric inhibitory peptide (GIP). FEBS Lett 172:142–148
Moore RD (1983) Effects of insulin upon ion transport. Biochim Biophys Acta 737:1–49

Morata P, Vargas AM, Pita ML, Sanchez-Medina F (1982a) Hormonal effects on the liver glucose metabolism in rainbow trout (*Salmo gairdneri*). Comp Biochem Physiol 72B:543–545
Morata P, Vargas AM, Pita ML, Sanchez-Medina F (1982b) Involvement of gluconeogenesis in the hyperglycemia induced by glucagon, adrenaline and cyclic AMP in rainbow trout (*Salmo gairdneri*). Comp Biochem Physiol 73A:379–381
Morel A, Chang J-Y, Cohen P (1984) The complete amino acid sequence of anglerfish somatostatin-28-II. FEBS Lett 175:21–24
Morley J, Levin S, Pehlevanian M et al. (1979) The effects of thyrotropin-releasing hormone on the endocrine pancreas. Endocrinology 104:137–139
Morris R, Islam DS (1969) The effects of hormones and hormone inhibitors on blood sugar regulation and the follicles of Langerhans in ammocoete larvae. Gen Comp Endocrinol 12:81–90
Mössner J, Logsdon CD, Goldfine ID, Williams JA (1984) Regulation of pancreatic acinar cell insulin receptors by insulin. Am J Physiol 247:G15–G160
Muggeo M, van Obberghen E, Kahn CR et al. (1979) The insulin receptor and insulin of the Atlantic hagfish. Extraordinary conservation of binding specificity and negative cooperation in the most primitive vertebrate. Diabetes 28:175–181
Muglia L, Locker J (1984) Extrapancreatic insulin gene expression in the fetal rat. Proc Nat Acad Sci USA 81:3635–3639
Munford J, Greenwald L (1974) The hypoglycemic effects of external insulin on fish and frogs. J Exp Zool 190:341–346
Munger BL (1981) Morphological characterization of islet cell diversity. In: Cooperstein SJ, Watkins D (eds) The islets of Langerhans. Biochemistry, physiology and pathology, Chap 1. Academic Press, London, New York, pp 3–34
Munger BL, Caramia F, Lacy PE (1965) The ultrastructural basis for the identification of cell types in the pancreatic islets II. Rabbit, dog and opossum. Z Zellforsch 67:776–798
Murat JC, Plisetskaya EM, Soltitskaya LP (1979) Glucose 6-phosphatase activity in kidney of the river lamprey *Lampetra fluviatilis* L. Gen Comp Endocrinol 39:115–117
Murat JC, Plisetskaya EM, Woo NYS (1981) Endocrine control of nutrition in cyclostomes and fish. Comp Biochem Physiol 68A:149–158
Murlin JR, Clough HD, Gibbs CBF, Stokes AM (1923) Wässerige Pankreasextrakte I. Einfluß auf den Kohlehydratstoffwechsel pankreasextirpierter Tiere. J Biol Chem 56:253
Musgrave IF, Bachmann AW, Jackson RV, Gordon RD (1985) Increased plasma noradrenaline during low dose adrenaline infusion in resting man and during sympathetic stimulation. Chem Exp Pharm Physiol 12:285–289
Mutt V, Said SI (1974) Structure of the porcine vasoactive intestinal octacosapeptide. Eur J Biochem 42:581–589
Mutt V, Jorpes JE, Magnusson S (1970) Structure of porcine secretin. The amino acid sequence. Eur J Biochem 15:513–519
Naber SP, Hazelwood RL (1977) In vitro insulin release from chicken pancreas. Gen Comp Endocrinol 32:495–504
Nakamura M, Yamada K, Yokote M (1971) Ultrastructural aspects of the pancreatic islets in carps of spontaneous diabetes mellitus. Experientia 27:75–76
Nagasawa H, Kataoka H, Isogai A et al. (1984) Amino-terminal amino acid sequence of the silkworm prothoracicotropic hormone: homology with insulin. Science 226:1344–1345
Naithani VK, Steffens GJ, Tager HS et al. (1984) Isolation and amino-acid sequence determination of monkey insulin and proinsulin. Hoppe-Seyler's Z Physiol Chem 365:571–575
Nedergaard J, Lindberg O (1982) The brown fat cell. In: Bourne GH, Danielli JF, Jeon KW (eds) International review of cytology, vol 74. Academic Press, New York, London, pp 187–286
Nesher R, Abramovitch E, Cerasi E (1985) Correction of diabetic pattern of insulin release from islets of the spiny mouse (*Acomys cahiranus*) by glucose priming in vitro. Diabetologia 28:233–236
Neubert K (1927) Bau und Entwicklung des menschlichen Pankreas. Arch Entwicklungsmich Organ III:29–118
Ng TB, Wong CM (1985) Epidermal and nerve growth factors manifest antilipolytic and lipogenic activities in isolated rat adipocytes. Comp Biochem Physiol 81B:687–689
Niall HD (1982) The evolution of peptide hormones. Ann Rev Physiol 44:615–624
Nilaver G, Defendini R, Zimmerman EA et al. (1982) Motilin in the Purkinje cell of the cerebellum. Nature (London) 295:597–598

Nilsson S (1983) Autonomic nerve function in vertebrates. In: Farner DS et al. (eds) Zoophysiology, vol 13. Springer, Berlin, 253 pp
Nishi M (1910) Über Glykogenbildung in der Leber pankreasdiabetischer Schildkröten. Arch Exp Pathol Pharmakol 62:170
Nissley SP, Rechler MM, Moses AC et al. (1977) Proinsulin binds to a growth peptide receptor and stimulates DNA synthesis in chick embryo fibroblasts. Endocrinology 101:708–716
Noe BD, Spiess J, Rivier JE, Vale W (1979) Isolation and characterization of somatostatin from anglerfish pancreatic islets. Endocrinology 105:1410–1415
Noe BD, McDonald JK, Greiner F et al. (1986) Anglerfish islets contain NPY immunoreactive nerves and produce the NPY analog aPY. Peptides 7:147–154
O'Boyle RN, Beamish FWH (1977) Growth and intermediary metabolism of larval and metamorphosing stages of the landlocked sea lamprey, *Petromyzon marinus*. Env Biol Fish 42:219–235
O'Conner DT (1983) Chromogranin: widespread immunoreactivity in polypeptide hormone producing tissues and in serum. Regul Pept 6:263–280
O'Conner DT, Burton DS, Deftos LT (1983) Chromogranin A: immunohistology reveals its universal occurrence in normal polypeptide hormone producing endocrine glands. Life Sci 33: 1657–1663
O'Conner KJ, Baxter D (1985) The demonstration of insulin-like material in the honey bee, *Apis mellifera*. Comp Biochem Physiol 81B:755–760
O'Day DH, Horgen PA (eds) (1981) Sexual interactions in eukaryote microbes. Academic Press, New York
Oertel WH, Mugnaini E, Tappaz ML, Wiese VK et al. (1982) Central GABAergic innervation of neurointermediate pituitary lobe: biochemical and immunohistochemical study in the rat. Proc Nat Acad Sci USA 79:675–679
O'Hare M (1983) Pancreatic polypeptide in diabetes mellitus. In: Mngola EN (ed) Diabetes 1982. Excepta Medica, Amsterdam, pp 517–523
O'Hea ER, Leveille GA (1969) Lipid biosynthesis and transport in the domestic chick (*Gallus domesticus*). Comp Biochem Physiol 30:149–159
Ohtani O (1983) Microcirculation of the pancreas: A correlative study of intravital microscopy with scanning electron microscopy of vascular corrosion casts. Arch Histol Japan 46:315–325
Okada Y, Taniguchi H, Shimada C (1976) High concentration of GABA and high glutamate decarboxylase activity in rat pancreatic islets and human isulinoma. Science 194:620–622
Ondo J, Mansky T, Wuttle W (1982) In vivo GABA release from the medial preoptic area of diestrus and ovariectomized rats. Exp Brain Res 46:69–72
O'Neal LW, Kipnis DM, Luse SA et al. (1968) Secretion of various endocrine substances by ACTH-secreting tumors – gastrin, melanotropin, norepinephrine, serotonin, parathormone, vasopressin, glucagon. Cancer 21:1219–1232
Oomura Y (1983) Glucose as a regulator of neuronal activity. Adv Metab Disorders 10:32–65
Oomura Y, Nijima A (1983) Chemosensitive neurons and neural control of pancreatic secretion. In: Mngola EN (ed) Diabetes 1982. Excerpta Medica, Amsterdam, pp 201–211
Orci L (1983) Cellular relationships in the islet of Langerhans: a regulatory perspective. In: Federlin K, Scholtolt J (eds) Importance of islets of Langerhans for modern endocrinology. Raven Press, New York, pp 11–26
Orci L (1985) The insulin factory: a tour of the plant surroundings and a visit to the assembly line. Diabetologia 28:528–546
Orci L, Perrelet A, Ravazzola M et al. (1973) A specialized membrane junction between nerve endings and B-cells in islets of Langerhans. Eur J Clin Invest 3:443–445
Orci L, Malaisse-Lagae F, Amherdt M et al. (1975) Cell contacts in human islets of Langerhans. J Clin Endocrinol Metabol 41:841–844
Orci L, Baetens D, Ravazzola M (1976) Ilots a polypeptide pancreatique (PP) et ilots a glucagon. Distribution topographique distincte dans la pancréas de rat. CR Acad Sci (Ser D) 283:12–1215
Osche G (1982) Rehapitulationsentwicklung und ihre Bedeutung für die Phylogenetik – Wann gilt die „Biogenetische Grundregel"? Verh Naturwiss Ver Hamburg (NF) 25:5–31
Östberg Y, van Noorden S, Pearse AGE (1975) Cytochemical, immunofluorescence, and ultrastructural investigations on polypeptide hormone localization in the islet parenchyma and bile duct mucosa of a cyclostome, *Myxine glutinosa*. Gen Comp Endocrinol 25:274–291

Östberg Y, van Noorden S, Pearse AGE, Thomas N (1976) Cytochemical, immunofluorescence, and ultrastructural investigations on polypeptide hormone containing cells in the intestinal mucosa of a cyclostome, *Myxine glutinosa*. Gen Comp Endocrinol 28:213–227

Ottolenghi C, Puviani AC, Baruffaldi A, Brighenti L (1982) In vivo effects of insulin on carbohydrate metabolism of catfish (*Ictalurus melas*). Comp Biochem Physiol 72A:35–41

Owada K, Yamada C, Kobayashi H (1985) Immunohistochemical investigations of urotensins in the caudal spinal cord of four species of elasmobranchs and the lamprey, *Lampetra japonica*. Cell Tissue Res 242:527–530

Oyama H, Martin J, Sussman K et al. (1981) The biological activity of catfish pancreatic somatostatin. Regul Pept 1:387–396

Oyama H, Gabbay K, Loo SW et al. (1982) Radioimmunoassay for catfish pancreatic somatostatin-22. Regul Pept 3:383–396

Padgaonkar AS, Rangneker PV (1975) Insulin tolerance tests in the normal, pancreatectomized, hypophysectomized and hypophysopancreatectomized water-snake, *Natrix piscator* (Russell). J Anim Morphol Physiol 22:38–42

Palmer JP, Porte D (1983) Neural control of glucagon secretion. In: Lefébvre PJ (ed) Glucagon II, Chap 30. Springer, Berlin New York Tokyo, pp 115–132

Palokangas R, Vihko V, Nuuja I (1973) The effects of cold and glucagon on lipolysis, glycogenolysis and oxygen consumption in young chicks. Comp Biochem Physiol 45:489–496

Pandol SJ, Sutlif VE, Jones SW, Charlton CG et al. (1983) Action of natural glucagon on pancreatic acini: due to contamination by previously undescribed secretagogues. Am J Physiol 245:G703–G710

Patel YC, Wheatley T, Ning C (1981) Multiple forms of immunoreactive somatostatin: comparison of distribution in neural and nonneural tissues and portal plasma of the rat. Endocrinology 109:1943–1949

Patel YC, Amherdt M, Orci L (1982) Quantitative electron microscopic autoradiography of insulin, glucagon, and somatostatin binding sites on islets. Science 217:1155–1156

Patent GJ (1970) Comparison of some hormonal effects on carbohydrate metabolism in an elasmobranch (*Squalus acanthus*) and a holocephalan (*Hydrolagus colliei*). Gen Comp Endocrinol 14:215–242

Patent GJ (1973) The chondrichthyean endocrine pancreas: What are its functions? Am Zool 13:639–651

Patent GJ, Epple A (1967) On the occurrence of two types of argyrophil cells in the pancreatic islets of the holocephalan fish, *Hydrolagus colliei*. Gen Comp Endocrinol 9:325–333

Patent GJ, Foà PP (1971) Radioimmunoassay of insulin in fishes, experiments in vivo and in vitro. Gen Comp Endocrinol 16:41–46

Patent GJ, Kechele PO, Carrano VT (1978) Nonconventional innervation of the pancreatic islets of the teleost fish, *Gillichthys mirabilis*. Cell Tissue Res 191:305–315

Patzelt C, Schiltz E (1984) Conversion of proglucagon in pancreatic alpha cells: the major end-products are glucagon and a single peptide, the major proglucagon fragment, that contains two glucagon-like sequences. Proc Nat Acad Sci USA 81:5007–5011

Patzelt C, Tager HS, Carroll RJ, Steiner DF (1980) Identification of pro-somatostatin in pancreatic islets. Proc Nat Acad Sci USA 77:2410–2414

Paulin C, Dubois PM (1978) Immunohistochemical identification and localization of pancreatic polypeptide cells in the pancreas and gastrointestinal tract of the human fetus and adult man. Cell Tissue Res 188:251–257

Pearse AGE (1968) Common cytochemical and ultrastructural characteristics of cells producing polypeptide hormones (the APUD series) and their relevance to thyroid and ultimobranchial C cells and calcitonin. Proc R Soc Lond 170:71–80

Pearse AGE (1969) The cytochemistry and ultrastructure of polypeptide hormone producing cells of the APUD series and the embryologic, physiologic and pathologic implications of the concept. J Histochem Cytochem 17:303–313

Pearse AGE (1977) The diffuse neuroendocrine system and the APUD concept: related 'endocrine' peptides in brain, intestine, pituitary, placenta, and anuran cutaneous glands. Med Biol 55:115–125

Pearse AGE (1979) The APUD concept and its relationship to the neuropeptides. In: Gotto AM, Pech EJ, Boyd AE (eds) Brain peptides: a new endocrinology. Elsevier/North Holland Biomedical Press, Amsterdam, New York, pp 89–101

Pearse AGE (1980) APUD: concept, tumours, molecular markers and amyloid. Mikroskopie 36:257–269

Pearse AGE (1982) Islet cell precursors are neurons. Nature 295:96–97

Pearse AGE (1984) The diffuse neuroendocrine system. Front Horm Res 12:1–7

Pearse AGE, Polak JM (1971) Neural crest origin of the endocrine polypeptide (APUD) cells of the gastrointestinal tract and pancreas. Gut 12:783–788

Pek SB, Spangler RS (1983) Hormones in the control of glucagon secretion. In: Lefébvre PJ (ed) Glucagon II, Chap 29. Springer, Berlin New York Tokyo, pp 99–114

Penhos JC, Ramey E (1973) Studies on the endocrine pancreas of amphibians and reptiles. Am Zool 13:667–698

Penhos JC, Krahl ME (1962) Insulin stimulus of leucine incorporation into frog liver protein. Am J Physiol 203:687–689

Penhos JC, Krahl ME (1963) Stimulation of leucine incorporation into perfused liver protein by insulin. Am J Physiol 204:140–142

Penhos JC, Wu CH, Reitman M et al. (1967a) Effects of several hormones after total pancreatectomy in alligators. Gen Comp Endocrinol 8:32–43

Penhos JC, Uno B, Houssay BA (1967b) Glucose and lipid metabolism in the toad's perfused liver. Gen Comp Endocrinol 8:297–304

Penman E, Lowry PJ, Wass JAH et al. (1980) Molecular forms of somatostatin in normal subjects and in patients with pancreatic somatostatinoma. Clin Endocrinol 12:611–620

Perez-Castillo A, Blázquez E (1980a) Tissue distribution of glucagon, glucagonlike immunoreactivity, and insulin in the rat. Am J Physiol 238:E258–E266

Perez-Castillo A, Blázquez E (1980b) Synthesis and release of glucagon by human salivary glands. Diabetologia 19:123–129

Peterson JD, Steiner DF, Emdin SO, Falkmer S (1975) The amino acid sequence of the insulin from a primitive vertebrate, the Atlantic hagfish (*Myxine glutinosa*). J Biol Chem 250:5183–5191

Petrusz P, Merchenthaler I, Maderdrut JL et al. (1983) Corticotropin-releasing factor (CRF)-like immunoreactivity in the vertebrate endocrine pancreas. Proc Nat Acad Sci USA 80:1721–1725

Peuler KJ, Johnson GA (1977) Simultaneous single isotope radioenzymatic assay of plasma norepinephrine, epinephrine and dopamine. Life Sci 21:625–636

Peyron JF, Samson M, van Obberghen E et al. (1985) Appearance of a functional insulin receptor during rabbit embryogenesis. Diabetologia 28:369–372

Phillippe M (1983) Fetal catecholamines. Am J Obst Gynecol 146:840–855

Phillips JW, Hird FJR (1977a) Gluconeogenesis in vertebrate livers. Comp Biochem Physiol 57B:127–131

Phillips JW, Hird FJR (1977b) Ketogenesis in vertebrate livers. Comp Biochem Physiol 57B:133–138

Pictet R, Rutter WJ (1972) Development of the embryonic endocrine pancreas. In: Freinkel N, Steiner DF (eds) Endocrine pancreas. Williams and Wilkins, Baltimore, Handbook of physiology, sect 7, vol 1, pp 25–66

Pictet RL, Clark WR, Williams RH, Rutter WJ (1972) An ultrastructural analysis of the developing embryonic pancreas. Dev Biol 29:436–467

Plisetskaya E (1975) Hormonal regulation of carbohydrate metabolism in lower vertebrates. Acad Sci USSR, Leningrad, 215 pp

Plisetskaya E (1980) Fatty acids levels in blood of cyclostomes and fish. Environ Biol Fishes 5:273–290

Plisetskaya E (1985) Some aspects of hormonal regulation of metabolism in agnathans. In: Foreman RE, Gorbman A, Dodd JM, Olsson R (eds) Evolutionary biology of primitive fishes. Plenum Press, New York London, pp 339–361

Plisetskaya E, Prozorovskaya MP (1971) Catecholamines in the blood and cardiac muscle of the lamprey, *Lampetra fluviatilis* during insulin hypoglycaemia. Zh Evol Biokhim Fiziol 7:101–103

Plisetskaya E, Bondareva VM (1972) The effect of acute insulin insufficiency on glycemic level and glycogen content in organs of some marine teleosts. Fiziol Zh USSR 58:1152–1157

Plisetskaya E, Kuzmina (1972) Glycogen levels in organs of cyclostomata and pisces. (In Russian) Woprosy Ichtiologii 12:335–343

Plisetskaya E, Leibush BN (1972) Insulin-like activity and immunoreactive insulin in the blood of the lamprey, *Lampetra fluviatilis*. J Evol Biochem Physiol 8:499–505

Plisetskaya E, Leibush BN (1974) Radioimmunoassay of insulin in lower vertebrates. Zh Evol Biokhim Fiziol 10:623–625

Plisetskaya E, Gorbman A (1982) The secretion of thyroid hormones and their role in regulation of metabolism in cyclostomes. Gen Comp Endocrinol 46:407–408

Plisetskaya E, Joosse J (1985) Hormonal regulation of carbohydrate metabolism in molluscs. In: Lofts B, Holmes WN (eds) Current Trends in Comparative Endocrinology. Hong Kong University Press, pp 1077–1079

Plisetskaya E, Leibush BN, Bondareva V (1976) The secretion of insulin and its role in cyclostomes and fishes. In: Grillo TAI, Leibson L, Epple A (eds) The evolution of pancreatic islets. Pergamon Press, Oxford, pp 251–269

Plisetskaya E, Kazakov VK, Solbitskaya L, Leibson LA (1978a) Insulin-producing cells in the gut of freshwater bivalve molluscs *Andonta cygnea* and *Unio pictorum* and the role of insulin in the regulation of their carbohydrate metabolism. Gen Comp Endocrinol 35:133–145

Plisetskaya E, Soltitskaya L, Rusacov YI (1978b) Production and role of insulin in the regulation of carbohydrate metabolism in freshwater and marine bivalve molluscs. In: Gaillard PJ, Poer HH (eds) Comparative endocrinology. Elsevier/North Holland Biomedical Press, Amsterdam, pp 449–453

Plisetskaya E, Dickhoff WW, Gorbman A (1983a) Plasma thyroid hormones in cyclostomes: do they have a role in regulation of glycemic levels? Gen Comp Endocrinol 49:97–107

Plisetskaya E, Sower SA, Gorbman A (1983b) The effects of insulin insufficiency on plasma thyroid hormones and some metabolic constituents in pacific hagfish, *Eptatretus stouti*. Gen Comp Endocrinol 49:315–319

Plisetskaya E, Woo NYS, Murat JC (1983c) Thyroid hormones in cyclostomes and fish and their role in regulation of intermediary metabolism. A review. Comp Biochem Physiol 74A:179–187

Plisetskaya E, Sower SA, Gorbman A (1984a) Profiles of plasma thyroid hormones and insulin during the upstream spawning migration of the sea lamprey, *Petromyzon marinus*. Gen Comp Endocrinol 53:451

Plisetskaya E, Rich AA, Dickhoff WW, Gorbman A (1984b) A study of triiodothyronine-catecholamine interactions: their effect on plasma fatty acids in Pacific hagfish *Eptatretus stouti*. Comp Biochem Physiol 78A:313–330

Plisetskaya E, Pollock HG, Rouse JB et al. (1985) Characterization of coho salmon (*Oncorhynchus kisutch*) insulin. Regul Pept 11:105–116

Plisetskaya E, Dickhoff WW, Paquette TL, Gorbman A (1986a) The assay of salmon insulin by homologous radioimmunoassay. Fish Physiol Biochem 1:37–43

Plisetskaya E, Pollock HG, Rouse JB et al. (1986b) Isolation and structures of coho salmon (*Oncorhynchus kisutch*) glucagon and glucagon-like peptide. Regul Pept 14:57–67

Plisetskaya E, Pollock HG, Rouse JB et al. (1986c) Characterization of coho salmon (*Oncorhynchus kisutch*) islet somatostatins. Gen Comp Endocrinol 63:252–263

Polak JM, Bloom SR (1982) Regulatory peptides and neuron specific enolase in the respiratory tract of man and other mammals. Exp Lung Res 3:313–358

Polak JM, Bloom SR (1984) Distribution of regulatory peptide-containing nerves. Front Horm Res 12:95–107

Polak JM, Buchan AMJ (1980) Motilin – cellular origin in human gut. J Histochem Cytochem 28:618

Polak JM, Grimelius L, Pearse A et al. (1975) Growth-hormone release-inhibiting hormone in gastrointestinal and pancreatic D cells. Lancet i:1220–1222

Pollock HG, Kimmel JR (1975) Chicken glucagon: isolation and amino acid sequence studies. J Biol Chem 250:9377–9380

Pollock HG, Kimmel JR (1981) Immunoassay for avian pancreatic polypeptide and applications in chickens. Gen Comp Endocrinol 45:386–394

Polyakova TI, Plisetskaya EM (1976) The effect of glucose on gut epithelium cells in the lancelot *Branchiostoma lanceolatum*. J Evol Biochem Physiol 12:184–186

Porte D Jr, Woods SC (1983) Neural regulation of islet hormones and its role in energy balance and stress hyperglycemia. In: Ellenberg M, Rifkin H (eds) Diabetes mellitus, theory and practice, Chap 13. Medical Examination Publishing Co, Inc, pp 267–294

Potter IC (1980) Ecology of larval and metamorphosing lampreys. Can J Fish Aquat Sci 37:1641–1657

Potter IC (1986) The distinctive characters of southern hemisphere lampreys (Geotriidae and Mordaciidae). In: Uyeno T, Ardi R, Taniuchi T, Matsuura K (eds) Indo-Pacific fish biology. The Ichthyological Society of Japan, Tokyo, pp 9–19

Pour P, Sayed S, Sayed G (1982) Hyperplastic, preneoplastic and neoplastic lesions found in 83 human pancreases. Am J Clin Pathol 77:137–152

Powley TL, Laughton W (1981) Neural pathways involved in hypothalamic integration of autonomic responses. Diabetologia 20:378–387

Prado JL, Prado ES (1946) Teor em insulina do pancreas da *Bothrops jararaca*. Rev Brasil Biol 6:133–137

Przybylski RJ (1967) Cytodifferentiation of the chick pancreas I. Ultrastructure of the islet cells in the initiation of granule formation. Gen Comp Endocrinol 8:115–128

Putzke H-P, Said F (1975) Different secretory responses of periinsular and other acini and the rat pancreas after pilocarpine injection. Cell Tissue Res 161:133–143

Puro DJ, Agardh E (1984) Insulin-mediated regulation of neuronal maturation. Science 225:1170–1172

Quéré M, El May A, Bouzakoura C, Dubois MP, Dubois PM (1985) Evidence for somatostatin-like immunoreactivity of discrete cells of the anterior pituitary of camel (*Camelus dromedarius*). Gen Comp Endocrinol 60:187–195

Quickel KE, Feldman JM, Lebovitz HE (1971) Inhibition of insulin secretion by serotonin and dopamine: species variation. Endocrinology 89:1295–1302

Racagni G, Apud JA, Cocchi D, Locatelli V, Muller EE (1982) Minireview. GABAergic control of anterior pituitary hormone secretion. Life Sci 31:823–838

Rafn S, Wingstrand KG (1981) Structure of intestine, pancreas, and spleen of the Australian lungfish, *Neoceratodus forsteri* (Krefft). Zool Scripta 10:223–239

Raghu PK, Taborsky GJ, Paquette TL et al. (1984) Evidence for noncholinergic ganglionic neural stimulation of B cell secretion. Am J Physiol 247:E265–E270

Raheja K (1973) Comparison of diurnal variations in plasma glucose, cholesterol, triglyceride, insulin and in liver glycogen in younger and older chicks (*Gallus domesticus*). Comp Biochem Physiol 44A:1009–1014

Rahier J, Goebbels RM, Henquin JC (1983) Cellular composition of the human diabetic pancreas. Diabetologia 24:366–371

Rangneker PV, Padgaonkar AS (1972) Effects of total pancreatectomy in the snake, *Natrix piscator*, Russell. J Anim Morphol Physiol 19:63–76

Rappaport AM, Ohira S, Coddling JA et al. (1972) Effects on insulin output and on pancreatic blood flow of endogenous insulin infusion into an in situ isolated portion of the pancreas. Endocrinology 91:168–173

Ravazzola M, Orci L (1979) Anatomie immunocytochimique des granules a dans pancréas endocrine humain. CR Acad Sci Paris (Ser D) 289:1161–1163

Ravazzola M, Orci L (1980) Glucagon and glicentin immunoreactivity are topologically segregated in the a granule of the human pancreatic A cell. Nature (London) 284:66–67

Ravazzola M, Siperstein A, Moody AJ et al. (1979) Glicentin immunoreactive cells: their relationship to glucagon-producing cells. Endocrinology 105:499–508

Rawdon BB (1984) Gastrointestinal hormones in birds: morphological, chemical and developmental aspects. J Exp Zool 232:659–670

Rawdon BB, Andrew A (1979) An immunocytochemical study of the distribution of pancreatic endocrine cells in chicks, with special reference to the relationship between pancreatic polypeptide and somatostatin-immunoreactive cells. Histochemistry 59:189–197

Rawdon BB, Kramer B, Andrew A (1984) The distribution of endocrine cell progenitors in the gut of chick embryos. J Embryol Exp Morphol 82:131–145

Rayfield EJ, Yoon J-W (1981) Role of viruses in diabetes. In: Cooperstein SJ, Watkins D (eds) The islets of Langerhans. Biochemistry, physiology and pathology, Chap 16. Academic Press, London, New York, pp 427–451

Reaven GM (1984) Insulin secretion and insulin action in non-insulin-dependent diabetes mellitus: which defect is primary? Diabetes Care 7(Suppl 1):17–24

Redding TW, Schally AV (1984) Inhibition of growth of pancreatic carcinomas in animal models by analogs of hypothalamic hormones. Proc Nat Acad Sci USA 81:248–252

Reddy S, Bibby NJ, Elliott RB (1986) Immunolocalization of insulin, glucagon, pancreatic polypeptide, and somatostatin in the pancreatic islets of the possum, *Trichosurus vulpecula*. Gen Comp Endocrinol 64:157–162

Rehfeld JF, Heding LG, Hølst JJ (1973) Increased gut glucagon release as a pathogenetic factor in reactive hypoglycemia. Lancet i:116–118

Rehfeld JF, Larsson L-I, Goltermann NR et al. (1980) Neural regulation of pancreatic hormone secretion by the C-terminal tetrapeptide of CCK. Nature (London) 184:33–38

Reichlin S (1983a) Somatostatin. (First of two parts) New Eng J Med 309:1495–1501

Reichlin S (1983b) Somatostatin. (Second of two parts) New Eng J Med 309:1556–1563

Reinecke M (1981) Immunohistochemical localization of polypeptide hormones in endocrine cells of the digestive tract of *Branchiostoma lanceolatum*. Cell Tissue Res 219:445–456

Renaud M, Moon TW (1980) Starvation and the metabolism of hepatocytes isolated from the American eel, *Anguilla rostrata* Le Suerur. J Comp Physiol 135:127–137

Reusens-Billen B, Perlot X, Remacle C et al. (1984) Localization of GABA high-affinity binding sites in the pancreas of neonatal rat. Cell Tissue Res 235:503–508

Rhoten WB (1973a) Perifusion of saurian pancreatic islets and biphasic insulin release following glucose stimulation. Comp Biochem Physiol 45A:1001–1007

Rhoten WB (1973b) Insulin release from pancreatic islets of a lizard following glycine or arginine. J Herpetol 7:207–210

Rhoten WB (1974a) Sensitivity of saurian pancreatic islets to glucose. Am J Physiol 227:993–997

Rhoten WB (1974b) Effects of cytochalasin B and other agents on insulin secretion from saurian islets. Comp Biochem Physiol 47A:959–969

Rhoten WB (1976) Glucagon levels in pancreatic extracts and plasma of the lizard. Am J Anat 147:131–137

Rhoten WB (1978) Effects of glucose on glucagon secretion by the anolian splenic pancreas. Proc Soc Exp Biol Med 157:180–183

Rhoten WB (1983) Biosynthesis of a proinsulin-like molecule in the pancreas of a lizard. Gen Comp Endocrinol 51:163–174

Rhoten WB (1984) Immunocytochemical localization of four hormones in the pancreas of the garter snake, *Thamnophis sirtalis*. Anat Rec 208:233–242

Rhoten WB, Hall CE (1981) Four hormones in the pancreas of the lizard *Anolis carolinensis*. Anat Rec 199:89–97

Rhoten WB, Hall CE (1982) An immunocytochemical study of the cytogenesis of pancreatic endocrine cells in the lizard, *Anolis carolinensis*. Am J Anat 163:181–193

Rhoten WB, Smith PH (1978) Localization of four polypeptide hormones in the saurian pancreas. Am J Anat 151:595–601

Rizza RA, Haymond MW, Cryer P, Gerich JE (1979) Differential effects of epinephrine on glucose production and disposal in man. Am J Physiol 237:E356–E362

Rizza RA, Cryer PE, Haymond MW, Gerich JE (1980a) Adrenergic mechanisms for the effects of epinephrine on glucose production and clearance in man. J Clin Invest 65:682–689

Rizza RA, Haymond MW, Miles JM et al. (1980b) Effect of α-adrenergic stimulation and its blockade on glucose turnover in man. Am J Physiol 238:E467–E472

Robbins MS, Grouse LH, Sorensen RL, Elde RP (1981) Effect of muscimol on glucose-stimulated somatostatin and insulin release from the isolated, perfused rat pancreas. Diabetes 30:168–171

Robertson RP (1981) Interrelationships between prostaglandins and biogenic amines during modulation of beta-cell function. In: Cooperstein SJ, Watkins D (eds) The islets of Langerhans. Biochemistry, physiology and pathology, Chap 8. Academic Press, London, New York, pp 173–188

Robinson AM, Williamson DH (1980) Physiological roles of ketone bodies as substrates and signals in mammalian tissues. Physiol Rev 60:143–187

Rohner-Jeanrenaud F, Jeanrenaud B (1985) Involvement of the cholinergic system in insulin and glucagon oversecretion of genetic preobesity. Endocrinology 116:830–834

Rohner-Jeanrenaud F, Bobbioni E, Ionescu E et al. (1983) Central nervous system regulation of insulin secretion. Adn Metabol Disorders 10:193–220

Rombout JHWM, Rademakers LHPM, van Hees JP (1979) Pancreatic endocrine cells of *Barbus conchonius* (Teleostei, Cyprinidae), and their relation to the enteroendocrine cells. Cell Tissue Res 203:9–23

Rombout JHWM, Taverne-Thiele JJ (1982) An immunocytochemical and electron-microscopical study of endocrine cells in the gut and pancreas of a stomachless teleost fish, *Barbus chonchonius* (Cyprinidae). Cell Tissue Res 227:577–593

Ronner P, Scarpa A (1982) Isolated perfused Brockmann body as a model for studying pancreatic endocrine secretion. Am J Physiol 243:E352–E359

Ronner P, Scarpa A (1984) Difference in glucose dependence of insulin and somatostatin release. Am J Physiol 246:E506–E509

Rooth P, Grankvist K, Täljedal I-B (1985) In vivo fluorescence microscopy of blood flow in mouse pancreatic islets: Adrenergic effects in lean and obese-hyperglycemic mice. Microvasc Res 30:176–184

Rosen DE, Forey PL, Gardiner BG, Patterson C (1981) Lungfishes, tetrapods, paleontology, and plesiomorphy. Bull Amer Mus Nat Hist 167:159–276

Rosenzweig JL, LeRoith D, Lesniak MA et al. (1983) Two distinct insulins in the guinea pig: the broad relevance of these findings to evolution of peptide hormones. Fed Proc 42:2608–2614

Rosenzweig JL, LeRoith D, Lesniak MA, Yip CC et al. (1985) Two distinct insulin-related molecules in the guinea pig: immunological and biochemical characterization of insulin-like immunoactivity from extrapancreatic tissues of the guinea pig. Diabetologia 28:237–243

Roth J, Bonner-Weir S, Norman AW, Orci L (1982) Immunocytochemistry of vitamin D-dependent calcium binding protein in chick pancreas: exclusive localization in B-cells. Endocrinology 110:2216–2218

Rothwell NJ, Stock MJ (1985) Biological distribution and significance of brown adipose tissue. Comp Biochem Physiol 82A:745–751

Rovainen CM, Lemcoe GE, Peterson A (1971) Structure and chemistry of glucose-producing cells in the lamprey. Brain Res 30:99–118

Rubenstein AH, Pottenger LA, Mako M et al. (1972) The metabolism of proinsulin and insulin by the liver. J Clin Invest 51:912–921

Rubenstein AH, Kuzuya H, Horwitz DL (1977a) Clinical significance of circulating C-peptide in diabetes mellitus and hypoglycemic disorders. Arch Int Med 137:625–632

Rubenstein AH, Steiner DF, Horwitz DL et al. (1977b) Clinical significance of circulating proinsulin and C-peptide. Rec Prog Hormone Res 33:435–475

Russek M, Racotta R (1980) A possible role of adrenaline and glucagon in the control of food intake. Front Horm Res 6:120–137

Rutter WJ (1980) The development of the endocrine and exocrine pancreas. In: Fitzgerald PJ, Morrison AM (eds) The pancreas, Chap 2. Williams and Wilkins, Baltimore, London, pp 30–38

Ryle AP, Sanger F, Smith LF, Kitae R (1955) The disulfide bonds of insulin. Biochem J 60:541–545

Sacks H, Terry LC (1981) Clearance of immunoreactive somatostatin by perfused rat liver. J Clin Invest 67:419A

Saffrey MJ, Marcus N, Jessen KR, Burnstock G (1983) Distribution of neurons with high-affinity uptake for GABA in the myenteric plexus of the guinea pig, rat and chicken. Cell Tissue Res 234:231–235

Sakamoto C, Goldfine ID, Williams JA (1984) The somatostatin receptor on isolated acinar cell plasma membranes. J Biol Chem 259:9623–9627

Saltiel AR, Fox JA, Sherline P, Cuatracasas P (1986) Insulin-stimulated hydrolysis of a novel glycolipid generates modulation of cAMP phosphodiesterase. Science 233:967–972

Samols E, Bonner-Weir S, Weir GC (1986) Intra-islet insulin-glucagon-somatostatin relationships. Clin Endocrinol Metabol 15:33–58

Samols E, Weir GC, Bonner-Weir S (1983) Intraislet insulin-glucagon-somatostatin relationships. In: Lefébvre PJ (Ted) Glucagon II, Chap 31. Springer, Berlin New York Tokyo, pp 133–173

Samsel J (1973) Acides aminés et sécrétion des glucagons chez le Canard. These de 3éme cycle, Université Strasbourg, 274 pp

Sanders B (1983a) Insulin-like peptides in the lobster *Homarus americanus* I. Insulin immunoreactivity. Gen Comp Endocrinol 50:366–373

Sanders B (1983b) Insulin-like peptides in the lobster *Homarus americanus* II. Insulin-like biological activity. Gen Comp Endocrinol 50:374–377

Sanders B (1983c) Insulin-like peptides in the lobster *Homarus americanus* III. No glucostatic role. Gen Comp Endocrinol 50:378–382

Sandor T, Mehdi AZ (1979) Steroids and evolution. In: Barrington EJW (ed) Hormones and evolution. Academic Press, New York, pp 1–72
Sarson DL, Wood SM, Kansal PC, Bloom SR (1984) Glucose-dependent insulinotropic polypeptide augmentation of insulin. Physiology or pharmacology? Diabetes 33:389–393
Sasaki Y, Takahashi H, Aso H, Ohneda A, Weekes TEC (1982) Effects of cold exposure on insulin and glucagon secretion in sheep. Endocrinology 111:2070–2076
Sato T, Herman L (1981) Stereological analysis of normal rabbit pancreatic islets. Am J Anat 161:71–84
Savina MV, Derkerchev EF (1983) Switch on and switch off phenomenon of liver gluconeogenic function in the lamprey (*Lampetra fluviatilis* L.) under the influence of season and temperature. Comp Biochem Physiol 75B:531–539
Schade DS, Eaton RP (1983) Hormonal interrelationships. In: Ellenberg M, Rifkin H (eds) Diabetes mellitus, theory and practice, Chap 12. Medical Examination Publishing Co, Inc, pp 255–266
Schaller HC, Bodenmüller H, Kemmner W (1982) Structure and function of morphogenetic substances from Hydra. Verh Dtsch Zool Ges 1982:81–90
Schilt J, Richoux JP, Dubois MP (1981) Demonstration of peptides immunologically related to vertebrate hormones in *Dugesia lugubris* (Turbellaria, Tricladida). Gen Comp Endocrinol 43: 331–335
Schirner H (1963) Das Pankreas von *Myxine glutinosa* and *Bdellostoma stouti*. Ein Beitrag zur Phylogenie des Pankreas. Nytt mag zool 11:5–18
Schlaghecke R, Blüm V (1981) Seasonal variation in insulin and glucagon concentrations of *Rana esculenta* (L.) Gen Comp Endocrinol 43:479–483
Schlüter KJ, André J, Enzmann F, Kerp L (1984) Insulin receptor binding in pork brain: different affinities of porcine and human insulin. Horm Metab Res 16:411–414
Schmechel D, Marangos PJ, Brightman M (1978) Neuron specific enolase is a molecular marker for peripheral and central neuroendocrine cells. Nature (London) 276:834–836
Schmidt WE, Siegel EG, Creutzfeldt W (1985) Glucagon-like peptide-1 but not glucagon-like peptide-2 stimulates insulin release from isolated rat pancreatic islets. Diabetologia 28:704–707
Schot LPC, Boer HH, Swaab DF, van Noorden S (1981) Immunocytochemical demonstration of peptidergic neurons in the central nervous system of the pond snail, *Lymnaca stagnalis* with antisera raised to biologically active peptides of vertebrates. Cell Tissue Res 216:273–291
Schuit FC, Pipeleers DG (1986) Differences in adrenergic recognition by pancreatic A and B cells. Science 232:875–877
Schusdziarra V (1980) Somatostatin – a regulatory modulator connecting nutrient entry and metabolism. Horm Metab Res 12:563–577
Schusdziarra V, Lawecki J, Ditschuneit HH et al. (1985) Effect of low-dose somatostatin infusion on pancreatic and gastric endocrine function in lean and obese nondiabetic human subjects. Diabetes 34:595–601
Schwabe C, Gowan LK, Reinig JW (1982) Evolution, relaxin and insulin: a new perspective. Ann New York Acad Sci 380:6–12
Schwabe C, LeRoith D, Thompson RP et al. (1983) Relaxin extracted from protozoa (*Tetrahymena pyriformis*). J Biol Chem 258:2778–2781
Schwartz TW (1980) Enteroinsular axis for pancreatic polypeptide. Front Horm Res 7:82–91
Schwartz TW, Tager HS (1981) Isolation and biogenesis of a new peptide from pancreatic islets. Nature 294:589–591
Schwartz TW, Gingerich RL, Tager HS (1980) Biosynthesis of pancreatic polypeptide. Identification of a precursor and a co-synthesized product. J Biol Chem 255:11494–11498
Schwartz TW, Holst JJ, Fahrenkrug J et al. (1978) Vagal, cholinergic regulation of pancreatic polypeptide secretion. J Clin Invest 61:781–789
Schwyzer R (1982) Peptides and the new endocrinology. Naturwissenschaften 69:150
Scow RO (1957) Total pancreatectomy in the rat: operation, effects and post-operative care. Endocrinology 60:359–367
Sedelmeier D (1985) Mode of action of the crustacean hyperglycemic hormone. Am Zool 25:223–232
Sekine Y, Yui R (1981) Immunohistochemical study of the pancreatic endocrine cells of the ray *Dasyatis akajei*. Arch Histol Jap 44:95–101
Seppälä M, Wahlström T, Leppaluto J (1979) Luteinizing hormone-releasing factor (LRF)-like immunoreactivity in rat pancreatic islet cells. Life Sci 25:1489–1496

Shen L-P, Rutter WJ (1984) Sequence of the human somatostatin I gene. Science 224:168–171
Shen L-P, Pictet RL, Rutter WJ (1982) Human somatostatin I: sequence of the cDNA. Proc Nat Acad Sci USA 79:4575–4579
Sheppard MN, Johnson JF, Cole GA et al. (1982) Neurone specific enolase (NSE) immunostaining: a useful tool for the light microscopical detection of endocrine cell hyperplasia in adult rats exposed to asbestos. Histochemistry 74:503–513
Sheridan VA, Plisetskaya E, Bern HA, Gorbman A (1985) Effects of somatostatin and urotensin II on lipid and carbohydrate metabolism in coho salmon. Fed Proc 44:632
Sherwin RS, Sacca L (1984) Effect of epinephrine on glucose metabolism in humans: contribution of the liver. Am J Physiol 247:E157–E165
Sherwin RS, Rosa G, Hendler RG, Felig P (1976) Effect of diabetes mellitus and insulin on the turnover and metabolic response to ketones in man. Diabetes 25:776–784
Shields D (1980) In vitro biosynthesis of fish islet pre-prosomatostatin: evidence of processing and segregation of a high molecular weight precursor. Proc Nat Acad Sci USA 77:4574–4578
Shikata H, Utsumi N, Hiramatsu M et al. (1984) Immunohistochemical localization of nerve growth factor and epidermal growth factor in guinea pig prostatic gland. Histochemistry 80:411–413
Shima K, Morishita S, Sawazaki N et al. (1977) Failure of exogenous insulin to inhibit insulin secretion in man. Horm Metab Res 9:441–443
Shimazu T (1983) Reciprocal innervation of the liver: its significance in metabolic control. Adv Metab Disorders 10:355–384
Shoelson S, Haneda M, Blix P et al. (1983) Three mutant insulins in man. Nature (London) 302:540–543
Shulkes A, Hardy KJ (1982) Ontogeny of circulating gastrin and pancreatic polypeptide in the foetal sheep. Acta Endocrinol 100:565–572
Siigur E, Neuman T, Gärve V et al. (1985) Isolation and characterization of nerve growth factor from *Vipera lebetina* (snake) venom. Comp Biochem Physiol 81B:211–215
Silva P, Stoff JS, Leone DR, Epstein FH (1985) Mode of action of somatostatin to inhibit secretion by shark rectal gland. Am J Physiol 249:R329–R334
Sirek A (1969) Pancreatectomy and diabetes. In: Pfeiffer EF (ed) Handbuch des Diabetes. JF Lehmann, München, pp 727–743
Sitbon G, Mialhe P (1978) Pancreatic hormones and plasma glucose: regulation mechanisms in the goose under physiological conditions III. Inhibitory effect of insulin on glucagon secretion. Horm Metab Res 10:473–477
Sitbon G, Mialhe P (1979) Pancreatic hormones and plasma glucose: regulation mechanisms in the goose under physiological conditions. IV – Effects of food ingestion and fasting on pancreatic hormones and gut GLI. Horm Metab Res 11:123–129
Sitbon G, Strosser MT, Gross R et al. (1980) Endocrine factors in intermediary metabolism with special reference to pancreatic hormones. In: Epple A, Stetson MT (eds) Avian endocrinology. Academic Press, New York, pp 251–270
Sive AA, Vinik AI, Levitt N et al. (1980) Adrenergic modulation of human pancreatic polypeptide (hPP) release. Gastroenterology 79:665–672
Siwe SA (1926) Pankreasstudien. Gegenbaurs Morphol Jahrb 57:84–307
Smith CL (1954) The relation between seasonal hyperglycemia and thyroid activity in the frog *Rana temporaria*. J Endocrinol 10:184–191
Smith LF (1966) Species variation in the amino acid sequence of insulin. Amer J Med 40: 662–666
Smith PH (1974) Pancreatic islets of the coturnix quail. A light and electron microscopic study with special reference to the islet organ of the splenic lobe. Anat Rec 178:567–586
Smith PH (1975) Structural modification of Schwann cells in the pancreatic islets of the dog. Am J Anat 144:513–517
Smith PH, Davis BJ (1983) Morphological aspects of pancreatic islet innervation. J Auton Nerv Syst 9:53–66
Smith PH, Madson KL (1981) Interactions between autonomic nerves and endocrine cells of the gastroenteropancreatic system. Diabetologia 20:314–324
Smith PH, Patel DG (1984) Immunochemical studies of the insulin-like material in parotid gland of rats. Diabetes 33:661–666

Smith PH, Marchant FW, Johnson DG et al. (1977) Immunocytochemical localization of gastric inhibitory polypeptide-like material within A-cells of the endocrine pancreas. Am J Anat 149:585–590
Sodoyez-Goffaux F, Sodoyez JC, DeVos CJ (1985) Insulin receptors in the gastrointestinal tract of the rat fetus: quantitative autoradiographic studies. Diabetologia 28:45–50
Solcia E, Buffa R, Capella C et al. (1980) Immunohistochemical and ultrastructural characterization of gut cells producing GIP, GLI, glucagon, secretin and PP-like materials. Front Horm Res 7:7–12
Solcia E, Capella C, Usellini L et al. (1985) The PP cell. In: Volk BW, Arquilla ER (eds) The diabetic pancreas, second edition. Plenum Press, New York, pp 107–115
Soret MG, Dulin WE (1981) Animal models of spontaneous diabetes. In: Cooperstein SJ, Watkins D (eds) The islets of Langerhans. Biochemistry, physiology and pathology, Chap 14. Academic Press, London New York, pp 357–385
Sorokin AV, Petrenko OI, Kavsan VM et al. (1982) Nucleotide sequence analysis of the cloned salmon preproinsulin cDNA. Gene 20:367–376
Sower SA, Plisetskaya E, Gorbman A (1985) Changes in plasma steroid and thyroid hormones, IR insulin and protein during final maturation and spawning of the sea lamprey, *Petromyzon marinus*. Gen Comp Endocrinol 58:259–269
Sperling MA, Ganguli S, Leslie N, Landt K (1984) Fetal-perinatal catecholamine secretion: role in perinatal glucose homeostasis. Am J Physiol 247:E69–E74
Sporn MB, Todaro GJ (1980) Autocrine secretion and malignat transformation of cells. N Eng J Med 303:878–880
Ssobolew LW (1902) Zur normalen und pathologischen Morphologie der inneren Sekretion der Bauchspeicheldrüse. Virchows Arch 168:91–128
Stagner JI, Samols E (1985a) Perturbation of insulin oscillations by nerve blockade in the in vitro canine pancreas. Am J Physiol 248:E516–E522
Stagner JI, Samols E (1985b) Role of intrapancreatic ganglia in regulation of periodic insular secretions. Am J Physiol 248:E522–E530
Stanchfield JE, Yager JD (1979) Insulin effects on protein synthesis and secretion in primary cultures of amphibian hepatocytes. J Cell Physiol 100:279–290
Staub A, Sinn L, Behrens OK (1955) Purification and crystallization of glucagon. J Biol Chem 214:619–632
Stefan Y, Falkmer S (1980) Identification of four endocrine cell types in the pancreas of *Cottus scorpius* (teleostei) by immunofluorescence and electron microscopy. Gen Comp Endocrinol 42:171–178
Stefan Y, Ravazzola M, Orci L (1981) Primitive islets contain two populations of cells with differing glucagon immunoreactivity. Diabetes 30:192–195
Stefan Y, Orci L, Malaisse-Lagae F et al. (1982a) Quantitation of endocrine cell content in the pancreas of nondiabetic and diabetic humans. Diabetes 31:694–700
Stefan Y, Ravazolla M, Grasso S et al. (1982b) Glicentin precedes glucagon in the developing human pancreas. Endocrinology 110:2189–2191
Steffens AB, Strubbe JH (1983) CNS regulation of glucagon secretion. Adv Metabol Disorders 10:221–257
Steinberg JP, Leitner JW, Draznin B, Sussman KE (1984) Calmodulin and cyclic AMP. Possible different sites of action of these two regulatory agents in exocytotic hormone release. Diabetes 33:339–345
Steiner DF, Oyer PE (1967) The biosynthesis of insulin and a probable precursor of insulin by a human islet cell adenoma. Proc Nat Acad Sci USA 73:473–480
Steiner DF, Chan SJ, Docherty K. et al. (1984) Evolution of polypeptide hormones and their precursor processing mechanisms. In: Falkmer S, Hakånson R, Sundler F (eds) Evolution and tumour pathology of the neuroendocrine system, Chap 12. Elsevier, Amsterdam New York, pp 203–223
Sterne J, Hirsch C, Pele MF (1968) Disorders of carbohydrate metabolism following feeding of glucose alone in the fish *Carassius aurathus*. Pathol Biol 16:639
Stevens EVJ, Husbands DR (1985) Insulin-dependent production of low-molecular-weight compounds that modify key enzymes in metabolism. Comp Biochem Physiol 81B:1–8
Stevenson RW (1983) Further evidence for non-pancreatic insulin immunoreactivity in guinea pig brain. Horm Metab Res 15:526–529

Stewart JK, Goodner CJ, Koerker DJ et al. (1978) Evidence for a biological role of somatostatin in the Pacific hagfish, *Eptatretus stouti*. Gen Comp Endocrinol 36:408–414
Stjernquist M, Emson P, Owman Ch et al. (1983) Neuropeptide Y in the female reproductive tract of the rat. Distribution of nerve fibers and motor effects. Neurosci Lett 39:279–284
Strahan R, Maclean JL (1969) A pancreas-like organ in the larva of the lamprey, *Mordacia mordax*. Aust J Sci 32:54–55
Strosser JT, DiScala D, Koch B, Mialhe P (1983a) Inhibitory effect and mode of action of somatostatin on lipolysis in chicken adipocytes. Biochim Biophys Acta 763:191–196
Strosser MT, Foltzer C, Cohen L, Mialhe P (1983b) Evidence for an indirect effect of somatostatin on glucagon secretion via inhibition of free fatty acids release in the duck. Horm Metab Res 15:279–283
Strosser MT, Harvey S, Foltzer C, Mialhe P (1984) Comparative effects of somatostatin-28 and somatostatin-14 on basal growth hormone and pancreatic function in immature ducks (*Anas platyrhynchos*). Gen Comp Endocrinol 56:265–270
Stuart CA, Brosnan PG, Furlanetto RW (1984) The somatomedin-C receptor of human placenta. Metabolism 33:90–96
Sundby F (1976) Species variations in the primary structure of glucagon. Metabolism 25:1319–1321
Sundby F, Markussen J (1971a) Isolation, crystallization and amino acid composition of rat glucagon. Horm Metab Res 3:184–187
Sundby F, Markussen J (1971b) Rabbit glucagon: isolation, crystallization and amino acid composition. Horm Metab Res 4:56
Sundby F, Markussen J, Danho W (1974) Camel glucagon: isolation, crystallization and amino acid composition. Horm Metab Res 6:425
Sundby F, Frandsen EK, Thomsen J, Kristiansen K, Brunfeldt K (1972) Crystallization and amino acid sequence of duck glucagon. FEBS Lett 26:289–293
Sundler F, Håkanson R (1984) Gastro-entero-pancreatic endocrine cells in higher mammals, with special reference to their ontogeny in the pig. In: Falkmer S, Håkanson S, Sundler F (eds) Evolution and tumour pathology of the neuroendocrine system. Elsevier, New York, pp 111–135
Sundler F, Håkanson R, Alumets J, Walles B (1977) Neuronal localization of pancreatic polypeptide (PP) and vasoactive intestinal peptide (VIP) in the earthworm (*Lumbricus terrestris*). Brain Res Bull 2:61–65
Sundler F, Alumets J, Håkanson R et al. (1978) Peptidergic (VIP) nerves in the pancreas. Histochemistry 55:173–176
Sundler F, Moghimzadeh E, Håkanson R et al. (1983) Nerve fibers in the gut and pancreas of the rat displaying neuropeptide-Y immunoreactivity. Cell Tissue Res 230:487–493
Suryawanshi SA, Rangneker PV (1971) Effects of administration of glucagon and insulin in the lizard, *Uromastix hardwickii*, Gray. J Anim Morphol Physiol 18:134–144
Susa JB, Wildness JA, Hentz R, Liu F et al. (1984a) Somatomedins and insulin in diabetic pregnancies: effects on fetal macrosomia in the human and rhesus monkey. J Clin Endocrinol Metabol 58:1099–1105
Susa JB, Neave C, Sehgal P et al. (1984b) Chronic hyperinsulinemia in the fetal rhesus monkey. Effects of physiologic hyperinsulinemia on fetal growth and composition. Diabetes 33:656–660
Suttie JM, Gluckman PD, Butter JH et al. (1985) Insulin-like growth factor (IGF-I). Antler-stimulating hormone? Endocrinology 116:846–848
Svennevig J-L (1967) Entwicklung des Inselorgans bei der Hausente, die Entstehung der dunklen und der hellen Inseln. Z Mikrosk-Anat Forsch 76:568–584
Syed Ali S (1982) Vascular pattern in the pancreas of the cat. Cell Tissue Res 223:221–234
Syed Ali S (1984) Angioarchitecture of the pancreas of the cat. Cell Tissue Res 235:675–682
Syed Ali S (1986) Microvasculature of the principal islets in the scorpion fish, *Myxocephalus scorpius*. Arch Histol Jap 48:363–371
Szabo S (1983) Somatostatin stimulates clearance and hepatic uptake of colloidal carbon in the rat. Life Sci 33:1975–1980
Szecowka J, Tendler D, Efendic S (1983) Effects of PHI on hormonal secretion from perfused rat pancreas. Am J Physiol 245:E313–E317
Taborsky GJ (1983) Evidence of a paracrine role for pancreatic somatostatin in vivo. Am J Physiol 245:E598–E603

Tager HS (1984) Glucagon-containing and glucagon-related peptides: evolutionary, structural and biosynthetic considerations. In: Falkmer S, Håkanson R, Sundler F (eds) Evolution and tumour pathology of the neuroendocrine system Chap 16. Elsevier, Amsterdam, New York, pp 285–311

Tager HS, Hohenbrocken M, Markese J, Kinerstein RJ (1980) Identification and localization of glucagon related peptides in rat brain. Proc Nat Acad Sci USA 77:6229–6233

Tager HS, Given B, Baldwin D et al. (1979) A structurally abnormal insulin causing human diabetes. Nature (London) 281:122–125

Täljedal I-B (1981) On insulin secretion. Diabetologia 21:1–7

Tamborlane WV, Hintz RL, Bergman M et al. (1981) Insulin-infusion-pump treatment of diabetes. Influence of improved metabolic control on plasma somatomedin levels. N Eng J Med 305:303–307

Tan K, Tsiolakis D, Marks V (1985) Effect of glucagon antibodies on plasma glucose, insulin and somatostatin in the fasting and fed rat. Diabetologia 28:435–440

Tanaka J, Shiosaka S, Tsubouchi H et al. (1983) Immunoreactive glucagon in the vascular walls of the rat. Life Sci 33:1599–1604

Taniguchi H, Okada Y, Hosoya Y, Baba S (1980a) Comparison of uptake of α-aminobutyric acid by pancreatic islets and by substantia nigra. Biomed Res 1(Suppl):175–179

Taniguchi H, Yoshioka M, Tsutou A et al. (1980b) Effect of α-aminobutyric acid on somatostatin and insulin content of rat cultured islets. Biomed Res 1(Suppl):180–182

Taniguchi M, Yoshioka M, Ejirl K et al. (1982a) Suppressor of somatostatin release and increase of somatostatin content in pancreatic islets by GABA. In: Okada Y, Roberts E (eds) Problems in GABA research from brain to bacteria, vol 565. Excerpta Medica, Amsterdam, pp 406–412

Tannenbaum GS, Ling N, Brazeau P (1982) Somatostatin-28 is longer acting and more selective than somatostatin-14 on pituitary and pancreatic hormone release. Endocrinology 111:101–107

Taparel D, Esteve JP, Susini C et al. (1983) Binding of somatostatin to guinea-pig pancreatic membranes: regulation of ions. Biochem Biophys Res Comm 115:827–833

Taparel D, Susini Ch, Esteve JP et al. (1985) Somatostatin analogs: Correlation of receptor affinity with inhibition of cyclic AMP formation in pancreatic acinar cells. Peptides 6:109–114

Tasaka Y, Inoue S, Marumo K, Hirata Y (1984) Plasma responses of pancreatic polypeptide, glucagon and insulin in normal and alloxan diabetiod dogs and their regional levels in the pancreas. Acta Endocrinol 105:233–238

Tashima LS, Cahill GF (1964) Role of glucagon and insulin in the carbohydrate metabolism of the toadfish. Excerpta Med Int Congr Ser 74:140

Tashima L, Cahill GF Jr (1968) Effects of insulin in the toadfish, *Opsanus tau*. Gen Comp Endocrinol 11:262–271

Taylor TL, Vaillant CR (1983) Pancreatic polypeptide-like material in nerves and endocrine cells of the rat. Peptides 4:245–253

Teitelman G, Joh TH, Reis DJ (1981a) Transformation of catecholaminergic precursors into glucagon (A) cells in mouse embryonic pancreas. Proc Nat Acad Sci USA 78:5225–5229

Teitelman G, Joh TH, Reis DJ (1981b) Linkage of the brain-skin-gut axis: islet cells originate from dopaminergic precursors. Peptides 2:157–168

Teller JK, Pilc L (1985) Insulin in insects: analysis of immunoreactivity in tissue extracts. Comp Biochem Physiol 81B:493–497

Teller JK, Rosinski G, Pilc L et al. (1983) The presence of insulin-like hormone in heads and midguts of *Tenebrio molitor* L. (Coleoptera) larvae. Comp Biochem Physiol 74A:463–465

Theodorsson-Norheim E, Öberg K, Rosell S, Boström H (1983) Neurotensin-like immunoreactivity in plasma and tumor tissue from patients with endocrine tumors of the pancreas and gut. Gastroenterology 85:881–889

Thim L, Moody AJ (1981) The primary structure of porcine glicentin (proglucagon). Regul Pept 2:139–150

Thomas NW (1975) Observations on the cell types present in the principal islet of the dab *Limanda limanda*. Gen Comp Endocrinol 26:496–503

Thomas TB (1937) Cellular components of the mammalian islets of Langerhans. Am J Anat 62:31–57

Thomas TB (1940) Islet tissue in the pancreas of the elasmobranchii. Anat Rec 76:1–18

Thorpe A (1976) Studies on the role of insulin in teleost metabolism. In: Grillo TAI, Leibson L, Epple A (eds) The evolution of pancreatic islets. Pergamon Press, Oxford, New York, Toronto, Sydney, Paris, Frankfurt, pp 271–284

Thorpe A, Ince BW (1974) The effects of pancreatic hormones, catecholamines, and glucose loading on blood metabolites in the northern pike (*Esox lucius*). Gen Comp Endocrinol 23:29–44

Thorpe A, Ince BW (1976) Plasma insulin levels in teleosts determined by a charcoal-separation radioimmunoassay technique. Gen Comp Endocrinol 30:332–339

Thorpe A, Duve H (1984) Insulin- and glucagon-like peptides in insects and molluscs. Molec Physiol 5:235–260

Thorpe A, Duve H (1985) Studies on the release of pancreatic hormones in cyclostomes and fishes in vitro. In: Lofts B, Holmes WN (eds) Current trends in comparative endocrinology. Hong Kong University Press, Hong Kong, pp 1055–1059

Tilzey JF, Waights V, Holmes R (1985a) The development of a homologous teleost insulin radioimmunoassay and its use in the study of adrenaline on insulin secretion from isolated pancreatic islet tissue of the rainbow trout, *Salmo gairdneri* R. Comp Biochem Physiol 81A:821–825

Tilzey JF, Waights V, Holmes R (1985b) Adrenergic control of insulin release from isolated islet tissue in the rainbow trout, *Salmo gairdneri* R. Gen Comp Endocrinol 59:460–467

Timson SM, Polak JM, Wharton J et al. (1979) Bombesin-like immunoreactivity in the avian gut and its localization to a distinct cell type. Histochemistry 61:213–221

Tominaga M, Ebitani I, Marubdshi S et al. (1981) Species differences of glucagon-like materials in the brain. Life Sci 29:1577–1581

Tomita T, Pollock HG (1981) Four pancreatic endocrine cells in the bullfrog (*Rana catesbeiana*). Gen Comp Endocrinol 45:355–363

Tomita T, Doull V, Pollock HG, Kimmel JR (1985) Regional distribution of pancreatic polypeptide and other hormones in chicken pancreas: reciprocal relationship between pancreatic polypeptide and glucagon. Gen Comp Endocrinol 58:303–310

Tompkins CV, Brandenburg D, Jones RH, Sonksen PH (1981) Mechanism of action of insulin and insulin analogues. Diabetologia 20:94–101

Tran VT, Beal MF, Martin JB (1985) Two types of somatostatin receptors differentiated by cyclic somatostatin analogs. Science 228:492–495

Trandaburu T (1976) Intrinsic innervation, monoamines and acetylcholinesterase activity in the pancreatic islets of some poikilothermic vertebrates and birds. In: Grillo TAI, Leibson L, Epple A (eds) The evolution of pancreatic islets. Pergamon Press, Oxford, pp 121–130

Trandaburu T, Calugareanu L (1966) Light and electron microscopic investigation of the endocrine pancreas of the grass-snake [*Natrix n. natrix* (L.)]. Z Zellforsch 97:212–225

Trenkle A, Hopkins K (1971) Immunological investigation of an insulin-like substance in the chicken egg. Gen Comp Endocrinol 16:493–497

Trimble ER, Bruzzone R, Gjinovci A, Renold AE (1985) Activity of the insulo-acinar axis in the isolated perfused rat pancreas. Endocrinology 117:1246–1252

Tsuda K, Seino Y, Mori K et al. (1983) Hyperfunction of the entero-PP axis in non-insulin diabetes mellitus. Horm Metab Res 15:581–585

Turner SD, Hazelwood RL (1974) Biliary excretion of insulin in the adult chicken. Comp Biochem Physiol 47A:303–313

Tyler JM, Kajinuma H (1972) Influence of beta-adrenergic and cholinergic agents in vivo on pancreatic glucagon and insulin secretion. Diabetes 21:332A

Umminger BL, Bair RD (1973) Role of islet tissue in the cold-induced hyperglycemia of the killifish, *Fundulus heteroclitus*. Exp Zool 183:65–70

Underhill LH, Rosenzweig JL, Roth JJ (1982) Insulin and insulin receptors in the nervous system of mammals. Front Horm Res 10:96–110

Underwood LE, D'Ercole AJ, Clemmons DR, van Wyk JJ (1986) Paracrine functions of somatomedins. Clin Endocrinol Metabol 15:59–77

Unger RH (1983) Insulin-glucagon relationships in the defense against hypoglycemia. Diabetes 32:575–583

Unger RH (1985) Glucagon physiology and pathophysiology in the light of new advances. Diabetologia 28:574–578

Unger RH, Eisentraut AM (1969) Enteroinsular axis. Arch Int Med 123:261–266

Unger RH, Foster DH (1985) Diabetes mellitus. In: Wilson JD, Foster DW (eds) Textbook of endocrinology. WB Saunders Co. Philadelphia, pp 1018–1080

Unger RH, Grundy S (1985) Hyperglycaemia as an inducer as well as a consequence of impaired islet cell function and insulin resistance: implications for the management of diabetes. Diabetologia 28:119–121

Unger RH, Orci L (1981 a) Glucagon and the A-cell. Physiology and pathophysiology (first of two parts). N Engl J Med 304:1518–1524

Unger RH, Orci L (1981 b) Glucagon and the A-cell. Physiology and pathophysiology (second of two parts). N Engl J Med 304:1575–1580

Unger RH, Eisentraut A, McCall MS et al. (1959) Glucagon antibodies and their use for immunoassay for glucagon. Proc Soc Exp Biol Med 102:621–623

Uttenthal LO, Ghiglione M, George SK, Bloom SR (1984) Effect of glucagon-like peptide-1 (GLP-1) on pancreatic juice output in the rat. Dig Dis Sci 29:91S(abstract)

Uvnäs-Wallensten K (1981) Peptides in metabolic autonomic nerves. Diabetologia 20(Suppl):337–342

Vale W, Ling N, Rivier J et al. (1976) Anatomic and phylogenetic distribution of somatostatin. Metabolism 25:1491–1494

Van Campenhout E, Cornelis G (1954) Les îlots endocrines du pancréas des oiseaux. CR Assoc Anat 79:462–466

Van Houten M, Posner BI (1981) Cellular basis of direct insulin action in the central nervous system. Diabetologia 20(Suppl):255–267

Van Houten M, Posner BI (1983) Circumventricular organs: receptors and mediators of direct peptide hormone action on brain. Adv Metabol Disorders 10:269–289

Van Noorden S (1984) The neuroendocrine system in protostomian and deuterostomian invertebrates and lower vertebrates. In: Falkmer S, Håkanson R, Sundler F (eds) Evolution and tumour pathology of the neuroendocrine system Chap 2. Elsevier, Amsterdam, New York, pp 7–38

Van Noorden S, Falkmer S (1980) Gut-islet endocrinology – some evolutionary aspects. Invest Cell Pathol 3:21–35

Van Noorden S, Patent GJ (1978) Localization of pancreatic polypeptide (PP)-like immunoreactivity in the pancreatic islets of some teleost fishes. Cell Tiss Res 188:521–525

Van Noorden S, Patent GJ (1980) Vasoactive intestinal polypeptide-like immunoreactivity in nerves of the pancreatic islet of the teleost fish, *Gillichthys mirabilis*. Cell Tiss Res 212:139–146

Van Noorden S, Pearse AGE (1976) The localization of immunoreactivity of insulin, glucagon and somatostatin in the gut of *Amphioxus* (*Branchiostoma*) *lanceolatus*. In: Grillo TAI, Leibson L, Epple A (eds) The evolution of pancreatic islets. Pergamon Press, Oxford, pp 163–178

Van Noorden S, Greenberg J, Pearse AGE (1972) Cytochemical and immunofluorescence investigations on polypeptide hormone localization in the pancreas and gut of the larval lamprey. Gen Comp Endocrinol 19:192–199

Varndell IM, Polak JM, Allen JM et al. (1984) Neuropeptide tyrosine (NPY) immunoreactivity in norepinephrine-containing cells and nerves of the mammalian adrenal gland. Endocrinology 114:1460–1462

Varndell IM, Sikri KL, Hennessy RJ et al. (1986) Somatostatin-containing D cells exhibit immunoreactivity for rat somatostatin cryptic peptide in six mammalian species. An electron microscopical study. Cell Tiss Res 246:197–204

Venturini G, Silei A, Palladini G, Carolei A, Margotta V (1984) Aminergic neurotransmitters and adenylate cyclase in hydra. Comp Biochem Physiol 78A:345–348

Vernon RG, Clegg RA, Flint DJ (1985) Adaptations of adipose tissue metabolism and number of insulin receptors in pregnant sheep. Biochem Physiol 81B:90–913

Vincent SR, Skirboll L, Hökfelt T et al. (1982) Coexistence of somatostatin- and avian pancreatic polypeptide (APP-) like immunoreactivity in some forebrain neurons. Neurosci 7:439–446

Vincent SR, Hökfelt T, Wu J-Y et al. (1983) Immunohistochemical studies of the GABA system in the pancreas. Neuroendocrinology 36:197–204

Viveros OH, Diliberto EJ, Daniels AJ (1983) Biochemical and functional evidence for the cosecretion of multiple messengers from single and multiple compartments. Fed Proc 42:2923–2928

Volk BW, Wellmann KF (1985 a) Historical review. In: Volk BW, Arquilla ER (eds) The diabetic pancreas, second edition. Plenum Press, New York, pp 1–16

Volk BW, Wellmann KF (1985b) Histology, cell types, and functional correlations of islets of Langerhans. In: Volk BW, Arquilla ER (eds) The diabetic pancreas, second edition. Plenum Press, New York, pp 81–106
Volk BW, Wellmann KF (1985c) Type II (idiopathic) diabetes (non-insulin-dependent). In: Volk BW, Arquilla ER (eds) The diabetic pancreas, second edition. Plenum Press, New York, pp 233–263
Volk BW, Wellmann KF (1985d) Pathogenetic considerations of type II diabetes. In: Volk BW, Arquilla ER (eds) The diabetic pancreas, second edition. Plenum Press, New York, pp 265–273
Volk BW, Wellmann KF (1985e) Hormonal diabetes secondary to extrapancreatic endocrinopathies. In: Volk BW, Arquilla ER (eds) The diabetic pancreas, second edition. Plenum Press, New York, pp 275–297
Vranic M, Pek S, Kawamori R (1974) Increased "glucagon immunoreactivity" in plasma of totally depancreatized dogs. Diabetes 23:205–212
Vranic M, Gauthier C, Bilinsky D et al. (1984) Catecholamine responses and their interactions with other glucoregulatory hormones. Am J Physiol 247:E145–E156
Wagner GF, McKeown BA (1981) Immunocytochemical localization of hormone-producing cells within the pancreatic islets of the rainbow trout (*Salmo gairdneri*). Cell Tissue Res 221:181–192
Wagner GF, McKeown BA (1982) Changes in plasma insulin and carbohydrate metabolism of zinc-stressed rainbow trout, *Salmo gairdneri*. Can J Zool 60:2079–2084
Wahren J, Felig P, Hagenfeldt L (1976) Effect of protein ingestion on splanchnic and leg metabolism in normal man and in patients with diabetes mellitus. J Clin Invest 57:987–999
Walton MJ, Cowey CB (1979) Gluconeogenesis by isolated hepatocytes from rainbow trout (*Salmo gairdneri*). Comp Biochem Physiol 62B:75–79
Wang Y-Q, Plisetskaya E, Baskin D, Gorbman A (1986) Immunocytochemical study of the pancreatic islets of the Pacific salmon, *Oncorhynchus kisutch*. Zool Sci 3:123–129
Watanabe T (1983) Ultrastructure of the chicken pancreatic islets with special reference to neural control. In: Mikami S, Komma K, Wada M (eds) Avian endocrinology. Springer, Berlin Heidelberg New York, pp 95–104
Watkins WB, Bruni JF, Yen SSC (1980) β-endorphin and somatostatin in the pancreatic D-cell. Colocalization by immunocytochemistry. J Histochem Cytochem 28:1170–1174
Weaver C (1980) Localization of parasympathetic preganglionic cell bodies innervating the pancreas within the vagal nucleus and nucleus ambiguus of the rat brain stem: evidence of dual innervation based on the retrograde axonal transport of horseradish peroxidase. J Autonom Nerv Syst 2:61–69
Weigle DS, Koerker DJ, Goodner CJ (1984) Pulsatile glucagon delivery enhances glucose production by perifused rat hepatocytes. Am J Physiol 247:E564–E568
Weir GC (1981) Glucagon in normal physiology and diabetes mellitus. In: Brownlee M (ed) Diabetes mellitus, vol I, Chap 4. Garland STPM Press, New York, London, pp 207–259
Weir GC (1983) Cyclic nucleotides in the control of glucagon secretion. In: Lefébvre PJ (ed) Glucagon II, Chap 27. Springer, Berlin Heidelberg New York, pp 75–82
Weir GC, Goltos PC, Sternberg EP, Patel YC (1976) High concentration of somatostatin immunoreactivity in chicken pancreas. Diabetologia 12:129–132
Weir GC, Samols E, Loo S et al. (1979) Somatostatin and pancreatic polypeptide secretion: effects of glucagon, insulin, and arginine. Diabetes 28:35–40
Weiss G (1984) Relaxin. Ann Rev Physiol 46:43–52
Welsch U, Dilly PN (1980) Elektronenmikroskopische Beobachtungen am Epithel des Verdauungstraktes der Hemichordaten. Ein Beitrag zur Evolution des Darmtraktes niederer Deuterostomier (Hemi- und Protochordaten) und seiner nervösen und hormonalen Steuerung. Zool Jahrb Anat 104:25–39
Welsch UN, Storch VN (1972) The fine structure of the endocrine pancreatic cells of *Ichthyophis kohtaoensis* (Gymnophiona, Amphibia). Arch Histol Japon 34:73–85
Werner H, Fridkin M, Aviv D, Koch Y (1985) Immunoreactive and bioactive somatostatin-like material is present in tobacco (*Nicotiana tabacum*). Peptides 6:797–802
Werther GA, Joffe S, Artal R, Sperling MA (1984) Opiates modulate insulin action in vivo in dogs. Diabetologia 26:65–69
Weyhenmeyer JA, Fellows RE (1983) Presence of immunoreactive insulin in neurons cultured from fetal rat brain. Cell Mol Neurobiol 3:81–86
White AW, Harrop CJF (1975) The islets of Langerhans of the pancreas of macropodid marsupials: a comparison with eutherian species. Aust J Zool 23:309–319

Williams JA, Goldfine ID (1985) The insulin-pancreatic acinar axis. Diabetes 34:980–986
Wilson S, Falkmer S (1965) Starfish insulin. Can J Biochem 43:1615–1624
Winborn WB, Sheridan PJ, McGill HC (1983) Estrogen receptors in the islets of Langerhans of baboons. Cell Tissue Res 230:219–223
Windbladh L (1976a) Pancreatic islets of some myxinoid cyclostomes. In: Grillo TAI, Leibson L, Epple A (eds) The evolution of pancreatic islets. Pergamon Press, Oxford New York, pp 179–188
Windbladh L (1976b) Follicles in the endocrine pancreas of some myxinoid species. Acta Zool 57:7–11
Winkler H, Westhead E (1980) The molecular organization of adrenal chromaffin granules. Neurosci 5:1803
Wollheim CB, Sharp GWB (1981) Regulation of insulin release by calcium. Physiol Rev 61:914–973
Wood SM, Polak JM, Bloom SR (1983) Neuropeptides in the control of the islets of Langerhans. Adv Metab Disorders 10:401–420
Woods SC, Porte D Jr (1978) The central nervous system, pancreatic hormones, feeding and obesity. Adv Metabol Disorders 9:313–331
Woods SC, Smith PH, Porte D Jr (1981) The role of the nervous system in metabolic regulation and its effects on diabetes and obesity. In: Brownlee M (ed) Diabetes mellitus, vol III, Chap 5. Garland STPM Press, New York London, pp 209–271
Wool IG, Castles JJ, Leader DP, Fox A (1972) Insulin and the function of muscle ribosomes. In: Steiner DF, Freinkel N (eds) Endocrine pancreas. Baltimore, Handbook of Physiology, section 7, vol I, pp 385–394
Wright G (1986) Immunocytochemical demonstration of growth hormone, prolactin and somatostatin-like immunoreactivities in the brain of larval, young adult and upstream migrant adult sea lamprey, *Petromyzon marinus*. Cell Tiss Res 246:23–31
Wriston JC (1984) Comparative biochemistry of the guinea pig: a partial checklist. Comp Biochem Physiol 77B:253–278
Wu P, Penman E, Coy DH, Rees LH (1983) Evidence for direct production of somatostatin-14 from a larger precursor than somatostatin-28 in a phaeochromocytoma. Regul Pept 5:219–233
Wurster DH, Miller MR (1960) Studies on the blood glucose and pancreatic islets of the salamander, *Taricha torosa*. Comp Biochem Physiol 1:101–109
Yaglov VV (1976) Evolutionary morphology and classification of pancreatic acinar-islet cells. In: Grillo TAI, Leibson L, Epple A (eds) The evolution of pancreatic islets. Pergamon Press, Oxford New York, pp 113–120
Yalow RS, Eng J (1981) Peptide hormones in strange places – are they there? Peptides 2:17–23
Yalow RS, Eng J (1983) Insulin in the central nervous system. Adv Metabol Disorders 10:341–354
Yamada H (1951) The postbranchial gut of *Lampetra planeri* Bloch, with special reference to its vascular system. Okayama Igakkai-Zasshi 63:1–52
Yamada J, Kitamura N, Yamashita T (1983) Avian gastrointestinal endocrine cells. In: Mikami S, Homma K, Wada M (eds) Avian endocrinology. Japan Scientific Societies Press, Tokyo, pp 67–79
Yanni M (1964) Effect of insulin and excess glucose on the carbohydrates, water and fat contents of the tissues of *Clarius lazera*. Z Vergl Physiol 48:624–631
Yokote M (1970a) Sekoke disease, spontaneous diabetes in carp, *Cyprinus carpio*, found in fish farms. I. Pathological study. Bull Freshwater Fish Res Lab 20:39–72
Yokote M (1972b) Sekoke disease, spontaneous diabetes in carp found in fish farms. III. Response to mammalian insulin. Bull Japanese Soc Sci Fish 36:1219–1223
Young FG (1963) Pancreatic hormones: insulin. In: Von Euler US, Heller H (eds) Comparative endocrinology, Chap 10. Academic Press, New York, pp 371–409
Young WS, Kuhar MJ, Roth JJ, Brownstein MJ (1980) Radiohistochemical localization of insulin receptors in the adult and developing rat brain. Neuropeptides 1:15–22
Young JB, Rosa RM, Landsberg L (1984) Dissociation of sympathetic nervous system and adrenal medullary responses. Am J Physiol 247:E35–E40
Youson JH (1981a) The alimentary canal. In: Hardisty MW, Potter IC (eds) The biology of lampreys, vol 3. Academic Press, London New York, pp 95–189
Youson JH (1981b) The liver. In: Hardisty MW, Potter IC (eds) The biology of lampreys, vol 3. Academic Press, London New York, pp 263–332

Zamnit VA, Newsholme EA (1979) Activities of enzymes of fat and ketone-body metabolism and effects of starvation on blood concentrations of glucose and fat fuels in teleost and elasmobranch fish. Biochem J 184:313–322
Zapf J, Rinderknecht E, Humbel RE, Froesch ER (1981) Insulin-like growth factors from human serum. In: Brownlee M (ed) Diabetes mellitus, vol I, Chap 5. Garland STPM Press, New York London, pp 261–295
Zelnik PR, Hornsey DJ, Hardisty MW (1977) Insulin and glucagon-like immunoreactivity in the liver lamprey (*Lampetra fluviatilis*). Gen Comp Endocrinol 33:53–60
Zierler K, Rogus EM, Scherer RW, Wu F-S (1985) Insulin action on membrane potential and glucose uptake: effects of high potassium. Am J Physiol 249:E17–E25
Zimmerman AE, Yip CC (1974) Guinea pig insulin I. Purification and physical properties. J Biol Chem 249:4021–4025
Zimmerman AE, Moule ML, Yip CC (1974) Guinea pig insulin. II Biological activity. J Biol Chem 249:4026–4029
Zimmerman KW (1927) C. Die Speicheldrüsen der Mundhöhle und die Bauchspeicheldrüse. In: Von Mollendorf W (ed) Handbuch der Mikroskopischen Anatomie des Menschen vol V/1. Springer, Berlin, pp 214–244
Zingg HH, Patel YC (1983) Processing of somatostatin-28 to somatostatin-14 by rat hypothalamic synaptosomal membranes. Life Sci 33:1241–1247
Zwilling E (1948) Insulin hypoglycemia in chick embryos. Proc Soc Exp Biol Med 67:192

Subject and Species Index